Lecture Notes in Bioinformatics 10545

Subseries of Lecture Notes in Computer Science

More information about this series at http://www.springer.com/series/5381

Jérôme Feret · Heinz Koeppl (Eds.)

Computational Methods in Systems Biology

15th International Conference, CMSB 2017
Darmstadt, Germany, September 27–29, 2017
Proceedings

 Springer

Editors
Jérôme Feret 🅭
Inria & École normale supérieure
Paris
France

Heinz Koeppl
Technische Universität Darmstadt
Darmstadt
Germany

ISSN 0302-9743   ISSN 1611-3349   (electronic)
Lecture Notes in Bioinformatics
ISBN 978-3-319-67470-4   ISBN 978-3-319-67471-1   (eBook)
DOI 10.1007/978-3-319-67471-1

Library of Congress Control Number: 2017952847

LNCS Sublibrary: SL8 – Bioinformatics

Printed on acid-free paper

This Springer imprint is published by Springer Nature
The registered company is Springer International Publishing AG
The registered company address is: Gewerbestrasse 11, 6330 Cham, Switzerland

Preface

This volume contains the papers presented at CMSB 2017, the 15th Conference on Computational Methods in Systems Biology held on September 27–29, 2017 at the Technische Universität Darmstadt (Germany).

The CMSB annual conference series, initiated in 2003, provides a unique discussion forum for computer scientists, biologists, mathematicians, engineers, and physicists interested in a system-level understanding of biological processes. Topics of interest include formalisms for modeling biological processes; models and their biological applications; frameworks for model verification, validation, analysis, and simulation of biological systems; high-performance computational systems biology and parallel implementations; model inference from experimental data; model integration from biological databases; multi-scale modeling and analysis methods; and computational approaches for synthetic biology. Case studies in systems and synthetic biology are especially encouraged.

There were 30 regular submissions, 5 tool-paper submissions, 1 presentation-only submission (for works already published in a journal), and 5 poster submissions. Each regular submission, tool-paper submission, and presentation-only submission was reviewed by 3 Program Committee members. The committee decided to accept 15 regular papers, 4 tool papers, the presentation, and 4 submitted posters. To complement the contributed papers, we also included in the program one tutorial on Biocham, by François Fages and Sylvain Solyman, and four invited lectures by: Michael Brenner (Harvard University, USA), Russ Harmer (CNRS and École normale supérieure de Lyon, France), Stefan Grill (Technische Universität Dresden, Germany), and Philipp Hennig (Max Planck Institute for Intelligent Systems, Tübingen, Germany).

As program co-chairs, we have many people to thank. We are extremely grateful to the members of the Program Committee and the external reviewers for their peer reviews and the valuable feedback they provided to the authors. We thank also the authors of the accepted papers for revising their papers according to the suggestions of the Program Committee and for their responsiveness in providing the camera-ready copies within the deadline. Our special thanks goes to François Fages and Ezio Bartocci, and all the members of the CMSB Steering Committee, for their advice on organizing and running the conference. We acknowledge the support of the EasyChair conference system during the reviewing process and the production of these proceedings. We also thank Springer. Our gratitude also goes to the tool track chair, Pierre Boutillier, and also to Christine Cramer, for her help and support before, during, and after the conference. It is our pleasant duty to acknowledge the financial support of our sponsors Analytikjena, Dispendix, the Profile Area Internet and Digitization of the Technische Universität Darmstadt, IBM Research Zurich, IEEE, LOEWE CompuGene, Nikon, and the support of the Bioinspired Communication Systems Lab at the

Technische Universität Darmstadt, where this year's event was hosted. Finally, we would like to thank all the participants of the conference. It was the quality of their presentations and their contribution to the discussions that made the meeting a scientific success.

September 2017

Jérôme Feret
Heinz Koeppl

Organization

Steering Committee

Ezio Bartocci	TU Wien, Austria
Finn Drablos	NTNU, Norway
François Fages	Inria, France
David Harel	Weizmann Institute of Science, Israel
Monika Heiner	Brandenburg University of Technology, Germany
Pietro Liò	University of Cambridge, UK
Tommaso Mazza	IRCCS Casa Sollievo della Sofferenza - Mendel, Italy
Satoru Miyano	University of Tokyo, Japan
Nicola Paoletti	Stony Brook University, USA
Gordon Plotkin	University of Edinburgh, UK
Corrado Priami	CoSBi/Microsoft Research, University of Trento, Italy
Carolyn Talcott	SRI International, USA
Adelinde Uhrmacher	University of Rostock, Germany

Program Committee Co-chairs

Jérôme Feret	Inria/ENS, France
Heinz Koeppl	Technische Universität Darmstadt, Germany

Tools Track Chair

Pierre Boutillier	Harvard Medical School, USA

Local Organization Chair

Heinz Koeppl	Technische Universität Darmstadt, Germany

Program Committee

John Albeck	University of California, Davis, USA
Ezio Bartocci	TU Wien, Austria
Guillaume Beslon	INSA-Lyon/Inria, France
Luca Bortolussi	University of Trieste, Italy
Jérémie Bourdon	Nantes University, France
Pierre Boutillier	Harvard Medical School, USA
Luca Cardelli	Microsoft Research, Cambridge, UK
Vincent Danos	CNRS/ENS, France and University of Edinburgh, UK
Joëlle Despeyroux	Inria Sophia Antipolis, France
James Faeder	University of Pittsburgh, USA

François Fages	Inria Paris-Saclay, France
Jérôme Feret	Inria/ENS, France
Tomas Gedeon	Montana State University, USA
Manoj Gopalkrishnan	Indian Institute of Technology Bombay, India
Calin Guet	IST, Austria
Monika Heiner	Brandenburg Technical University, Germany
Heinz Koeppl	Technische Universität Darmstadt, Germany
Jean Krivine	IRIF/Paris Diderot University, France
Reinhard Laubenbacher	University of Connecticut Health Center, USA
Axel Legay	IRISA/Inria, France
Stefan Legewie	IMB Mainz, Germany
Pietro Liò	University of Cambridge, UK
Oded Maler	CNRS-VERIMAG, France
Nicola Paoletti	Stony Brook University, USA
Loïc Paulevé	CNRS/LRI, France
Tatjana Petrov	IST, Austria
Ovidiu Radulescu	University of Montpellier 2, France
Marc Riedel	University of Minnesota, USA
Olivier Roux	LS2N and Ecole Centrale de Nantes, France
David Šafránek	Masaryk University, Czech Republic
Guido Sanguinetti	University of Edinburgh, UK
Carolyn Talcott	SRI International, USA
Fabian J. Theis	Helmholtz Zentrum München, Germany
Adelinde Uhrmacher	Universität Rostock, Germany
Verena Wolf	Saarland University, Germany
Jean Yang	Carnegie Mellon University, Pittsburgh, USA
Paolo Zuliani	Newcastle University, UK

Tool Evaluation Committee

Pierre Boutillier	Harvard Medical School, USA
Martin Demko	Masaryk University, Czech Republic
Maud Fagny	Harvard T.H. Chan School of Public Health and Dana Farber Cancer Institute, USA
Marieke Kuijjer	Harvard T.H. Chan School of Public Health and Dana Farber Cancer Institute, USA
Sébastien Légaré	ENS Lyon, France
David Šafránek	Masaryk University, Czech Republic
Andrea Vandin	IMT School for Advanced Studies Lucca, Italy
Alejandro F. Villaverde	IIM-CSIC, Vigo, Spain
Yarden Katz	Harvard Medical School, USA

Additional Reviewers

Backenköhler, Michael
Behr, Nicolas
Bhattacharyya, Arnab
Doty, David
Fippo Fitime, Louis
Hajnal, Matej
Helms, Tobias
Herajy, Mostafa
Husson, Adrien
Krüger, Thilo

Lück, Alexander
Magnin, Morgan
Mens, Irini-Eleftheria
Revell, Jeremy
Ripka, Lorenz
Šalagovič, Jakub
Selyunin, Konstantin
Soliman, Sylvain
Ulus, Dogan

Contents

Tool Papers

Posters

Invited Paper

Bio-Curation for Cellular Signalling: The KAMI Project

Russ Harmer$^{(\boxtimes)}$, Yves-Stan Le Cornec, Sébastien Légaré,
and Ievgeniia Oshurko

Université de Lyon, CNRS – ENS Lyon – Université Claude Bernard Lyon 1, LIP,
Lyon, France
`russell.harmer@ens-lyon.fr`

Abstract. The general question of what constitutes bio-curation for rule-based modelling of cellular signalling is posed. A general approach to the problem is presented, based on rewriting in hierarchies of graphs, together with a specific instantiation of the methodology that addresses our particular bio-curation problem. The current state of the ongoing development of the KAMI (Knowledge Aggregator & Model Instantiator) bio-curation tool, based on this approach, is detailed along with our plans for future development.

1 The Bio-curation Problem

In multi-cellular organisms, tissue development, maintenance and repair are largely coordinated via decentralized *signalling*: cells send signals—usually small proteins such as hormones, growth factors or cytokines—to be received by other cells through the agency of dedicated receptor proteins embedded in their external membranes. Reception of a signal is typically transduced across the external membrane by a conformational change of the receptor protein which, in consequence, triggers various intra-cellular signalling 'pathways' [9].

Despite their name, these latter do not exist physically, as actual pathways in the cell, but rather as metaphors for the cascaded activation of enzymes that perform post-translational modifications (PTMs)—most commonly phosphorylation and dephosphorylation—in order to control the assembly and disassembly of protein complexes. The metaphorical 'destination' of a pathway is the cell's DNA and the 'journey' ends in the modulation of gene expression as effected by the assembly or disassembly of complexes of transcription factors that bind directly to the DNA.

This intrinsic signalling system can be perturbed by modifications to a cell's DNA—mutations or gene ablation, duplication or rearrangement—that 'reroute', 'block' or 'short-cut' its pathways; and by pharmacological interventions intended to counteract such pathological changes.

Even in the absence of such extrinsic perturbations, different cells may respond differently to the same signal. In particular, different cell *types*—which express different repertoires of proteins—need not express the same receptors

J. Feret and H. Koeppl (Eds.): CMSB 2017, LNBI 10545, pp. 3–19, 2017.
DOI: 10.1007/978-3-319-67471-1_1

so that the 'starting point' of a pathway may be present in some cases yet absent in others. More generally, the intricate choreography of protein-protein interactions (PPIs)—bindings, unbindings and PTMs—that we conceptualize as pathways clearly depends on the gene expression profile of the cell (including its expression levels): a 'highway' in one cell may be a 'country lane' in another.

1.1 Modelling Pathways

Considerable work has been done, *e.g.* [14, 18, 19], to determine statistical 'models from data', highly specific to the *context* of a particular cell type. Although able to recapitulate successfully the principal highways known to operate in that context, such models (unsurprisingly) tend to have limited predictive power in other contexts. Indeed, this kind of work never intended, nor claimed, to seek such predictive power; on the contrary, it was exploiting extreme contextuality to provide deeper insight into the workings of particular cells. However, it also illustrates very clearly the difficulty of trying to model directly in terms of pathways: such models have an inherently holistic nature and, realistically, can only be built by unbiased, statistical learning methods.

Our approach, as initially advocated in [5], adopts a different stance: we step down a level, instead seeking a *de-contextualized* representation of the PPIs that underlie pathways; then provide the means to *re-instantiate* automatically that knowledge in any context in the form of an *executable* model [2]. We then attempt to reconstruct the biologist's notion of pathway either by the extraction of a (suitably post-processed) *causal trace* from a (stochastic) simulation of the model [4,5]; or by direct construction of such a causal trace through static analysis of the model [15].

This factorization of the modelling process allows us to focus attention on *bio-curation*: the construction of the de-contextualized representation of PPIs. The consequences of this knowledge in any particular cell context will be revealed by the automatic generation of an executable model and subsequent analysis. This contrasts with most modelling methodologies that require the modeller first to understand sufficiently the very system they are seeking to model; instead, we aim to enable an *exploratory* form of modelling as 'tool for discovery' in order to investigate how a single 'roadmap' of PPIs can be deployed, in varying (normal or pathological) contexts, to exhibit distinct cell type-specific signalling.

However, our approach poses certain constraints on what constitutes an appropriate executable model. The principal requirement is that the model provides a notion of *execution trace* based on discrete *events*, *i.e.* occurrences of PPIs, from which *causal traces* can be extracted, cf. Mazurkiewicz traces [17]. This immediately rules out ODE models. More subtly, although Mazurkiewicz's theory applies to reaction-based models—formulated either in terms of Petri nets or multi-set rewriting—the resulting causal traces contain a great deal of *spurious* causality since a single PPI is typically encoded as a family of reactions.

For example, suppose a protein B can independently bind proteins A and C to form a complex ABC via intermediates AB or BC. In the event that an A and B first react to form AB, via the reaction $A, B \rightarrow AB$, a spurious causality would be identified to the subsequent $AB, C \rightarrow ABC$ event. Indeed, the *independence* of B's bindings to A and C are expressed by the fact that the system also admits $A, BC \rightarrow ABC$ and $B, C \rightarrow BC$. If these latter reactions were removed from the system, this would imply a sequential assembly of ABC and the above causality would no longer be spurious. This mismatch between the level of representation and the desired notion of causality vastly complicates—and compromises the scalability of—the use of reaction-based models for our purposes.

This mismatch can be alleviated through the use of models based on graph rewriting, an approach known as *rule-based modelling*, exemplified by the BioNet-Gen[1] [13] and Kappa[2] [5] languages. In this setting, a PPI is represented by a single graph rewriting rule and the above issue of spurious causality no longer arise: the protein B would have two binding *sites*, one for A and one for C, and the rule 'A binds B' would not mention the binding site for C (and vice versa). More generally, Mazurkiewicz traces can be generalized to such graph rewriting settings [1,4,12] although questions still remain as to the most appropriate notion(s) of causal trace in the context of reversible systems[3].

Kappa provides three notions of causal trace: an *uncompressed* trace that may contain many uninformative 'do-undo' event pairs; a *weakly compressed* trace that employs heuristics to eliminate such 'do-undo's; and a *strongly compressed* trace that further quotients by conflating all instances, *i.e.* individual proteins, of each agent, *i.e.* type of protein [4,15]. The latter two notions correspond closely, in many cases, to the intuitive notions of pathway employed by biologists.

1.2   Representing PPIs

The protein-centric representation of Kappa—as opposed to the complex-centric representation of reaction-based models—fixes, at least to a good first approximation, the mismatch with the desired notion of causality. However, for the purposes of providing a de-contextualized representation of PPIs, it has some serious shortcomings. The principal difficulty comes from the fact that, although one Kappa rule corresponds to one PPI, in practice many PPIs share a single *mechanism*. If we wish to update our knowledge about such a mechanism, this necessitates identifying, and then making 'the same' change to, every Kappa rule corresponding to that mechanism. The significance of this problem became apparent during the first author's development (in 2007–08) of a Kappa model of the erbB signalling network, as partially documented in [5], and led directly to the work on MetaKappa [6,11].

MetaKappa provided a partial solution to this problem by enabling the definition of mechanisms as *generic* rules—that were automatically expanded into sets of underlying Kappa rules—shared by splice variants, loss-of-function mutants and even related genes. However, it was unable to treat the important case of gain-of-function mutants and, critically, the fact that mechanisms had to be

---

[1] http://bionetgen.org/index.php/Main_Page.
[2] http://dev.executableknowledge.org.
[3] Ioana Cristescu, private communication.

defined in MetaKappa implicitly required the modeller to have already in mind an intended set of underlying Kappa rules. In other words, a choice of generic rules expressed only one possible way of compressing a *known*, contextualized set of Kappa rules.

Let us now state explicitly our *bio-curation problem* for signalling. We are seeking to enable the de-contextualized representation of knowledge about PPIs: specifically, the known *necessary* conditions under which a PPI may take place. Furthermore, we need to be able to express this knowledge in such a way that a single *mechanism* corresponds to a single 'element' of our knowledge representation in order to avoid the 'update problem' above. In particular, a mechanism that is potentially shared by a family of splice variants and/or mutants of a given gene should correspond to a single element.

We also need to provide the means to *deploy* this knowledge in context via the automatic determination of which mechanisms give rise to which specific PPIs: a mechanism may not apply to a particular splice variant that lacks, for example, the necessary binding site; or a mutated protein may lose, or gain, the ability to participate in a given mechanism. Finally, this contextualized knowledge should then be automatically transformed into an executable model for detailed analysis.

1.3   Plan of the Paper

In Sect. 2, we present briefly our `ReGraph`[4] Python library which provides the underlying graph rewriting machinery necessary for our bio-curation tool `KAMI`[5] and discuss its use to support a de-contextualized representation of PPIs. In Sect. 3, we discuss the front-end—which performs semi-automatic update of this knowledge—and back-end of `KAMI`—which automatically instantiates this knowledge into an executable Kappa model. We conclude with a discussion of perspectives for future development of `KAMI` in Sect. 4.

2   KAMI's Knowledge Representation

2.1   The `ReGraph` Library

In previous work [2], the first author presented a theoretical framework for graph-based knowledge representation specifically tailored to the needs of representing PPIs for the purposes of rule-based modelling. In this setting, one first defines a so-called *meta-model*, a particular graph intended to define the kinds of entities that can exist: genes, features of genes (regions, key residues, modifiable states) and actions (binding, unbinding and state modification). The meta-model is then used to *type* a second graph called the 'pre-model', but which we rename as *action graph* in this paper, which defines the specific genes, features and actions that occur in a model. By typing, we mean the existence of a homomorphism

---

[4] https://github.com/Kappa-Dev/ReGraph.
[5] https://github.com/Kappa-Dev/KAMI.

from the action graph to the meta-model [7,12]. Finally, the action graph types a collection of *nuggets* that represent the PPIs in the model. A *model* thus comprises an action graph typing a collection of nuggets.

This framework supports *sesqui-push-out* graph rewriting [3,12] so it can express adding, deleting, cloning and merging of nodes and edges. An *update* of knowledge about a PPI can thus be expressed as an appropriate step of graph rewriting. An important technical point in this approach is that PPIs—themselves graph rewriting rules—are reified as *graphs*. This enables updates of PPIs to be written as ordinary graph rewriting rules even though, conceptually, they should be thought of as second-order rules that rewrite rules, cf. [16]. This is a particularity of our meta-model and clearly the generic framework could also be used in completely different domains—with or without the need to 'reduce' second-order to first-order rewriting. However, the rather *ad hoc* nature of the graphs used—simple graphs with two kinds of directed edges where nodes can have attributes—imposes unnecessary limitations on applicability of the framework.

We address this by adopting a more general theoretical framework based on simple directed graphs where nodes *and* edges have attributes that can be assigned sets of values. This still provides all the structure necessary to support sesqui-push-out rewriting but provides greater flexibility; in particular, different kinds of edges can be expressed by the use of edge attributes.

The well-known Python library `networkX`[6] provides exactly this class of graphs; as such, we chose to build our `ReGraph` library for (sesqui-push-out) graph rewriting on top of `networkX`. The `ReGraph` library also provides support for typing *hierarchies*: collections of graphs connected by (i) typing homomorphisms that form a forest or, more generally, a DAG (provided all typing paths between two graphs coincide); and (ii) binary *relations* in the form of spans of typing homomorphisms.

The notion of typing immediately extends to rewriting rules and, given a rule and a graph G typed by T, the result of rewriting G remains typed by T [12]. Conversely, if we rewrite T, we can restore typing by *propagating* the rewrite to G: if a node/edge is deleted or cloned in T, we delete or clone all nodes/edges typed by it in G [12]. This allows us to update an entire hierarchy upon rewriting of one of its constituent graphs: we propagate the rewrite to all other graphs typed—directly or transitively—by the rewritten graph and restore all typing homomorphisms. This is exploited by the back-end of KAMI for knowledge instantiation; see Sect. 3.2.

The notion of typing can be refined by placing *constraints* on the in- or out-degree of certain nodes: a constraint in T must be satisfied by all graphs G typed—directly or transitively—by T. This is used to express domain-specific semantic constraints in the front-end of KAMI; see Sect. 3.1.

---

[6] https://networkx.github.io.

2.2   The Meta-model

The heart of KAMI is an instance of ReGraph with a particular hierarchy, rooted in a *meta-model*, that includes—in addition to the action graph and nugget graphs—background knowledge in the form of (i) domain-specific PPI templates, *e.g.* 'phosphorylation', used to perform semantic checks or auto-completion; and (ii) definitions of gene products, *e.g.* splice variants and mutants, used to instantiate knowledge into specific contexts.

The meta-model, shown in Fig. 1, remains more or less unchanged from that originally proposed in [2]. The principal difference lies in two new nodes, defining *tests* of binding status, that were previously encoded in a rather opaque fashion; these allow nuggets to express conditions that are tested, but not modified, by the graph rewriting rules they reify. The 'source' and 'target' nodes, which played a purely formal rôle in [2], have been replaced by a single kind of *site* which should be thought of as representing a template of a physical binding site that can occur in multiple genes. As before, there are two kinds of arrows—distinguished by attributes: dotted arrows represent a *belongs to* relation, *i.e.* hierarchical structuring of actors; while solid arrows relate actions and actors.

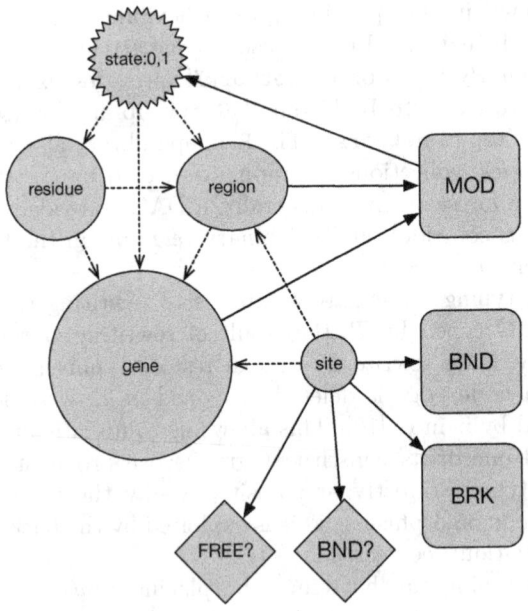

Fig. 1. The meta-model of KAMI

The meta-model also defines some standard meta-data as attributes:

– for *genes*, a string-valued attribute for the UniProt[7] accession number;

---

[7] http://www.uniprot.org/uniprot/.

- for *residues*, an attribute **aa** with values in the set of twenty one-letter codes for amino acids;
- for the dotted arrow *from residues to genes*, a positive integer-valued attribute **pos** for its position in the sequence;
- for all actions, a positive real-valued attribute **rc** for its rate constant;
- for *MOD* actions, a {0, 1}-valued attribute **val** specifying the value written by the modification.

Note that a *state* is simply an attribute whose value can be modified by actions from *within* the system; as such, in order to be able to express such a MOD action, it must be reified explicitly as a node.

2.3   Action Graphs

An instance of KAMI's hierarchy contains two action graphs: one that is built up during the development of a model; and a second that frames the built-in domain-specific background knowledge. In ontological terms, where the meta-model defines *general* concepts—genes, actions, &c.—the action graphs define which entities *actually* exist: the specific genes, actions, &c. under consideration; and the entities—binding domains, PTM states, &c.—for which the system has background knowledge.

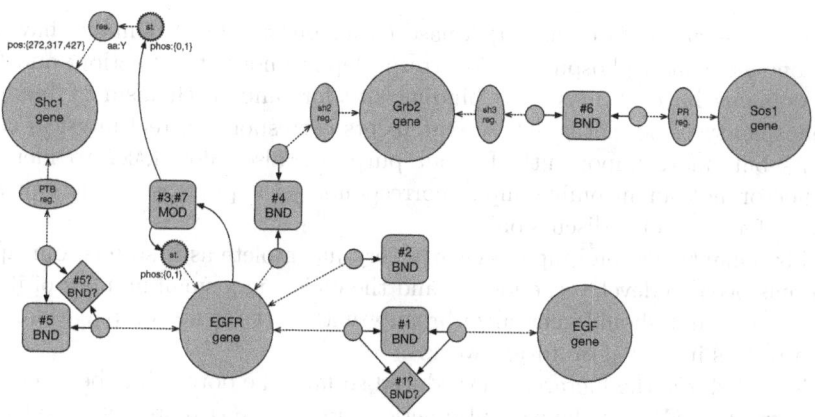

Fig. 2. An example of action graph

Figure 2 shows a typical (small) example of the first kind of action graph. It defines five actual *genes*, in the sense that those five nodes are typed by the *gene* node of the meta-model, each of which defines a type—*Shc1*, *Grb2*, *EGFR*, *EGF* and *Sos1*—that can be used by nugget graphs. The other nodes also have this dual typing aspect which occurs in any graph which is neither a sink nor a source node of its hierarchy.

The current *semantic* action graph of KAMI is shown in Fig. 3. It defines three types of regions—*kinase* domains, *phosphatase* domains and *SH2* domains—and other associated entities that will be referenced by semantic nuggets. These four domains participate in three kinds of actions—*phosphorylation, dephosphorylation* and *SH2–phospho-tyrosine motif binding*.

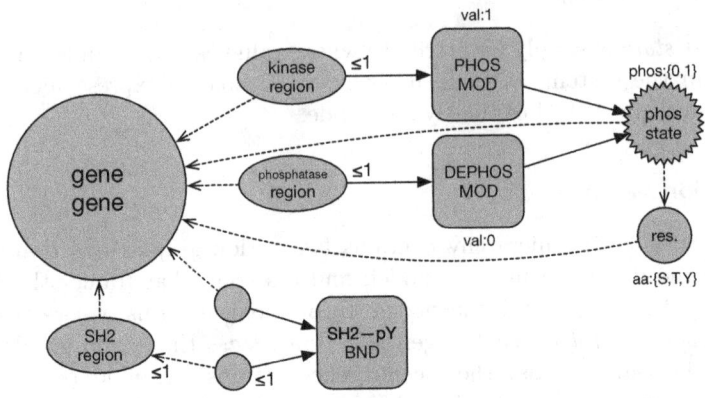

Fig. 3. The semantic action graph

The *constraints* state that (i) kinase (resp. phosphatase) domains have at most one associated phosphorylation (resp. dephosphorylation) action; and (ii) SH2 domains have at most one binding site for, and mechanism of binding to, phospho-tyrosine motifs. These statements correspond to real physical constraints but, more importantly for our purposes, also allow KAMI to identify whether or not an incoming input corresponds to a pre-existing action; see Sect. 3.1 for a detailed discussion.

This semantic action graph is clearly very incomplete as it stands; our approach has been to develop the ideas—and the code—in a small number of illustrative cases that should generalize broadly with little or no complication. We return to this in Sect. 4 on future work.

Figure 4 shows the hierarchy introduced so far. The dotted line between the action graph (AG) and the semantic action graph (SAG) represents a *relation* between the two graphs which, internally, corresponds to a span from the graph ● to AG and SAG: the typing from ● to AG picks out those nodes of AG that have been assigned a semantic attribution in SAG; and the typing from ● to SAG specifies that assignment. Note that, in order to be a valid hierarchy, the two paths from ● to the meta-model (MM) must commute.

In our example, the node *#3* of the AG is assigned to the *PHOS* node of the SAG and the (unique) state of *EGFR* is assigned to the *phos* state of the SAG; *EGFR* is also assigned to the *gene* node of the SAG. This means that node *#3 is* a phosphorylation and any domain-specific constraints—expressed in the SAG—of phosphorylation therefore apply. Additionally, node *#4* of the

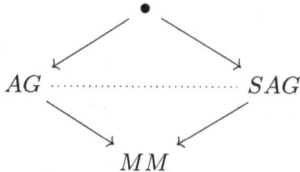

Fig. 4. The (partial) hierarchy of KAMI

AG is assigned to the *SH2–pY* node of the SAG and the region *sh2* of *Grb2* is assigned to the *SH2* node of the SAG; *Grb2* is also assigned to the *gene* node of the SAG. This means that node *sh2 is* an SH2 domain and node *#4 is* an SH2 domain–phospho-tyrosine binding.

2.4 Nuggets

An instance of KAMI's hierarchy may contain many nuggets, representing specific (families of) PPIs, typed by the action graph. It also contains a built-in—but modifiable—collection of *semantic* nuggets, typed by the semantic action graph, that provide *templates* for certain generic PPIs such as domain-domain or domain-motif bindings. These enable us to perform *semantic checks* that can reject *non-sense* nuggets.

Figure 5 shows an example of a nugget typed by the action graph of Fig. 2. Note how the nugget specifies all and only the (known) context—in this case, the test that a state of *EGFR* called *phos* has value 1 and that *Grb2* has a region *sh2*—necessary for this PPI to occur.

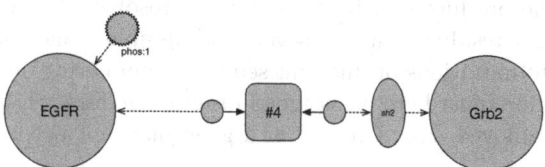

Fig. 5. An example of nugget

A nugget *N matches* a semantic nugget *SN* iff there is a span of injective homomorphisms $N \leftarrowtail \bullet \rightarrowtail SN$. A matching is *complete* iff the right leg $\bullet \rightarrowtail SN$ of the span is an isomorphism, *i.e.* there is an injective homomorphism $SN \rightarrowtail N$. For example, the nugget in Fig. 5 matches the semantic nugget in Fig. 6—which defines a template for SH2 domain–phospho-tyrosine binding—via the evident complete matching.

A given semantic action may have several associated semantic nuggets, *e.g.* Fig. 7 shows a more *refined* semantic nugget for SH2 domain–phospho-tyrosine binding. These two semantic nuggets are related by a span which also

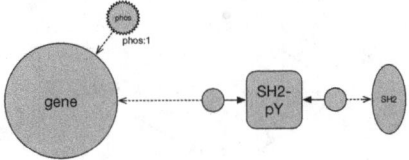

Fig. 6. An example of semantic nugget

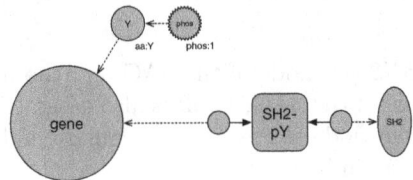

Fig. 7. Another example of semantic nugget

serves as a rewriting rule that can be applied to a nugget—provided (i) there is a complete matching to the LHS semantic nugget; and (ii) we supply a typing of the RHS into the action graph. This allows us to *upgrade* nuggets systematically once we have all extra needed details.

2.5   Protein Definitions

We represent *gene products*, *i.e.* proteins, as rewriting rules typed by the meta-model whose LHSs are injectively typed by the action graph, cf. complete matchings. A LHS comprises one gene and all features belonging to it; the RHS can have multiple gene products, each of which must resolve all disjunctive aspects of those features: a residue that has several admissible values of its **aa** or **pos** attributes—due to mutations or different sequence numbering due to splice variants or truncations—must here be assigned *exactly one* for each. Moreover, each feature may be removed, *e.g.* a region of a gene may not occur in some splice variants.

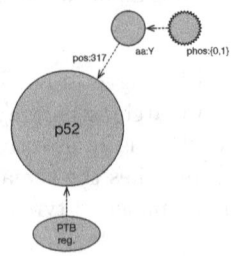

Fig. 8. Definition of a gene product

The gene *Shc1* has a residue with three admissible values for pos. We represent the *p52* splice variant, where pos = 317, of *Shc1* as in Fig. 8. We use these rewriting rules in Sect. 3.2 in the back-end of KAMI that generates Kappa models. The full current hierarchy of KAMI is shown in Fig. 9.

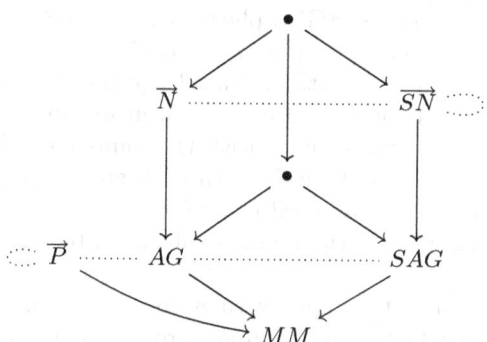

Fig. 9. The full 'hierarchy type' of KAMI including (semantic) nuggets and proteins

3 The KAMI Bio-curation Tool

In the previous section, we have seen how the generic framework of graph hierarchies, as provided by ReGraph, can be exploited to build a knowledge representation (KR) suitable for PPIs. Importantly, an update of the KR is defined by a step of graph rewriting defined in the terms of the KR's meta-model and, as such, has an intrinsic *semantic* character: an update expresses more than just a 'diff'; it is stated in terms of a *meaningful* change in an expert's knowledge about something in the KR.

The history of updates thus provides an *audit trail* that recapitulates, in properly semantic, domain-specific terms, the *modelling* process itself. In particular, it maintains a record of how knowledge was aggregated from various sources—principally scientific papers but also potentially from databases—thus providing some transparency and clarity—as well as support for model *maintenance* and future *update*—in the face of the fragmentary, dispersed nature of the primary bio-medical literature.

In this section, we describe the current front- and back-end to the KAMI bio-curation tool: the front-end takes input—either directly from the user via a GUI or through INDRA[8] statements[9]—and constructs, then applies, the appropriate step of graph rewriting. As we will explain, the system can exploit domain-specific background knowledge—in the form of semantic nuggets—to identify whether or not the input speaks of an interaction that already exists in the KR. We also very briefly describe the back-end of KAMI which takes a collection of

---

[8] https://github.com/sorgerlab/indra.
[9] http://indra.readthedocs.io/en/latest/modules/statements.html.

protein definitions and calculates the instantiation of nuggets to that collection of gene products, *i.e.* the *contextualization* of our representation to the 'cell type' defined by the given collection of proteins.

3.1 Knowledge Input and Aggregation

Given an (INDRA) input such as 'EGFR phosphorylates Shc1 on Y317' or 'Grb2's SH2 domain binds Shc1 phosphorylated on Y317', we need to compute the rewriting rule(s) required to insert this knowledge into KAMI's hierarchy. This problem is an instance of the standard problem in semantics—given an input, calculate its denotation—with a slight twist: the computed rules depend on the *current state* of the hierarchy. Indeed, given such an incoming input, depending on the current state, we may need to perform a significant update or there may be nothing to do at all as the input is subsumed by what the KR already contains.

The key task in computing update rules concerns identifying whether, or not, (i) each entity mentioned in the input already exists in the KR; and (ii) the (inter)action in question already exists in the KR. The first question can be resolved fairly easily using *grounding*: several standard names/IDs exist for genes (UniProt, HGNC, &c.) and regions/domains (PFAM, InterPro, &c.). The current version of KAMI takes inputs in the form of INDRA statements[10] which include such grounding information—at least for genes—as meta-data; however, it should be a straightforward task to obtain grounding in cases where INDRA does not provide it or, in the future, where we intend to use less pre-processed input formats.

KAMI contains a module, called the *gene anatomizer*, which takes a UniProt ID (or similar) and interrogates various databases (principally InterPro) to construct a representation of the gene and all its (significant) regions, including grounding information. By including all regions, not just those mentioned in an input, we often enable stronger inference during the construction of a rewriting rule: knowing that Grb2 has only one SH2 domain means that it *must* be the one referred to in the above input. Moreover, the anatomizer need only be run once on any given gene; the results are maintained in the action graph and can be reused freely.

The second *identification* problem, for interactions, has sharper teeth: to the best of our knowledge, no system of grounding for PPIs exists to date[11]. This problem cannot be solved automatically in general: even if an input speaks of '*A* binds *B*' and we already have a binding action between *A* and *B*, we *cannot* immediately infer that they refer to the same action as *A* and *B* may be able to bind in multiple ways. However, we can exploit background knowledge in some cases to establish that an input speaks of an existing interaction.

---

[10] We chose to use INDRA for now as it also provides us with import from BioPAX [8] and a number of NLP systems. However, there is no obstacle to providing direct import to KAMI from such sources; indeed, doing so would avoid losing certain kinds of information that are not represented in the current version of INDRA, *e.g.* regions.

[11] A notable side-effect of the KAMI project will be precisely to provide such a grounding.

For example, given an input of the form 'Grb2's SH2 domain binds Shc1 phosphorylated on Y317', KAMI would first construct a proto-nugget:

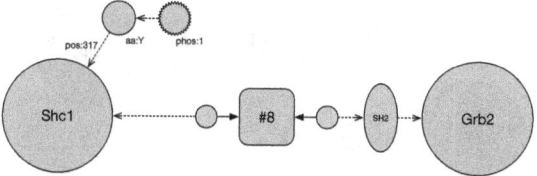

It would then use grounding meta-data to resolve *Shc1*, its residue Y317, the *phos* state of Y317, *Grb2* and its SH2 domain to existing nodes in the action graph. What about the remaining nodes—the two binding sites and the action? Given that the proto-nugget matches the semantic nugget of Fig. 7, its action is identified as an *SH2-pY* binding. The constraints imposed by the semantic action graph now require that the binding site of the SH2 domain and the *SH2-pY* action[12] be identified with those in the action graph already, giving rise to the following updated action graph (see Fig. 10).

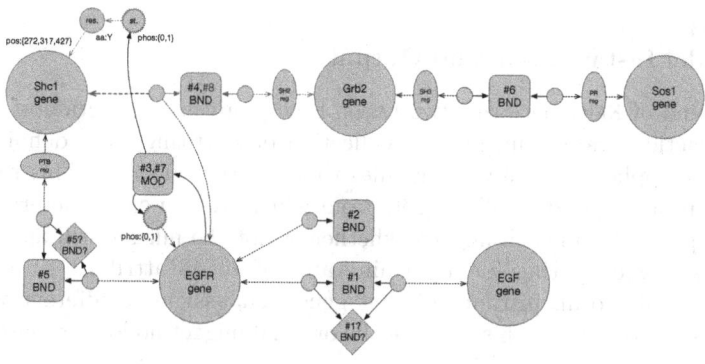

Fig. 10. The updated action graph

Moreover, the two nuggets for *Grb2*'s SH2 domain will also be merged, giving rise to a *disjunctive* nugget expressing '*Grb2*'s SH2 domain binds phosphorylated *EGFR* or *Shc1* phosphorylated on Y317'.

---

[12] This also implies that the second binding site must be identified with that belonging to *EGFR* as binding actions have at most two binding sites—a constraint, elided until now, enforced by the meta-model—*i.e.* this site is a template with an instance in *EGFR* and another in *Shc1*.

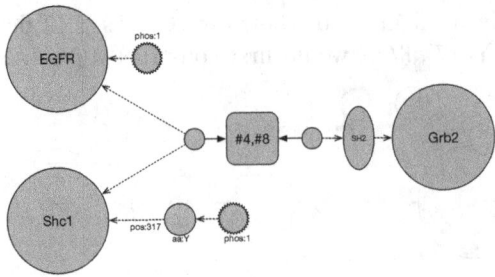

The ability to express such disjunctive statements means that a nugget corresponds to a shared *mechanism*—a family of PPIs—so any update, concerning *Grb2*, of a family member of this nugget—for example, that some mutation in the SH2 domain abrogates binding to *EGFR*—would apply at the level of the *mechanism*: *Grb2* binds *Shc1* in the same way as it binds *EGFR*; *therefore* the mutation also abrogates *Grb2*'s binding to *Shc1*. This solution of the update problem, discussed in the introduction, is a special case of what biologists call *by similarity* inference; but it occurs in KAMI not through logical 'inference' but by the merging of nodes.

3.2   Model Instantiation and Output

The back-end of KAMI performs two tasks. Firstly, given a collection of nuggets and their action graph, and given a collection of rewriting rules defining gene products, it applies those rewriting rules to the action graph[13]. This rewriting step is then propagated to all nuggets from which we can easily determine, for each gene product and each nugget, whether or not the nugget still applies. For example, a nugget testing for a certain value of an **aa** attribute of a residue would not apply to an instance of that gene that assigns a different value to that attribute. We detect this because the original nugget no longer matches the transformed one.

This effects a transformation from a gene- and mechanism-based level of representation to a protein- and PPI-based level: it *contextualizes* the knowledge with respect to the given collection of gene products. The second step now amounts to a standard parsing task: the contextualized knowledge is translated into Kappa. Each gene product defines a distinct agent type and the rules are read off by 'multiplying out' disjunctions, *e.g.* 'A_1 or A_2 binds B_1 or B_2' gives rise to four distinct rules.

4   Current and Future Work

We have presented an overview of the aims and functionality of our bio-curation tool KAMI with particular focus on the importance of capturing mechanisms,

---

[13] Unlike normal updates, this does not rewrite the action graph in-place; instead, it copies the relevant part of the action graph and rewrites *that* in-place.

not just individual PPIs, together with a curation procedure which exploits domain-specific background knowledge and intrinsically provides an audit trail documenting the curation process. The tool is based on solid theoretical foundations, discussed to some extent in [2,12], that will be further developed in the long version of the present paper.

The development of KAMI continues in earnest. The most immediate goals concern providing additional background knowledge, principally for the binding domains—PTB, SH3, WW, PDZ, &c.—and other enzymatic domains commonly implicated in signalling. This additional knowledge will already substantially increase the ability of the front-end to aggregate effectively through the merging of nodes. However, a further powerful source of background knowledge concerns closely related genes or, better, *conserved regions* of genes that typically share mechanisms. This could be captured by the merging of *region* nodes; in this way, we would extend the power of the system to identify automatically potential merging to a far wider class of (binding) actions.

In the longer term, we intend to broaden KAMI's current, very much mechanistically-oriented representation to incorporate *phenomenological* aspects. These will come in essentially two kinds: phenomenological *states*, such as 'activation' of an enzymatic domain; and *actions* that typically express the overall effect of an entire cascade of mechanistic actions. In a way somewhat analogous to the refinement of semantic templates outlined above, the tool must be able to support the gradual refinement of phenomenological knowledge about signalling—of which there is a great deal in the bio-medical literature—into its mechanistic 'implementation'.

In this way, we hope that KAMI can become an authentic 'tool for discovery' that provides automated support for the *book-keeping* aspects of curation, allowing the expert user to focus on hypothesis testing and investigating the consequences of curated knowledge in various contexts.

Related work. Our work bears a superficial similarity to the INDRA project developed in the Sorger Lab at Harvard Medical School [10]. However, the level of representation employed by INDRA corresponds to that of rule-based modelling—their *agents* are specific gene products, so mutants must be treated as distinct agents; and *statements* have none of the disjunctive flavour of nuggets—and therefore fails to solve the 'update problem'.

Indeed, INDRA sets out to solve a different problem: its aim is not the decontextualization of knowledge but the (semi-)automation of model construction. In line with this, INDRA does not seek a transparent and semantically rigorous curation procedure; instead it invests in a battery of techniques—some based on background knowledge, others on heuristics—to infer conflicts and other relationships between INDRA statements. The outcome of this *assembly* procedure is an executable model, either ODEs or rule-based, but whose provenance and built-in assumptions remain rather opaque since no meaningful audit trail can be provided.

Acknowledgements. This work was sponsored by the Defense Advanced Research Projects Agency (DARPA) and the U.S. Army Research Office under grant numbers W911NF-14-1-0367 and W911NF-15-1-0544. The views, opinions, and/or findings contained in this report are those of the authors and should not be interpreted as representing the official views or policies, either expressed or implied, of the Defense Advanced Research Projects Agency or the Department of Defense.

The first author thanks specially Walter Fontana for many discussions over the years related to this work. Thanks also to Pierre Boutillier, John Bachman and Ben Gyori; and to Adrien Basso-Blandin and Ismaïl Lahkim Bennani who worked on prototypes of KAMI and ReGraph respectively.

References

1. Baldan, P.: Modelling concurrent computations: from contextual Petri nets to graph grammars. Ph.D. thesis, Department of Computer Science, University of Pisa (2000)
2. Basso-Blandin, A., Fontana, W., Harmer, R.: A knowledge representation meta-model for rule-based modelling of signalling networks. EPTCS **204**, 47–59 (2016)
3. Corradini, A., Heindel, T., Hermann, F., König, B.: Sesqui-pushout rewriting. In: Corradini, A., Ehrig, H., Montanari, U., Ribeiro, L., Rozenberg, G. (eds.) ICGT 2006. LNCS, vol. 4178, pp. 30–45. Springer, Heidelberg (2006). doi:10.1007/11841883_4
4. Danos, V., Feret, J., Fontana, W., Harmer, R., Hayman, J., Krivine, J., Thompson-Walsh, C., Winskel, G.: Graphs, rewriting and pathway reconstruction for rule-based models. In: Foundations of Software Technology and Theoretical Computer Science (2012)
5. Danos, V., Feret, J., Fontana, W., Harmer, R., Krivine, J.: Rule-based modelling of cellular signalling. In: Caires, L., Vasconcelos, V.T. (eds.) CONCUR 2007. LNCS, vol. 4703, pp. 17–41. Springer, Heidelberg (2007). doi:10.1007/978-3-540-74407-8_3
6. Danos, V., Feret, J., Fontana, W., Harmer, R., Krivine, J.: Rule-based modelling and model perturbation. In: Priami, C., Back, R.-J., Petre, I. (eds.) Transactions on Computational Systems Biology XI. LNCS, vol. 5750, pp. 116–137. Springer, Heidelberg (2009). doi:10.1007/978-3-642-04186-0_6
7. Danos, V., Harmer, R., Winskel, G.: Constraining rule-based dynamics with types. MSCS **23**(2), 272–289 (2013)
8. Demir, E., et al.: The BioPAX community standard for pathway data sharing. Nat. Biotechnol. **28**(9), 935–942 (2010)
9. Gerhart, J.: 1998 Warkany lecture: signaling pathways in development. Teratology **60**(4), 226–239 (1999)
10. Gyori, B.M., Bachman, J.A., et al.: From word models to executable models of signaling networks using automated assembly. BioRxiv (2017)
11. Harmer, R.: Rule-based modelling and tunable resolution. EPTCS **9**, 65–72 (2009)
12. Harmer, R.: Rule-Based Meta-modelling for Bio-curation. Habilitation à Diriger des Recherches, ENS Lyon (2017)
13. Harris, L.A., et al.: BioNetGen 2.2: advances in rule-based modeling. Bioinformatics **32**(21), 3366–3368 (2016)
14. Janes, K.A., et al.: A systems model of signaling identifies a molecular basis set for cytokine-induced apoptosis. Science **310**(5754), 1646–1653 (2005)

15. Laurent, J.: Causal analysis of rule-based models of signaling pathways. Master's thesis, École Normale Supérieure, Paris, France (2015)

16. Machado, R., Ribeiro, L., Heckel, R.: Rule-based transformation of graph rewriting rules: towards higher-order graph grammars. Theoret. Comput. Sci. **594**, 1–23 (2015)

17. Mazurkiewicz, A.: Introduction to trace theory. In: The Book of Traces, pp. 3–41 (1995)

18. Molinelli, E.J., et al.: Perturbation biology: inferring signaling networks in cellular systems. PLoS Comput. Biol. **9**(12), e1003290 (2013)

19. Nelander, S., et al.: Models from experiments: combinatorial drug perturbations of cancer cells. Mol. Syst. Biol. **4**(1), 216 (2008)

Regular Papers

Quantitative Regular Expressions
for Arrhythmia Detection Algorithms

Houssam Abbas[1]([✉]), Alena Rodionova[2], Ezio Bartocci[2], Scott A. Smolka[3],
and Radu Grosu[2]

[1] Department of Electrical and Systems Engineering,
University of Pennsylvania, Philadelphia, USA
habbas@seas.upenn.edu

[2] Cyber-Physical Systems Group, Technische Universität Wien, Vienna, Austria
{alena.rodionova,ezio.bartocci,radu.grosu}@tuwien.ac.at

[3] Department of Computer Science, Stony Brook University, Stony Brook, USA
sas@cs.stonybrook.edu

Abstract. Motivated by the problem of verifying the correctness of
arrhythmia-detection algorithms, we present a formalization of these
algorithms in the language of *Quantitative Regular Expressions*. QREs
are a flexible formal language for specifying complex numerical queries
over data streams, with provable runtime and memory consumption
guarantees. The medical-device algorithms of interest include *peak detection* (where a peak in a cardiac signal indicates a heartbeat) and various
discriminators, each of which uses a feature of the cardiac signal to distinguish fatal from non-fatal arrhythmias. Expressing these algorithms'
desired output in current temporal logics, and implementing them via
monitor synthesis, is cumbersome, error-prone, computationally expen-
sive, and sometimes infeasible.

In contrast, we show that a range of peak detectors (in both the time
and wavelet domains) and various discriminators at the heart of today's
arrhythmia-detection devices are easily expressible in QREs. The fact
that one formalism (QREs) is used to describe the desired end-to-end
operation of an arrhythmia detector opens the way to formal analysis
and rigorous testing of these detectors' correctness and performance.
Such analysis could alleviate the regulatory burden on device developers
when modifying their algorithms. The performance of the peak-detection
QREs is demonstrated by running them on real patient data, on which
they yield results on par with those provided by a cardiologist.

Keywords: Peak Detection · Electrocardiograms · Arrythmia discrimination · ICDs · Quantitative Regular Expressions

1 Introduction

Medical devices blend signal processing (SP) algorithms with decision algorithms
such that the performance and correctness of the latter critically depends on
that of the former. As such, analyzing a device's decision making in isolation

© Springer International Publishing AG 2017
J. Feret and H. Koeppl (Eds.): CMSB 2017, LNBI 10545, pp. 23–39, 2017.
DOI: 10.1007/978-3-319-67471-1_2

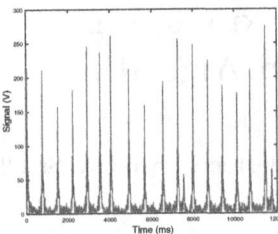 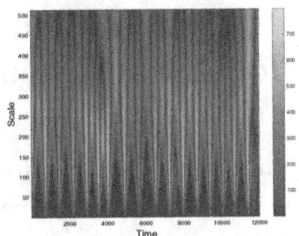

Fig. 1. Rectified EGM during normal rhythm (left) and its CWT spectrogram (right)

of SP offers at best an incomplete picture of the device's overall behavior. For example, an Implantable Cardioverter Defibrillator (ICD) will first perform *Peak Detection* (PD) on its input voltage signal, also known as an *electrogram* (see Fig. 1). The output of PD is a timed boolean signal where a 1 indicates a peak (local extremum) produced by a heartbeat, which is used by the downstream *discrimination algorithms* to differentiate between fatal and non-fatal rhythms. Over-sensing (too many false peaks detected) and under-sensing (too many true peaks missed) can be responsible for as much as 10% of an ICD's erroneous decisions [23], as they lead to inaccuracies in estimating the heart rate and in calculating important timing relations between the beats of the heart's chambers.

Motivated by the desire to verify ICD algorithms for cardiac arrhythmia discrimination, *we seek a unified formalism for expressing and analysing the PD and discrimination tasks commonly found in ICD algorithms.* A common approach would be to view these tasks as one of checking that the cardiac signal satisfies certain requirements, express these requirements in temporal logic, and obtain the algorithms by monitor synthesis. For example, PD evaluates to 1 if the signal (in an observation window) contains a peak, while the *V-Rate* discriminator evaluates to 1 if the average heart rate exceeds a certain threshold.

As discussed in Sect. 2, however, this approach quickly leads to a fracturing of the formalisms: PD algorithms and the various discriminators require different logics, and some simply cannot be expressed succinctly (if at all) in any logic available today. Thus, despite the increasingly sophisticated variety of temporal logics that have appeared in the literature [6,11], they are inadequate for expressing the operations of PD and discrimination succinctly. It should be noted that PD is an extremely common signal-processing primitive used in many domains, and forms of discrimination appear in several cardiac devices besides ICDs, such as Implantable Loop Recorders and pacemakers. Thus the observed limitations of temporal logics extend beyond just ICD algorithms.

PD and discrimination both require reasoning, and performing a wide range of numerical operations, over data streams, where the data stream is the incoming cardiac electrogram observed in real-time. For example, a commercial peak detector (demonstrated in Sect. 6) defines a peak as a value that exceeds a certain time-varying threshold, and the threshold is periodically re-initialized as a percentage of the previous peak's value. As another example, the *Onset* discriminator compares the average heart rate in two successive windows of fixed size. Thus, the desired formalism must enable value storage, time freezing, various arithmetic operations, and nested computations, *while remaining legible and succinct, and enabling compilation into efficient implementations.*

We therefore propose the use of Quantitative Regular Expressions (QREs) to describe (three different) peak detectors and a common subset of discriminators. QREs, described in Sect. 4, are a *declarative* formal language based on classical regular expressions for specifying *complex numerical queries on data streams* [1]. QREs' ability to interleave user-defined computation at any nesting level of the underlying regular expression gives them significant expressiveness. (Formally, QREs are equivalent to the streaming composition of regular functions [2]). QREs can also be compiled into runtime- and memory-efficient implementations, which is an important consideration for implanted medical devices.

To demonstrate the versatility and suitability of QREs for our task, we focus on PD in the rest of the paper, since it is a more involved than any single discriminator. Three different peak detectors are considered (Sect. 3): 1. detector WPM, which operates in the wavelet domain, 2. detector WPB, our own modification of WPM that sacrifices accuracy for runtime, and 3. detector MDT, which operates in the time domain, and is implemented in an ICD on the market today. For all three, a QRE description is derived (Sect. 5). The detectors' operations is illustrated by running them on real patient electrograms (Sect. 6).

In summary, our contributions are:

- We show that a common set of discriminators is easily encoded as QREs, and compare the QREs to their encoding in various temporal logics.
- We present two peak detectors based on a general wavelet-based characterization of peaks.
- We show that the wavelet-based peak detectors, along with a commercial time-domain peak detector found in current ICDs, are easily and clearly expressible in QREs.
- We implement the QREs for peak detection and demonstrate their capabilities on real patient data.

2 Challenges in Formalizing ICD Discrimination and Peak Detection

This section demonstrates the difficulties that arise when using temporal logic to express the discrimination and peak-detection tasks common to all arrhythmia-detection algorithms. Specifically: different discriminators require the use of different logics, whose expressive powers are not always comparable; the formulas quickly become unwieldy and error-prone; and the complexity of the monitor-synthesis algorithm, *when* it is available, rapidly increases due to nesting of freeze quantification. On the other hand, it will be shown that QREs are well-suited to these challenges: all tasks are expressible in the QRE formalism, the resulting expressions are simple direct encodings of the tasks, and their monitors are efficient. The syntax and semantics of the logics will be introduced informally as they are outside the scope of this paper.

An ICD *discriminator* takes in a finite discrete-timed signal $w : \{0,\ldots,T\} \to D$. (Signal w will also sometimes be treated as a finite string in D^* without causing confusion). The discriminator processes the signal w in a sliding-window fashion. When the window is centered at time instant t, the discriminator computes some feature of the signal (e.g., the average heart rate) and uses this feature to determine if the rhythm displays a potentially fatal arrhythmia in the current window (at time t). The ICD's overall Fatal vs Non-Fatal decision is made by combining the decisions from all discriminators.

In what follows, several discriminators that are found in the devices of major ICD manufacturers are described. Then for each discriminator, after discussing the challenges that arise in specifying the discriminator in temporal logic, a QRE is given that directly implements the discriminator. This will also serve as a soft introduction to QRE syntax. Fix a data domain D and a cost domain C. For now, we simplify things by viewing a QRE f as a regular expression r along with a way to assign costs to strings $w \in D^*$. If the string w matches the regular expression r, then the QRE maps it to $f(w) \in C$. If the string does not match, it is mapped to the undefined value $\bot$. The QRE's computations can use a fixed but arbitrary set of *operations* (e.g., addition, max, or insertion into a set). Operations can be thought of as arbitrary pieces of code.

The first example of discriminator checks whether the number of heartbeats in a one-minute time interval is between 120 and 150. This requires the use of a counting modality like that used in CTMTL [16]. If p denotes a heartbeat, then the following CTMTL formula evaluates to true exactly when the number of heartbeats lies in the desired range: $C_{[0,59]}^{\geq 120}p \wedge C_{[0,59]}^{\leq 150}p$.

This is equally easily expressed as a QRE: match 60 signal samples (at a 1 Hz sampling rate), and at every sample where p is true (this is a heartbeat), add 1 to the cost, otherwise add 0. Finally, check if the sum is in the range:

$$\mathtt{inrange}(iter_{60}-add(p?1 \text{ else } 0))$$

The second discriminator determines whether the heart rate increases by at least 20% when measured over consecutive and disjoint windows of 4 beats. In logic, this requires explicit clocks, such as those used in Explicit Clock Temporal Logic XCTL [14], since the beat-to-beat delay is variable. So let T denote the time state (which keeps track of time) and let the x_i's be rigid clock variables that store the times at which p becomes true. The following XCTL formula expresses the desired discriminator:

$$\Box(p \wedge (x_1 = T) \wedge \Diamond(p \wedge \ldots \Diamond(p \wedge (x_9 = T) \wedge [(x_5 - x_1) \cdot 0.8 \geq x_9 - x_5]) \ldots))$$

Note the need to explicitly mark the 9 heartbeats and nest the setting of clock variables 9-deep. This computation can be described in a QRE in a simpler, more concise manner. Just like the usual regular expressions, simpler QREs can be combined into more complex ones. We will now use the *split–op* combinator (see Fig. 2): given the input string $w = w_1w_2$ which is a concatenation of strings w_1

and w_2, and QREs f, g, $split{-}op(\mathsf{f},\mathsf{g})$ maps w to the cost value $op(\mathsf{f}(w_1),\mathsf{g}(w_2))$, where op is some operator (e.g., averaging). So let QRE fourBeats match four consecutive beats in the boolean signal w and let it compute the average cycle length of these 4 beats. Let $inc(x, y)$ be an operation that returns True whenever $0.8x \geq y$. Then QRE suddenOnset does the job:

$$
\begin{aligned}
\mathsf{suddenOnset} &:= && split{-}inc(\mathsf{fourBeats},\,\mathsf{fourBeats}) \\
\mathsf{fourBeats} &:= && iter_4{-}avg(\mathsf{intervalLength}) \\
\mathsf{intervalLength} &:= && split{-}left(\mathsf{countzeros},\,1) \quad /\!/ \; left(a, b) \text{ returns } a
\end{aligned}
$$

The third discriminator takes in a three-vaued signal $w : \mathbb{N} \to \{0, A, V\}$ where a 0 indicates no beat, an A indicates an atrial beat, and a V indicates a ventricular beat. One *simplified* version of this discriminator detects whether this pattern occurs in the current window: $V0^{a:b}A0^{c:d}V0^{e:f}A0^{g:h}V$. Here, a and b are integers, and $0^{a:b}$ indicates between a and b repetitions of 0. This can be expressed in discrete-time Metric Temporal Logic [15]. E.g. the prefix $V0^{a:b}A$ can be written as $w = V \implies X((w = 0)\,\mathcal{U}_{[a+1,b]}(w = A))$. And so on. This quickly becomes unwieldy as the pattern itself becomes lengthier and with more restrictions on the timing of the repetitions. On the other hand, this is trivially expressed as a (quantitative) regular expression.

Our final example comes from Peak Detection (PD), which takes in a real-valued signal $v : \mathbb{N} \to \mathbb{R}_{\geq 0}$. For *one component* of this PD, the objective is to detect when $v(t)$ exceeds a threshold value $h > 0$ which is reset as a function of the previous peak value. Thus the logic must remember the value of that peak. This necessitates *freeze quantification* of state variables, as used in Constraint LTL with Freeze Quantification CLTL$^{\downarrow}$ [10] ($\downarrow_{z\,=\,v}$ means that we freeze the variable z to the value of v):

$$
\Box(v > h \implies \downarrow_{BL\,=\,1} \Diamond\,(\varphi_{\text{local-max}} \implies h = 0.8z_2))
$$
$$
\varphi_{\text{local-max}} := \downarrow_{z_1\,=\,v} X(\downarrow_{z_2\,=\,v} X(z_2 > z_1 \wedge z_2 > v))
$$

The nesting of freeze quantifiers increases the chances of making errors when writing the specification and decreases its legibility. More generally, monitoring of nested freeze quantifiers complicates the monitors significantly and increases their runtimes. E.g., in [6] the authors show that the monitoring algorithm for STL with nested freeze quantifiers is exponential in the number of the nested freeze operators in the formula. This becomes more significant when dealing with the *full* PD, of which the above is one piece. On the other hand, we have implemented an even more complex PD as a QRE (Sect. 5.1).

The reader will recognize that the operations performed in these tasks are quite common, like averaging, variability, and state-dependent resetting of values, and can conceivably be used in numerous other applications.

This variety of logics required for these tasks, all of which are fundamental building blocks of ICD operation, means that a temporal logic-based approach to the problem is unlikely to yield a unifying view, whereas QREs clearly do. In the rest of the paper, the focus is placed on peak detection, as it is more complicated than discrimination, and offers a strong argument for the versatility and power of QREs in medical-device algorithms.

3   Peaks in the Wavelet Domain

Rather than confine ourselves to one particular peak detector, we first describe a general definition of peaks, following the classical work of Mallat and Huang [18]. Then two peak detectors based on this definition are presented. In Sect. 6, a third, commercially available, peak detector is also implemented.

3.1   Wavelet Representations

This definition operates in the wavelet domain, so a brief overview of wavelets is now provided. Readers familiar with wavelets may choose to skip this section. Formally, let $\{\Psi_s\}_{s>0}$ be a family of functions, called *wavelets*, which are obtained by scaling and dilating a so-called *mother wavelet* $\psi(t)$: $\Psi_s(t) = \frac{1}{\sqrt{s}}\psi\left(\frac{t}{s}\right)$. The *wavelet transform* W_x of signal $x : \mathbb{R}_+ \to \mathbb{R}$ is the two-parameter function:

$$W_x(s, t) = \int\limits_{-\infty}^{+\infty} x(\tau)\Psi_s(\tau - t)\, d\tau \tag{1}$$

An appropriate choice of ψ for peak detection is the nth derivative of a Gaussian, that is: $\psi(t) = \frac{d^n}{dt^n}G_{\mu,\sigma}(t)$. Equation (1) is known as a *Continuous Wavelet Transform* (CWT), and $W_x(s, t)$ is known as the *wavelet coefficient*.

Parameter s in the wavelet ψ_s is known as the *scale* of the analysis. It can be thought of as the analogue of frequency for Fourier analysis. A smaller value of s (in particular $s < 1$) *compresses* the mother wavelet as can be seen from the definition of Ψ_s, so that only values close to $x(t)$ influence the value of $W_x(s, t)$ (see Eq. (1)). Thus, at smaller scales, the wavelet coefficient $W_x(s, t)$ captures *local* variations of x around t, and these can be thought of as being the higher-frequency variations, i.e., variations that occur over a small amount of time. At larger scales (in particular $s > 1$), the mother wavelet is *dilated*, so that $W_x(s, t)$ is affected by values of x far from t as well. Thus, at larger scales, the wavelet coefficient captures variations of x over large periods of time.

Figure 1 shows a Normal Sinus Rhythm EGM and its CWT $|W_x(s, t)|$. The latter plot is known as a *spectrogram*. Time t runs along the x-axis and scale s runs along the y-axis. Brighter colors indicate larger values of coefficient magnitudes $|W_x(s, t)|$. It is possible to see that early in the signal, mid- to low-frequency content is present (bright colors mid- to top of spectrogram), followed by higher-frequency variation (brighter colors at smaller scales), and near the end of the signal, two frequencies are present: mid-range frequencies (the bright colors near the middle of the spectrogram), and very fast, low amplitude oscillations (the light blue near the bottom-right of the spectrogram).

3.2   Wavelet Characterization of Peaks

Consider the signal and its CWT spectrogram $|W_x(s,t)|$ shown in Fig. 1. The coefficient magnitude $|W_x(s,t)|$ is a measure of signal power at (s,t). At larger scales, one obtains an analysis of the low-frequency variations of the signal, which are unlikely to be peaks, as the latter are characterized by a rapid change in signal value. At smaller scales, one obtains an analysis of high-frequency components of the signal, which will include both peaks and noise. These remarks can be put on solid mathematical footing [19, Chap. 6]. **Therefore, for peak detection one must start by querying CWT coefficients that occur at an appropriately chosen scale $\bar{s}$.**

Given the fixed scale $\bar{s}$, the resulting $|W_x(\bar{s},t)|$ is a function of time. The next task is to find the *local maxima* of $|W_x(\bar{s},t)|$ as t varies. The times when local maxima occur are precisely the times when the energy of scale-$\bar{s}$ variations is locally concentrated. **Thus peak characterization further requires querying the local maxima at $\bar{s}$.**

Not all maxima are equally interesting; rather, only those with value above a threshold, since these are indicative of signal variations with large energy concentrated at $\bar{s}$. **Therefore, the specification only considers those local maxima with A value above a threshold $\bar{p}$.**

Maxima in the wavelet spectrogram are not isolated: as shown in [19, Theorem 6.6], when the wavelet ψ is the nth derivative of a Gaussian, the maxima belong to connected curves $s \mapsto \gamma(s)$ that are never interrupted as the scale decreases to 0. These *maxima lines* can be clearly seen in Fig. 1 as being the vertical lines of brighter color extending all the way to the bottom. Multiple maxima lines may converge to the same point $(0, t_c)$ in the spectrogram as $s \to 0$. A celebrated result of Mallat and Hwang [18] shows that *singularities* in the signal always occur at the convergence times t_c. For our purposes, a singularity is a time when the signal undergoes an abrupt change (specifically, the signal is poorly approximated by an $(n+1)^{th}$-degree polynomial at that change-point). **These convergence times are then the peak times that we seek.**

Although theoretically, the maxima lines are connected, in practice, signal discretization and numerical errors will cause some interruptions. Therefore, rather than require that the maxima lines be connected, we only require them to be (ϵ, δ)-connected. Given $\epsilon, \delta > 0$, an (ϵ, δ)-*connected curve* $\gamma(s)$ is one such that for any s in its domain, $|s - s'| < \epsilon \implies |\gamma(s) - \gamma(s')| < \delta$.

A succinct description of this *Wavelet Peaks with Maxima* (WPM) is then:

- (Characterization *WPM*) Given positive reals $\bar{s}, \bar{p}, \epsilon, \delta > 0$, a peak is said to occur at time t_0 if there exists a (ϵ, δ)-connected curve $s \mapsto \gamma(s)$ in the (s,t)-plane such that $\gamma(0) = t_0$, $|W_x(s, \gamma(s))|$ is a local maximum along the t-axis for every s in $[0, \bar{s}]$, and $|W_x(\bar{s}, \gamma(\bar{s}))| \geq \bar{p}$.

The choice of values $\bar{s}$, ϵ, δ and $\bar{p}$ depends on prior knowledge of the class of signals we are interested in. Such choices are pervasive and unavoidable in signal processing, as they reflect application domain knowledge. Such a specification is difficult, if not impossible, to express in temporal and time-frequency logics.

In the next section we show how WPM can be formalized using Quantitative Regular Expressions.

3.3 Blanking Characterization

For comparison, we modify WPM to obtain a peak characterization that is computationally cheaper but suffers some imprecision in peak-detection times. We call it *Wavelet Peaks with Blanking* (WPB). It says that one peak at the most can occur in a time window of size BL samples.

- (Characterization *WPB*) Given positive reals $\bar{s}, \bar{p} > 0$, a peak is said to occur at time t_0 if $|W_x(\bar{s}, t_0)|$ is a local maximum along t and $|W_x(\bar{s}, t_0)| > \bar{p}$, and there is no peak occurring anywhere in $(t_0, t_0 + BL]$.

Section 6 compares WPM and WPB on patient electrograms.

4   A QRE Primer

An examination of discrimination and PD (Sects. 2 and 3) shows the need for a language that: (1) Allows a rich set of numerical operations. (2) Allows matching of complex patterns in the signal, to select scales and frequencies at which interesting structures exist. (3) Supports the synthesis of time- and memory-efficient implementations. This led to the consideration of Quantitative Regular Expressions (QREs). A QRE is a symbolic regular expression over a data domain D, augmented with data costs from some cost domain C. A QRE views the signal as a *stream* $w \in D^*$ that comes in one data item at a time. As the Regular Expression (RE) matches the input stream, the cost of the QRE is evaluated.

Formally, consider a set of types $\mathcal{T} = \{T_1, T_2, \ldots, T_k\}$, a data domain $D \in \mathcal{T}$, a cost domain $C \in \mathcal{T}$, and a parameter set $X = (x_1, x_2, \ldots, x_k)$, where each x_i is of type T_i. Then a QRE f is a function

$$[\![f]\!]\colon D^* \to (T_1 \times T_2 \times \ldots \times T_k \to C) \cup \{\bot\}$$

where $\bot$ is the undefined value. Intuitively, if the input string $w \in D^*$ does not match the RE of f, then $[\![f]\!](w) = \bot$. Else, $[\![f]\!](w)$ is a function from $T_1 \times T_2 \times \ldots \times T_k$ to C. When a parameter valuation $\bar{v} \in T_1 \times \ldots \times T_k$ is given, this then further evaluates to a cost value in C, namely $[\![f]\!](w)(\bar{v})$. Figure 2 provides an overview of QREs and their combinators.

QREs can be compiled into efficient *evaluators* that process each data item in time (or memory) polynomial in the size of the QRE and proportional to the maximum time (or memory) needed to perform an *operation* on a set of cost terms, such as addition, least-squares, etc. The operations are selected from a set of operations *defined by the user. It is important to be aware that the choice of operations constitutes a trade-off between expressiveness (what can be computed) and complexity (more complicated operations cost more)*. See [1] for restrictions placed on the predicates and the symbolic regular expressions.

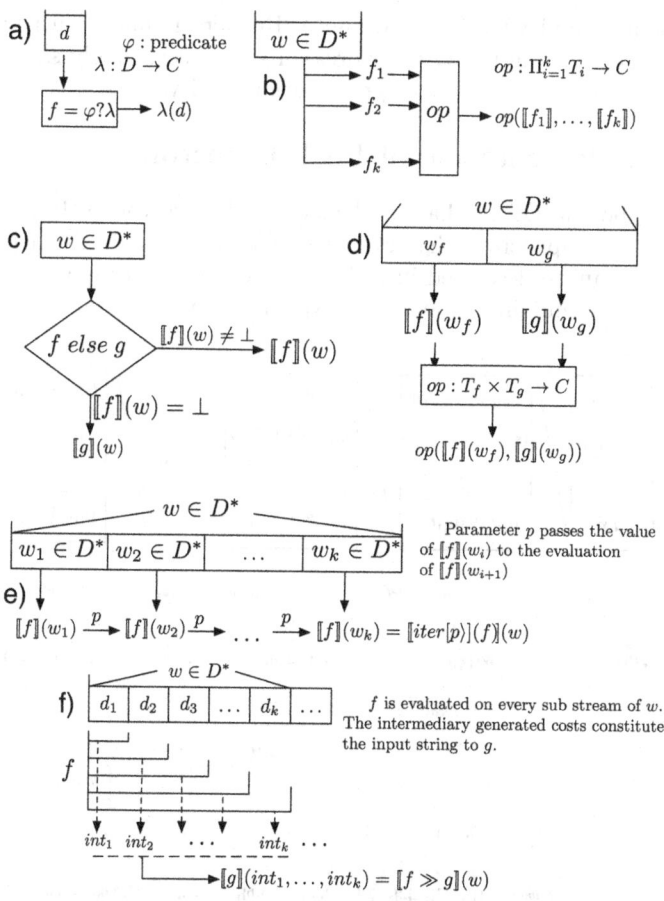

Fig. 2. QREs and their combinators. (a) Basic QRE $\varphi?\lambda$ matches one data item d and evaluates to $\lambda(d)$ if $\varphi(d)$ is True. (b) QRE $op(f_1, \ldots, f_k)$ evaluates the k QREs $f_1, \ldots, f_k$ on the same stream w and combines their outputs using operation op (e.g., addition). f_i outputs a value of type T_i. (c) QRE f else g evaluates to f if f matches the input stream; else it evaluates to g. (d) QRE $split-op(f, g)$ splits its input stream in two and evaluates f on the prefix and g on the suffix; the two results are then combined using operation op. (e) QRE $iter[p](f)$ iteratively applies f on substreams that match it, analogously to the Kleene-$*$ operation for REs. Results are passed between iterations using parameter p. (f) QRE $f \gg g$ feeds the output of QRE f into QRE g as f is being computed.

The declarative nature of QREs will be important when writing complex algorithms, without having to explicitly maintain state and low-level data flows. But as with any new language, QREs require some care in their usage. Space limitations preclude us from giving the formal definition of QREs. Instead, we will describe what each QRE does in the context of peak detection to give the reader a good idea of their ease of use and capabilities. Figure 2 illustrates how

QREs are defined and what they compute. Readers familiar with QREs will notice that, when writing the QRE expressions, we occasionally sacrifice strict syntactic correctness for the sake of presentation clarity.

5   QRE Implementation of Peak Detectors

We now describe the QREs that implement peak detectors WPM and WPB of Sect. 3.2. It is emphasized that even complicated procedures such as these two algorithms can be described in a declarative fashion using QREs, without resorting to a programming language or explicitly storing state, etc.

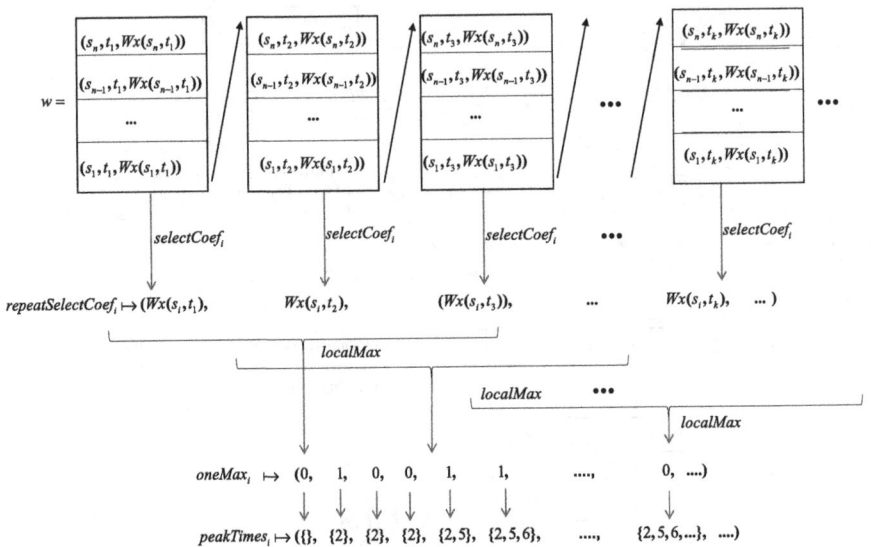

Fig. 3. QRE peakWPM

5.1   QRE for WPM

A numerical implementation of a CWT returns a discrete set of coefficients. Let $s_1 < s_2 < \ldots < s_n$ be the analysis scales and let $t_1, t_2, \ldots$ be the signal sampling times. Recall that a QRE views its input as a stream of incoming data items. A data item for WPM is $d = (s_i, t_j, |W_x(s_i, t_j)|) \in D := (\mathbb{R}_+)^3$. We use $d.s$ to refer to the first component of d, and $d.|W_x(s, t)|$ to refers to its last component. The input stream $w \in D^*$ is defined by the values from the spectrogram organized in a column-by-column fashion starting from the highest scale:

$$w = \underbrace{(s_n, t_1, |W_x(s_n, t_1)|), \ldots, (s_1, t_1, |W_x(s_1, t_1)|)}_{w_{t_1}} \ldots$$

$$\ldots \underbrace{(s_n, t_m, |W_x(s_n, t_m)|), \ldots, (s_1, t_m, |W_x(s_1, t_m)|)}_{w_{t_m}}$$

Let s_σ, $1 \leq \sigma \leq n$, the the scale that equals $\bar{s}$. Since the scales $s_i > s_\sigma$ are not relevant for peak detection (their frequency is too low), they should be discarded from w. Now, for each scale s_i, $i \leq \sigma$, we would like to find those local maxima of $|W_x(s_i, \cdot)|$ that are larger than threshold p_i[1]. We build the QRE peakWPM bottom-up as follows. In what follows, $i = 1, \ldots, \sigma$. See Fig. 3.

- QRE selectCoef$_i$ selects the wavelet coefficient magnitude at scale s_i from the incoming spectrogram column w_t. It must first wait for the entire column to arrive in a streaming fashion, so it matches n data items (recall there are n items in a column – see Fig. 3) and returns as cost $d.|W_x(s_i, t)|$.

$$\mathsf{selectCoef}_i := (d_n d_{n-1} \ldots d_1? \, d_i.|W_x(s_i, t)|).$$

- QRE repeatSelectCoef$_i$ applies selectCoef$_i$ to the latest column w_t. To do so, it splits its input stream in two: it executes selectCoef$_i$ on the last column, and ignores all columns that preceded it using $(d^n)^*$. It returns the selected coefficient $|W_x(s_i, t)|$ from the last column.

$$\mathsf{repeatSelectCoef}_i := split-right((d^n)^*, \mathsf{selectCoef}_i)$$

Combinator $split-right$ returns the result of operating on the right-hand side of the split, i.e. the suffix.

- QRE localMax$_i$ matches a string of real numbers of length at least 3: $r_1 \ldots r_{k-2} r_{k-1} r_k$. It returns the value of r_{k-1} if it is larger than r_k and r_{k-2}, and is above some pre-defined threshold p_i; otherwise, it returns 0. This will be used to detect local maxima in the spectrogram in a moving-window fashion. In detail:

$$\mathsf{localMax}_i := split-right(\mathbb{R}^*?0, \mathsf{LM}_3) \qquad (2)$$

localMax$_i$ splits the input string in two: the prefix is matched by $\mathbb{R}^*$ and is ignored. The suffix is matched by QRE LM$_3$: LM$_3$ matches a length-three string and simply returns 1 if the middle value is a local maximum that is above p_i, and returns zero, otherwise.

- QRE oneMax$_i$ feeds outputs of QRE repeatSelectCoef$_i$ to the QRE localMax$_i$.

$$\mathsf{oneMax}_i := \mathsf{repeatSelectCoef}_i \gg \mathsf{localMax}_i$$

Thus, oneMax$_i$ "sees" a string of coefficient magnitudes $|W_x(s_i, t_1)|$, $|W_x(s_i, t_2)|, \ldots$ generated by (streaming) repeatSelectCoef$_i$, and produces a 1 at the times of local maxima in this string.

- QRE peakTimes$_i$ collects the times of local maxima at scale s_i into one set.

$$\mathsf{peakTimes}_i := \mathsf{oneMax}_i \gg \mathsf{unionTimes}$$

It does so by passing the string of 1s and 0s produced by oneMax$_i$ to unionTimes. The latter counts the number of 0s separating the 1s and puts that in a set $\mathcal{M}_i$. Therefore, after k columns w_t have been seen, set $\mathcal{M}_i$ contains all local maxima at scale s_i which are above p_i in those k columns.

---

[1] $p_\sigma = \bar{p}$, $p_{i<\sigma} = 0$, since we threshold only the spectrogram values at scale $\bar{s}$. After this initial thresholding, tracing of maxima lines returns the peaks.

- QRE peakWPM is the final QRE. It combines results obtained from scales s_σ down to s_1:

$$\text{peakWPM} := conn_\delta(\text{peakTimes}_\sigma, ..., \text{peakTimes}_1)$$

Operator $conn_\delta$[2] checks if the local maxima times for each scale (produced by peakTimes_i) are within a δ of the maxima at the previous scale.

In summary, the complete QRE is given top-down by:

$$\text{peakWPM} := conn_\delta(\text{peakTimes}_\sigma, ..., \text{peakTimes}_1)$$
$$\text{peakTimes}_i := \text{oneMax}_i \gg \text{unionTimes}$$
$$\text{oneMax}_i := \text{repeatSelectCoef}_i \gg \text{localMax}_i$$
$$\text{localMax}_i := split-right(\mathbb{R}^*?0, \text{LM}_3)$$
$$\text{repeatSelectCoef}_i := split-right((d^n)^*, \text{selectCoef}_i)$$
$$\text{selectCoef}_i := (d_n \ldots d_1? \, d.|W_x(s_i, t)|)$$

5.2 QRE Implementation of WPB

Peak characterization WPB of Sect. 3.2 is implemented as QRE peakWPB. See Fig. 4. The input data stream is the same as before.

- QRE oneMax_σ (defined as before) produces a string of 1s and 0s, with the 1s indicating local maxima at scale $\bar{s} = s_\sigma$.
- QRE oneBL matches one blanking duration, starting with the maximum that initiates it. Namely, it matches a maximum (indicated by a 1), followed by a blanking period of length BL samples, followed by any-length string without maxima (indicated by 0^*): $\text{oneBL} := (1 \cdot (0|1)^{BL} \cdot 0^*)$

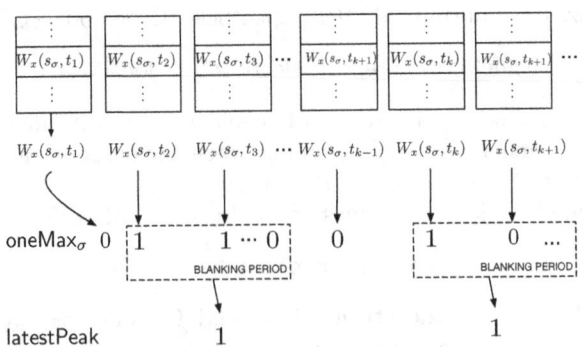

Fig. 4. QRE peakWPB

---

[2] Operator $conn_\delta$ can be defined recursively as follows: $conn_\delta(X, Y) = \{y \in Y : \exists x \in X : |x - y| \le \delta\}$, $conn_\delta(X_k, .., X_1) = conn_\delta(conn_\delta(X_k, .., X_2), X_1)$.

- QRE latestPeak will return a 1 at the time of the latest peak in the input signal: latestPeak $= split-right(\text{oneBL}^*?0, 1?1)$. It does so by matching all the blanking periods up to this point using oneBL* and ignoring them. It then matches the maximum (indicated by 1) at the end of the signal.
- QRE peakWPB feeds the string of 1s and 0s produced by oneMax$_\sigma$ to the QRE latestPeak: peakWPB $=$ oneMax$_\sigma \gg$ latestPeak

6   Experimental Results

We show the results of running peak detectors peakWPM and peakWPB on real patient data, obtained from a dataset of intra-cardiac electrograms. We also specified a peak detector available in a commercial ICD [22] as QRE peakMDT, and show the results for comparison purposes. The implementation uses an early version of the StreamQRE Java library [20]. Comparing the runtime and memory consumption of different algorithms (including algorithms programmed in QRE) in a consistent and reliable manner requires running a compiled version of the program on a particular hardware platform. No such compiler is available at the moment, so we don't report such performance numbers.

The results in this section should not be interpreted as definitively establishing the superiority of one peak detector over another, as this is not this paper's objective. Rather, the objective is to highlight the challenges involved in peak detection for cardiac signals, an essential signal-processing task in many medical devices. In particular, by highlighting how different detectors perform on different signals, it establishes the need for a formal (and empirical) understanding of their operation on classes of arrhythmias. This prompts the adoption of a formal description of peak detectors for further joint analysis with discrimination.

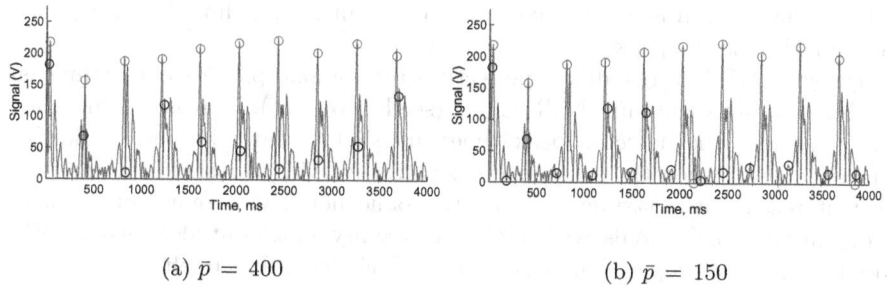

(a) $\bar{p} = 400$ (b) $\bar{p} = 150$

Fig. 5. peakWPM-detected peaks (red circles) and peakWPB-detected peaks (black circles) on a VT rhythm (Color figure online).

Figure 5 presents one rectified EGM signal of a Ventricular Tachycardia (VT) recorded from a patient. Circles (indicating detected time of peak) show the result of running peakWPM (red circles) and peakWPB (black circles). These results were obtained for $\bar{s} = 80$, $BL = 150$, and different values of $\bar{p}$. The first setting of $\bar{p}$ (Fig. 5 (a)) for both QREs was chosen to yield the best performance. This is akin to the way cardiologists set the parameters of commercial ICDs: they observe the signal, then set the parameters. We refer to this as the *nominal setting*. Ground-truth is obtained by having a cardiologist examine the signal and annotate the true peaks.

We first observe that the peaks detected by peakWPM match the ground-truth; i.e., the nominal performance of peakWPM yields perfect detection. This is not the case with peakWPB. Next, one can notice that the time precision of detected peaks with peakWPM is higher than with peakWPB due to maxima lines tracing down to the zero scale. Note also that the results of peakWPM are stable for various parameters settings. Improper thresholds $\bar{p}$ or scales $\bar{s}$ degrade the results only slightly (compare locations of red circles on Fig. 5 (a) with Fig. 5 (b)). By contrast, peakWPB detects additional false peaks (compare black circles in Figs. 5 (a) and (b)).

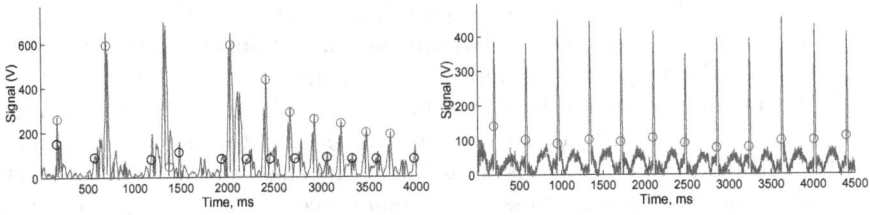

Fig. 6. WPM and peakMDT running on a VF rhythm (left) and peakMDT running on an NSR rhythm (right) (Color figure online).

Figure 6 (left) shows WPM (red circles) running on a Ventricular Fibrillation (VF) rhythm, which is a potentially fatal disorganized rhythm. Again, we note that WPM finds the peaks.

Detector MDT works almost perfectly with nominal parameters settings on any Normal Sinus Rhythm (NSR) signal (see Fig. 6 right). NSR is the "normal" heart rhythm. The detected peak times are slightly early because peakMDT declares a peak when the signal exceeds a time-varying threshold, rather than when it reaches its maximum. Using the same nominal parameters on more disorganized EGM signals with higher variability in amplitude, such as VF, does not produce proper results; see the black circles in Fig. 6 left.

7   Related Work

Signal Temporal Logic (STL) [17] was designed for the specification of temporal, real-time properties over real-valued signals and has been used in many applications including the differentiation of medical signals [4,7]. In [6], STL was augmented with a signal value *freeze operator* that allows one to express oscillatory patterns, but it is not possible to use it to discriminate oscillations within a particular frequency range. The spectrogram of a signal can be represented as a 2D map (from time and scale to amplitude) and one may think to employ a spatial-temporal logic such as SpaTeL [13] or Signal Spatio-Temporal Logic (SSTL) [21] on spectrograms. However, both of their underlying spatial models, graph structures for SSTL and quadtrees for SpaTeL, are not appropriate for this purpose. Logics for describing frequency and temporal properties have been proposed, including Time-Frequency Logic (TFL) in [11] and the approach in [8]. TFL is not sufficiently expressive for peak detection because it lacks the necessary mechanisms to quantify over variables or to freeze their values. Timed regular expressions [3, 24, 25] extend regular expressions by clocks and are expressively equivalent to timed automata, but cannot express the computations required for the tasks covered in this paper. Even the recent work proposed in [12] on measuring signals with timed patterns is not of help in our application, since it does not handle, neither in the specification nor in the measurement, the notion of local minima/maxima that is necessary for peak detection. Furthermore, the operator of measure is separated by the specification of the pattern to match.

SRV [9] is a stream runtime *verification* language that requires explicit encoding of relations between input and output streams, which is an awkward way of encoding the complex tasks of this paper. Moreover, unlike Boolean SRVs [5], QREs allow multiple unrestricted data types in intermediary computations and a number of their questions are decidable for these arbitrary types.

8   Conclusions and Future Work

The tasks of discrimination and peak detection, fundamental to arrhythmia-discrimination algorithms, are easily and succinctly expressible in QREs. One obvious limitation of QREs is that they only allow regular matching, though this is somewhat mitigated by the ability to chain QREs (though the streaming combinator ≫) to achieve more complex tasks. One advantage of programming in QREs is that it automatically provides us with a base implementation, whose time and memory complexity is independent of the stream length.

As future work, it will be interesting to compile a QRE into C or assembly code to measure and compare actual performance on a given hardware platform. Also, just like an RE has an equivalent machine model (DFA), a QRE has an equivalent machine model in terms of a deterministic finite-state transducer [1]. This points to an analysis of a QRE's correctness and efficiency beyond testing. Two lines of inquiry along these lines are promising in the context of medical devices.

Probabilistic analysis. Assume a probabilistic model of the QRE's input strings. For medical devices, such a model might be learned from data. We may then perform a statistical analysis of the output of the QRE under such an input model. In particular, we may estimate how long it takes the ICD to detect a fatal arrhythmia, or the probability of an incorrect detection by the ICD.

Energy calculations. We may compute the energy consumption of an algorithm that is expressed as a QRE, by viewing consumption as another quantity computed by the QRE. Alternatively, we may label the transitions of the underlying DFA by "energy terms", and leverage analysis techniques of weighted automata to analyze the energy consumption. Energy considerations are crucial to implanted medical devices that must rely on a battery, and which require surgery to replace a depleted battery.

Acknowledgments. The authors would like to thank Konstantinos Mamouras for insightful discussions about QREs and for providing the QRE Java library we used in this paper. This work is supported in part by AFOSR Grant FA9550-14-1-0261 and NSF Grants IIS-1447549, CNS-1446832, CNS-1446664, CNS-1445770, and CNS-1445770, and Austrian National Research Network grants S 11405-N23 and S 11412-N23 (RiSE/SHiNE of the FWF) and the ICT COST Action IC1402 Runtime Verification beyond Monitoring.

References

1. Alur, R., Fisman, D., Raghothaman, M.: Regular programming for quantitative properties of data streams. In: Thiemann, P. (ed.) ESOP 2016. LNCS, vol. 9632, pp. 15–40. Springer, Heidelberg (2016). doi:10.1007/978-3-662-49498-1_2
2. Alur, R., Freilich, A., Raghothaman, M.: Regular combinators for string transformations. In: Proceedings of the Joint Meeting of the Twenty-Third EACSL Annual Conference on Computer Science Logic (CSL) and the Twenty-Ninth Annual ACM/IEEE Symposium on Logic in Computer Science (LICS), CSL-LICS 2014, pp. 9:1–9:10. ACM, New York (2014)
3. Asarin, E., Caspi, P., Maler, O.: Timed regular expressions. J. ACM **49**(2), 172–206 (2002)
4. Bartocci, E., Bortolussi, L., Sanguinetti, G.: Data-driven statistical learning of temporal logic properties. In: Legay, A., Bozga, M. (eds.) FORMATS 2014. LNCS, vol. 8711, pp. 23–37. Springer, Cham (2014). doi:10.1007/978-3-319-10512-3_3
5. Bozzelli, L., Sánchez, C.: Foundations of boolean stream runtime verification. In: Bonakdarpour, B., Smolka, S.A. (eds.) RV 2014. LNCS, vol. 8734, pp. 64–79. Springer, Cham (2014). doi:10.1007/978-3-319-11164-3_6
6. Brim, L., Dluhos, P., Safránek, D., Vejpustek, T.: STL-*: extending signal temporal logic with signal-value freezing operator. Inf. Comput. **236**, 52–67 (2014)
7. Bufo, S., Bartocci, E., Sanguinetti, G., Borelli, M., Lucangelo, U., Bortolussi, L.: Temporal logic based monitoring of assisted ventilation in intensive care patients. In: Margaria, T., Steffen, B. (eds.) ISoLA 2014. LNCS, vol. 8803, pp. 391–403. Springer, Heidelberg (2014). doi:10.1007/978-3-662-45231-8_30
8. Chakarov, A., Sankaranarayanan, S., Fainekos, G.: Combining time and frequency domain specifications for periodic signals. In: Khurshid, S., Sen, K. (eds.) RV 2011. LNCS, vol. 7186, pp. 294–309. Springer, Heidelberg (2012). doi:10.1007/978-3-642-29860-8_22
9. D'Angelo, B., Sankaranarayanan, S., Sánchez, C., Robinson, W., Finkbeiner, B., Sipma, H.B., Mehrotra, S., Manna, Z.: LOLA: runtime monitoring of synchronous systems. In: Proceedings of the 12th International Symposium of Temporal Representation and Reasoning (TIME 2005), pp. 166–174. IEEE Computer Society Press (2005)

10. Demri, S., Lazic, R., Nowak, D.: On the freeze quantifier in constraint LTL: decidability and complexity. Inf. Comput. **205**(1), 2–24 (2007)
11. Donzé, A., Maler, O., Bartocci, E., Nickovic, D., Grosu, R., Smolka, S.A.: On temporal logic and signal processing. In: Chakraborty, S., Mukund, M. (eds.) ATVA 2012. LNCS, vol. 7561, pp. 92–106. Springer, Heidelberg (2012). doi:10.1007/978-3-642-33386-6_9
12. Ferrère, T., Maler, O., Ničković, D., Ulus, D.: Measuring with timed patterns. In: Kroening, D., Păsăreanu, C.S. (eds.) CAV 2015. LNCS, vol. 9207, pp. 322–337. Springer, Cham (2015). doi:10.1007/978-3-319-21668-3_19
13. Haghighi, I., Jones, A., Kong, Z., Bartocci, E., Grosu, R., Belta, C.: Spatel: a novel spatial-temporal logic and its applications to networked systems. In: Proceedings of HSCC 2015: The 18th International Conference on Hybrid Systems: Computation and Control, pp. 189–198. ACM (2015)
14. Harel, E., Lichtenstein, E., Pnueli, A.: Explicit clock temporal logic. IEEE (1990)
15. Koymans, R.: Specifying real-time properties with metric temporal logic. Real-Time Syst. **2**(4), 255–299 (1990)
16. Krishna, S.N., Madnani, K., Pandya, P.K.: Metric temporal logic with counting. In: Jacobs, B., Löding, C. (eds.) FoSSaCS 2016. LNCS, vol. 9634, pp. 335–352. Springer, Heidelberg (2016). doi:10.1007/978-3-662-49630-5_20
17. Maler, O., Nickovic, D.: Monitoring temporal properties of continuous signals. In: Lakhnech, Y., Yovine, S. (eds.) FORMATS/FTRTFT -2004. LNCS, vol. 3253, pp. 152–166. Springer, Heidelberg (2004). doi:10.1007/978-3-540-30206-3_12
18. Mallat, S., Hwang, W.L.: Singularity detection and processing with wavelets. IEEE Trans. Inf. Theor. **38**(2), 617–643 (1992)
19. Mallat, S.G.: A Wavelet Tour of Signal Processing, Third Edition: The Sparse Way. Academic Press, Amsterdam (2008)
20. Mamouras, K., Raghothaman, M., Alur, R., Ives, Z., Khanna, S.: StreamQRE: modular specification and efficient evaluation of quantitative queries over streaming data. In: Proceedings of 38th ACM SIGPLAN Conference on Programming Language Design and Implementation, pp. 693–708 (2017)
21. Nenzi, L., Bortolussi, L., Ciancia, V., Loreti, M., Massink, M.: Qualitative and quantitative monitoring of spatio-temporal properties. In: Bartocci, E., Majumdar, R. (eds.) RV 2015. LNCS, vol. 9333, pp. 21–37. Springer, Cham (2015). doi:10.1007/978-3-319-23820-3_2
22. Stroobandt, R.X., Barold, S.S., Sinnaeve, A.F.: Implantable Cardioverter - Defibrillators Step by Step. Wiley, Hoboken (2009)
23. Swerdlow, C.D., Asirvatham, S.J., Ellenbogen, K.A., Friedman, P.A.: Troubleshooting implanted cardioverter defibrillator sensing problems I. Circ. Arrhythm. Electrophysiol. **7**(6), 1237–1261 (2014)
24. Ulus, D.: MONTRE: a tool for monitoring timed regular expressions. In: Majumdar, R., Kunĉak, V. (eds.) CAV 2017. LNCS, vol. 10426. Springer, Cham (2017). doi:10.1007/978-3-319-63387-9_16
25. Ulus, D., Ferrère, T., Asarin, E., Maler, O.: Timed pattern matching. In: Legay, A., Bozga, M. (eds.) FORMATS 2014. LNCS, vol. 8711, pp. 222–236. Springer, Cham (2014). doi:10.1007/978-3-319-10512-3_16

Detecting Attractors in Biological Models with Uncertain Parameters

Jiří Barnat, Nikola Beneš[✉], Luboš Brim, Martin Demko, Matej Hajnal, Samuel Pastva, and David Šafránek

Systems Biology Laboratory, Faculty of Informatics, Masaryk University, Botanická 68a, 602 00 Brno, Czech Republic
{barnat,xbenes3,brim,xdemko,xhajnal,xpastva,safranek}@fi.muni.cz

Abstract. Complex behaviour arising in biological systems is typically characterised by various kinds of attractors. An important problem in this area is to determine these attractors. Biological systems are usually described by highly parametrised dynamical models that can be represented as parametrised graphs typically constructed as discrete abstractions of continuous-time models. In such models, attractors are observed in the form of terminal strongly connected components (tSCCs). In this paper, we introduce a novel method for detecting tSCCs in parametrised graphs. The method is supplied with a parallel algorithm and evaluated on discrete abstractions of several non-linear biological models.

1 Introduction

Biological systems as understood in systems biology are considered to be complex dynamical systems with a large extent of non-linear interactions. Interactions among systems components have the form of negative or positive feedback, the interplay of which can cause hardly predictable or even chaotic behaviour to emerge. In general, long-term systems behaviour may be significantly affected by the coexistence of dozens of complex and concurrent flows of information. For example, the irreversible decision processes observed in cell-cycle [24] or tissue development [18] arise from feedback loops that allow the cell to stabilise in several significantly different states each implying a unique phenotype.

Some of the problems related to the study of systems dynamics, which initially appear extremely complicated, can be greatly simplified if we concentrate on their *long-term behaviour*, i.e. what happens eventually. This idea finds its mathematical expression in the concept of an *attractor*.

Attractors can be seen as a special type of a portrait in the phase space. Points in a phase space represent the value of each of the system's variables at each moment of time. As the system changes over time, the data points make up a trajectory. Trajectories can be arranged into a phase portrait. Certain phase portraits then display attractor(s) as the long-term stable sets of points of the

This work has been supported by the Czech Science Foundation grant GA15-11089S and by the Czech National Infrastructure grant LM2015055.

J. Feret and H. Koeppl (Eds.): CMSB 2017, LNBI 10545, pp. 40–56, 2017.
DOI: 10.1007/978-3-319-67471-1_3

dynamical system, i.e. the locations in the phase portrait towards which the system's dynamics are attracted after transient phenomena have died down.

Attractors can reveal important information about the causal elements operative in a system, e.g. that the variables are non-linearly related to one another and so forth. That is why a quantitative and qualitative study of the geometrical, topological, and other properties of the attractors can yield deep insights into the system's dynamics.

Complex behaviour arising in biological systems is thus typically characterised by various kinds of attractors. An important problem of systems analysis is to determine the number and position of attractors. Biological systems are usually described by highly parametrised differential equations that can be approximated and abstracted by discrete systems [3,11,16]. In discrete systems, the most typical attractors can be observed in the form of terminal strongly connected components (tSCCs) [23]. We use this fact to provide an *efficient parallel* algorithm for automatised detection of such attractors in discrete models and in discrete abstractions of continuous models of dynamical systems. Alternatively, we could use a *general* method based on model checking for identifying non-trivial phase portraits in systems dynamics [4]. That general method needs to employ a hybrid temporal logic for which the algorithm is significantly more computationally demanding. This is a motivation to focus on a specific method targeting attractors.

Our Contribution. We introduce a novel approach for detection of attractors in parametrised systems in which attractors are understood as tSCCs in graphs representing the systems dynamics. We provide a parallel efficient algorithm for detecting tSCCs in parametrised graphs. We supply the method with a set of heuristics improving expected computation times. We evaluate the algorithm on several non-linear biological models. We additionally provide efficient algorithm variants for the simpler problems of tSCC counting and for deciding the question whether the parametrised graph has at least a given number tSCC.

Related Work. The existing solutions to attractor detection in non-linear continuous models are typically based on numerical methods working in two-dimensional systems while higher-dimensional systems remain a challenge [15]. In the case of discrete models, the problem is directly reduced to identification of SCCs which can be done efficiently for higher-dimensional systems in the non-parametrised case [8,9,17]. Owing to the fact that parameter space of a biological system explodes combinatorially with the arity of component influences, parameter uncertainty results in enormously large sets of parameter values. To that end, attractor detection in parametrised models remains to be a grand challenge in general.

On the technical side, our algorithm adapts the known parallel algorithms [1] to the parametrised setting and adds the possibility to accelerate the computation if only the number of tSCCs is requested without the need to enumerate the attractors. Moreover, this paper shows that exploiting the projection of parameters to the system dynamics gives an advantage of significantly faster computation than that achievable with a naïve execution of Tarjan's algorithm for SCC decomposition [25] scanning all parameter values one-by-one.

2   Methods

In the following, we will consider parametrised graphs which are directed graphs with self-loops allowed and edges labelled by parameters taken from a given parameter set.

Definition 1. *A* parametrised graph *is a triple* $G = (V, E, \mathbb{P})$ *where V is a finite set of* vertices, $\mathbb{P}$ *is a set of* parameter valuations *and* $E \subseteq V \times \mathbb{P} \times V$ *is the set of parametric edges. We write* $u \xrightarrow{p} v$ *instead of* $(u, p, v) \in E$ *when E is clear from the context. For a set of parameters* $P \subseteq \mathbb{P}$, *we also write* $u \xrightarrow{P} v$ *to denote that* $P = \{p \in \mathbb{P} \mid u \xrightarrow{p} v\}$. *For every* $p \in \mathbb{P}$, *the* restriction *of G on p is the graph* $G_p = (V, E_p)$ *where* $E_p = \{(u, v) \mid (u, p, v) \in E\}$.

Note that the $\xrightarrow{P}$ notation allows us to see a parametrised graph as an edge-labelled graph whose edges are labelled by sets of parameter valuations. The sets of parameter valuations can be possibly encoded in a symbolic way (say, using interval representation or formulae of a suitable logic).

In order to define our main problem, we now define the attractors of a graph. In general dynamical systems theory an attractor [20] is the smallest set of states (points in the phase space) invariant under the dynamics. Here we consider a discrete abstraction of a dynamical system in the form of a parametrised graph in which the dynamics is represented using paths. The respective abstraction of the notion of an attractor coincides with the notion of a terminal strongly connected component (tSCC) of a graph. We will thus use the notions of an attractor and a tSCC interchangeably.

Definition 2. *Let* $G = (V, E)$ *be a directed graph. We say that a vertex* $t \in V$ *is* reachable *from a vertex* $s \in V$ *if* $(s, t) \in E^*$ *where E^* denotes the reflexive and transitive closure of E.*

A set of vertices $C \subseteq V$ *is* strongly connected, *if for any two vertices* $u, v \in C$, *we have that v is reachable from u. A* strongly connected component (SCC) *is a* maximal *strongly connected set* $C \subseteq V$, *i.e. such that no C' with* $C \subsetneq C' \subseteq V$ *is strongly connected. A strongly connected component C is* trivial *if C is made of a single vertex c and* $(c, c) \notin E$, *and is* non-trivial *otherwise. Furthermore, C is called* terminal (tSCC) *if* $(C \times (V \setminus C)) \cap E = \emptyset$.

In graph theory, tSCCs are also called knots [7,13], with the minor difference that some authors require a knot to have at least two vertices.

We are now ready to state the algorithmic problem we are interested in.

Problem 1 (tSCCs Detecting Problem). Let $G = (V, E, \mathbb{P})$ be a parametrised graph. Our goal is to enumerate, for every parameter valuation $p \in \mathbb{P}$, all tSCCs in the graph G_p, the restriction of G on p.

We may also sometimes be interested in certain simpler versions of the problem. In the *counting* version, we are only interested in the number of tSCCs, i.e. we want to compute the function $c : \mathbb{P} \rightarrow \mathbb{N}$ that assigns to each parameter valuation p the number of tSCCs in G_p. In the *threshold* version, we are given a threshold number of tSCCs and want to partition the set of parameter valuations $\mathbb{P}$ into those for which G_p contains at least the given number of tSCCs and those for which it does not. Finally, in the *existential threshold* version, we only aim to decide whether there exists a parameter valuation p for which G_p contains at least the given number of tSCCs.

2.1   Algorithm

Our goal is to develop a parallel (shared-memory or distributed-memory) algorithm for solving the tSCCs Detecting Problem. A simple *sequential* solution to the problem is to use any reasonable SCC decomposition algorithm (e.g. Tarjan's [25]) and enumerate the terminal components in the residual graph. However, all known optimal sequential SCC decomposition algorithms use the depth-first search algorithm, which is suspected to be non-parallelisable [22]. There are known parallel SCC decomposition algorithms; for a survey we refer to [1]. Our approach here, however, is based on the observation that we do not have to compute all the SCCs in order to enumerate the terminal ones. Furthermore, instead of scanning through all parameter valuations and solving the problem for every one of them separately our approach deals with sets of parameter valuations directly. This makes our algorithm suitable for use in connection with various kinds of symbolic set representations.

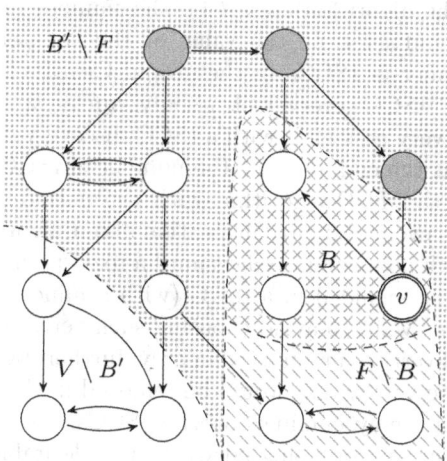

Fig. 1. Illustration of the non-parametrised version of our algorithm.

The main idea of the Terminal Component Detection (TCD) algorithm lies in repeated reachability, which is known to be easily parallelisable. To explain the idea we start with a non-parametrised version of the algorithm first. The following explication is illustrated in Fig. 1. Let us assume a given (non-parametrised) graph $G = (V, E)$. We initialise a tSCC counter to 1: clearly, every graph has at least one tSCC. We choose an arbitrary vertex $v \in V$ (denoted by the double circle in the illustration) and compute all vertices reachable from v; let us call the resulting set of vertices F. We further compute the set of all vertices backwards-reachable from v inside F; we call the resulting set B. Finally, we compute all vertices backwards-reachable from any vertex of F; let us call this set B'.

There are several observations to be made at this point. Clearly, B is an SCC of the graph and moreover, it is a terminal SCC iff $F \setminus B$ is empty. Furthermore, $B' \setminus F$ contains no tSCCs: all vertices in $B' \setminus F$ have a path to a vertex in F. We are thus in one of the following situations:

- $F \setminus B = \emptyset$, $V \setminus B' = \emptyset$: There are no further tSCCs and the algorithm ends.
- $F \setminus B \neq \emptyset$, $V \setminus B' = \emptyset$: We recursively run the algorithm in $F \setminus B$.
- $F \setminus B = \emptyset$, $V \setminus B' \neq \emptyset$: We have found one tSCC (namely, B) and there is at least one tSCC in $V \setminus B'$. We thus increase the counter by one and recursively run the algorithm in $V \setminus B'$.
- $F \setminus B \neq \emptyset$, $V \setminus B' \neq \emptyset$: Observe that no tSCC may intersect both $F \setminus B$ and $V \setminus B'$. We thus increase the counter by one (we know that there is one more tSCC) and recursively run the algorithm twice: in $F \setminus B$ and in $V \setminus B'$.

Note that when running the algorithm recursively twice, the two subgraphs are independent (there is no path from $F \setminus B$ to $V \setminus B'$ or vice versa) and the tasks can thus be run in parallel. The correctness of the algorithm is based on the following invariant. Let t be the number of concurrently running tasks (i.e. invocations of the algorithm), let d be the number of discovered tSCCs (see item 3 above), and let c be the value of the counter. The invariant is $t + d = c \leq$ the number of tSCCs in the graph. Clearly, the algorithm eventually discovers every tSCC: every task is only run on a subgraph known to contain a tSCC and every task ends with a tSCC discovery or a recursive run of another task. Thus, at the end, $d = c =$ the number of tSCCs in the graph.

We also enhance the TCD algorithm with the notion of *trimming* in the manner of [19]. Before every recursive invocation of our algorithm, we iteratively remove all vertices with no incoming edges until a fixed-point is reached. Clearly, such vertices cannot be included in a tSCC (with a minor technical exception, see below) and we thus want to avoid choosing such vertices as starting points. In Fig. 1, the removed vertices are marked in grey; furthermore, the $V \setminus B'$ part of the graph contains one vertex that would be removed in the next recursive run.

Note that trimming may eliminate trivial tSCCs, i.e. tSCCs consisting of a single vertex without any outgoing edges. If it is desirable to include trivial tSCCs in the output, we simply modify the trimming algorithm to check whether the eliminated vertices are tSCCs.

2.2 Parametrised Algorithm

We now extend the basic idea with parameter valuations. We first need to modify the reachability procedure to take parameter valuations into account. To be able to formulate the algorithm, we need a notion of parametrised sets of vertices.

Formally, a parametrised set of vertices $\widehat{A}$ is a function $\widehat{A} : V \to 2^{\mathbb{P}}$. We say that v is present in $\widehat{A}$ for $p \in \mathbb{P}$ if $p \in \widehat{A}(v)$. Clearly, whenever $\widehat{A}(v) = \emptyset$, this means that the vertex v is not present in the parametrised set for any parameter valuation. We call a parametrised set $\widehat{A}$ *empty* if $\widehat{A}(v) = \emptyset$ for all vertices v.

To deal with parametrised sets, we use a generalisation of the standard set operations. All the operations are performed element-wise, i.e. the union of parametrised sets $\widehat{A} \cup \widehat{B}$ is defined as the parametrised set $\widehat{C}$ such that $\widehat{C}(v) = \widehat{A}(v) \cup \widehat{B}(v)$ for all v; similarly for intersection and set difference.

To define the forward and backward reachable sets, we first need a notion of p-reachability. We say that v is p-reachable from s in $G|_{\widehat{V}}$ if the restricted graph G_p contains a path from s to v that only includes vertices that are present in $\widehat{V}$ for p. We then define $\mathcal{R}_{\widehat{V}}(s,v) = \{p \mid v \text{ is } p\text{-reachable from } s \text{ in } G|_{\widehat{V}}\}$. The reachability sets are then defined as follows: $\texttt{cfwd}(\widehat{V}, \widehat{X})$ denotes the parametrised set of vertices forward reachable from $\widehat{X}$ inside $\widehat{V}$ and similarly for $\texttt{cbwd}$.

$$\texttt{cfwd}(\widehat{V}, \widehat{X}) = \widehat{A}, \text{ where } \widehat{A}(v) = \widehat{V}(v) \cap \bigcup_{s \in V} \left(\widehat{X}(s) \cap \mathcal{R}_{\widehat{V}}(s,v) \right)$$

$$\texttt{cbwd}(\widehat{V}, \widehat{X}) = \widehat{A}, \text{ where } \widehat{A}(v) = \widehat{V}(v) \cap \bigcup_{s \in V} \left(\widehat{X}(s) \cap \mathcal{R}_{\widehat{V}}(v,s) \right)$$

Both functions can be effectively computed using a fixed-point algorithm.

Algorithm 1 shows the resulting parametrised algorithm. The basic idea is the one described above, extended with parametrised sets. There is, however, one key difference. It is not sufficient to select just one vertex as the basis for the first (forward) reachability. The reason is that as the algorithm proceeds, the investigated parametrised set of vertices may contain vertices with various incomparable sets of associated parameter valuations. Therefore, in the main part of the algorithm we collect several starting vertices with disjoint parameter valuation sets that together cover all parameter valuations that are present in the currently explored parametrised set of vertices.

The problem is illustrated in Fig. 2. We start with a parametrised graph with four vertices and two parameter valuations, depicted by the red (empty) and blue (filled) coloured dots. In the first iteration of the algorithm, $\widehat{V}$ consists of all the vertices with both parameter valuations. We select a starting vertex (depicted by a double circle) and compute $\widehat{F}$, which is in this case equal to $\widehat{B}$ as well as $\widehat{B'}$. The parametrised set $\widehat{V} \setminus \widehat{B'}$ is non-empty, we thus increase the counter for both parameter valuations. Note that the counter also has to be parametrised.

In the next iteration of the algorithm, let us first assume that we would only select a single vertex (see the second row in Fig. 2). Let us thus select t and compute $\widehat{F}$ again. It is again equal to $\widehat{B}$ and $\widehat{B'}$; this means that $\widehat{V} \setminus \widehat{B'}$ is again non-empty. In this case, however, it would be an error to increase the counter for

```
1  procedure init(G = (V, E, ℙ))
2  │   count_p ← 1 for all p ∈ ℙ
3  │   V̂ ← [∀v ∈ V : v ↦ ℙ]
4  │   main(V̂)

5  procedure main(V̂)
6  │   trim V̂
7  │   P ← {p ∈ ℙ | ∃u : p ∈ V̂(u)}
8  │   Ŝ ← [∀v ∈ V : v ↦ ∅]
9  │   while P is not empty do
10 │   │   choose v such that V̂(v) ∩ P ≠ ∅
11 │   │   add v ↦ V̂(v) ∩ P to Ŝ
12 │   │   P ← P \ V̂(v)
13 │   F̂ ← cfwd(V̂, Ŝ)
14 │   B̂ ← cbwd(F̂, Ŝ)
15 │   run in parallel
16 │   │   worker 1
17 │   │   │   P ← {parameters appearing in B̂ and not appearing in F̂ \ B̂}
18 │   │   │   B̂ restricted to p is a tSCC for all p ∈ P
19 │   │   │   main(F̂ \ B̂) if F̂ \ B̂ is not empty
20 │   │   worker 2
21 │   │   │   B̂' ← cbwd(V̂, F̂)
22 │   │   │   count_p ← count_p + 1 for all p occurring in V̂ \ B̂'
23 │   │   │   main(V̂ \ B̂') if V̂ \ B̂' is not empty
```

Algorithm 1. Parallel algorithm for tSCC counting in parametrised graphs.

the red parameter valuation as there are, in fact, only two tSCCs for each of the parameter valuations. We thus need to keep track of the parameter valuations of the selected vertex and if they do not cover all parameter valuations, we select another one. In the case of the example, it is correct to choose both t and u as starting vertices (see the third row).

The example also illustrates that the choice of the vertex on line 10 may influence the performance of the algorithm. Had we chosen v in the second iteration of the algorithm, no other vertices would be necessary. It might, however, be not always possible to find one vertex that covers all parameter valuations in $\widehat{V}$. Another issue is that a wrong choice of starting vertices may slice the set of parameter valuations into too many small subsets. In Sect. 4.1 we discuss and evaluate two vertex selection heuristics, one based on the cardinality of the parameter valuation set and another that aims to choose vertices close to tSCCs.

It remains to describe how the parametrised TCD algorithm proceeds with trimming and keeping the counter. The parametrised trimming works as follows: For every vertex v in $\widehat{V}$, we compute the set of parameter valuations under which v has no incoming edges. If all the sets are empty, the trimming is done.

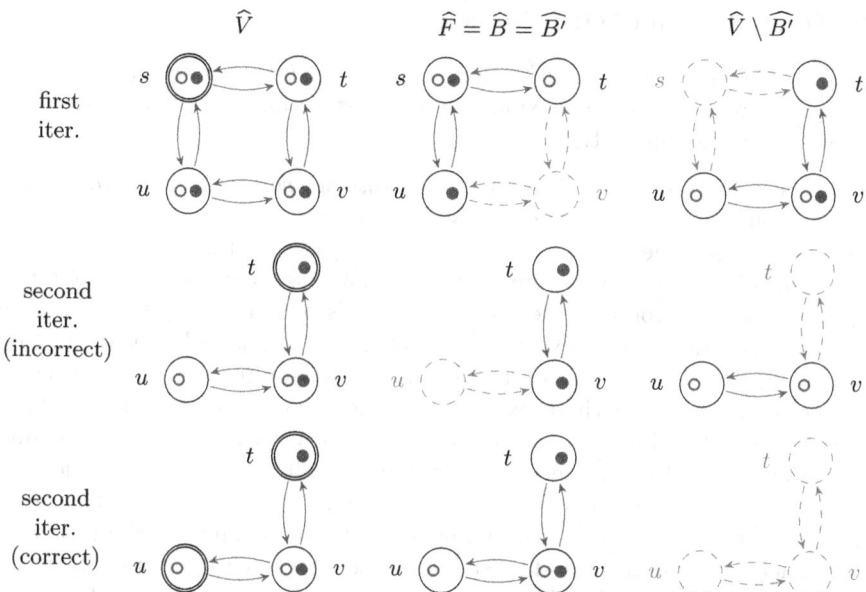

Fig. 2. Illustration of Algorithm 1. (Color Figure Online)

Otherwise, we remove the parameter valuations from $\widehat{V}$ and repeat the process. As for the counter, instead of a single number, we use a mapping $\mathbb{P} \rightarrow \mathbb{N}$ that assigns to each parameter valuation the number of tSCCs in its induced graph. The actual implementation of the counter depends on the (symbolic) representation of the parameter valuations and is discussed in Sect. 4.1.

Note that the algorithm as presented in Algorithm 1 solves both the tSCCs Detecting Problem and its counting version. If only the *counting* version is considered, we simply remove lines 17 and 18. Furthermore, if we are only interested in the *threshold* version of the problem, we may stop considering all parameter valuations p for which $\mathtt{count}_p$ has already reached the threshold. Moreover, in the *existential threshold* version, we stop the whole algorithm once any parameter valuation has reached the threshold.

3 Applications

We apply the method to several models used in systems biology. Since most of the existing and widely used models are represented by means of ordinary differential equations (ODEs), we employ the piece-wise multi-affine approximation [16] and rectangular abstraction procedures [2,11] to obtain a discrete representation of the systems dynamics in the form of a finite parametrised graph.

3.1   Discretisation of ODE Models

In this section, we briefly describe the format of the ODE models used and the subsequent procedures of approximation and abstraction that allow us to apply the method defined in Sect. 2.

Model. We consider $\mathbb{P} \subseteq \mathbb{R}^m_{\geq 0}$ as the *continuous parameter valuation space* of dimension m. A *biological model* $\mathcal{M}$ is given as a system of ODEs of the form $\dot{x} = f(x, \mu)$ where $x = (x_1, \ldots, x_n) \in \mathbb{R}^n_{\geq 0}$ is a vector of variables, $\mu = (\mu_1, \ldots, \mu_m)$ is a vector of parameters such that μ is evaluated in $\mathbb{P}$, and $f = (f_1, \ldots, f_n)$ is a vector where each component is a function constructed as a sum of reaction rates where every sum member is an affine or bi-linear function of x, or a sigmoidal function of x. An important requirement is that each f_i must be affine in μ and there exist no k, l such that $k \neq l$ and μ_k, μ_l both occur in some f_i. Moreover, we assume that every variable x_i has a bound denoted by x_{max_i}. In consequence, we require for all $p \in \mathbb{P}$ that no trajectory can exit the bounds. Formally, $\forall p \in \mathbb{P}, \forall i \in \{1, ..., n\} : (x_i = 0 \Rightarrow f_i(x, p) > 0) \wedge (x_i = max_i \Rightarrow f_i(x, p) < 0)$. Similarly to [2], we assume $\mathbb{P}$ includes *almost all* parameter valuations excluding singular cases for which some trajectory can slide along a threshold plane. In particular, any parameter valuation p for which some component of $f(x, p)$ can be zero on a boundary of some rectangle (as defined below) is not allowed. In consequence, a fixed point can appear only in a rectangle interior.

The restriction imposed on f covers mass action kinetics with stoichiometric coefficients not greater than one and any sigmoidal kinetics such as all significant variants of enzyme or Hill kinetics. Parameters must be independent and cannot appear in an exponent or a denominator of the kinetic function employed.

Approximation. To proceed with discretisation, the model $\dot{x} = f(x, \mu)$ has to satisfy the criterion that every f_i is piecewise multi-affine (PMA) in x. To transform the model into this form, we employ the approach defined in [16]. In particular, each sigmoidal function member in f_i is approximated with an optimal sequence of piecewise affine ramp functions. In this procedure, a finite number of thresholds is introduced for every component of x. The crucial factor of the approximation error is the number of piecewise affine segments. Though there is not yet a method that would somehow propagate the information on approximation error into the trajectories of the resulting PMA model, it has been shown on several case studies that the approximation does affect the system's vector field only negligibly [12, 16].

Abstraction. We employ the rectangular abstraction [3, 16]. We assume that we are given a set of thresholds $\{\theta^i_1, \ldots, \theta^i_{n_i}\}$ for each variable x_i satisfying $\theta^i_1 < \theta^i_2 < \cdots < \theta^i_{n_i}$. Each f_i is assumed to be multi-affine on each n-dimensional interval $[\theta^1_{j_1}, \theta^1_{j_1+1}] \times \cdots \times [\theta^n_{j_n}, \theta^n_{j_n+1}]$. We call these intervals rectangles. Each rectangle is uniquely identified via an n-tuple of numbers: $R(j_1, \ldots, j_n) = [\theta^1_{j_1}, \theta^1_{j_1+1}] \times \cdots \times [\theta^n_{j_n}, \theta^n_{j_n+1}]$, where the range of each j_i is $\{1, \ldots, n_i - 1\}$. We also define $VR(j_1, \ldots, j_n)$ to be the set of all vertices of $R(j_1, \ldots, j_n)$.

The abstraction results in a symbolic description of a parametrised graph, $G = (V, E, \mathbb{P})$ where $V = \{(j_1, \ldots, j_n) \mid \forall i : 1 \leq j_i < n_i\}$ such that each $v \in V$ represents the rectangle $R(v)$. The relation $u \xrightarrow{P} v$ is defined for a parameter valuations set $P \subseteq \mathbb{P}$ between any two nodes $u, v \in V$, $u \neq v$, for which $R(u) \cap R(v)$ forms an $(n-1)$-dimensional (hyper)rectangle ($R(u), R(v)$ are neighbouring in one dimension) and for which one of the following conditions holds:

- $\exists! j. v_j = u_j + 1$, $\forall i, i \neq j : v_i = u_i$ and for each $p \in P$ there exists $\hat{x} \in VR(u) \cap VR(v)$ satisfying $f_j(\hat{x}, p) > 0$;
- $\exists! j. v_j = u_j - 1$, $\forall i, i \neq j : v_i = u_i$ and for each $p \in P$ there exists $\hat{x} \in VR(u) \cap VR(v)$ satisfying $f_j(\hat{x}, p) < 0$.

Additionally, there is a self-loop defined for any $u \in V$ and a parameter valuations set $P \subseteq \mathbb{P}$ such that $\forall p \in P : \mathbf{0} \in hull\{f(\hat{x}, p) \mid \hat{x} \in VR(u)\}$.

Every edge is associated with a subset $P \subseteq \mathbb{P}$ of parameter values under which it is enabled. Finite number of thresholds implies finite number of distinct parameter sets that can appear on transitions in the model. Total number of parameter sets for an abstraction of model $\mathcal{M}$, denoted $|\mathbb{P}_{|\mathcal{M}}|$, is thus finite.

The rectangular abstraction approximates the existence of a fixed point in a rectangle. This is achieved conservatively by introducing reflexivity for every rectangle such that there is a zero vector included in the convex hull of all vertices of the rectangle. In other words, this is a necessary condition for the existence of a point where the derivatives in all coordinates are zero. In this setting, it has been shown that rectangular abstraction is conservative (overapproximation) with respect to almost all trajectories of the approximated (PMA) model [2].

The conservativeness of the abstraction and the consideration of only those parameter values for which the dynamics is bounded (cannot exit the interval $[\theta^i_{j_1}, \theta^i_{j_{n_i}}]$ for any $i \leq n$) together imply that *every tSCCs in the abstraction covers an attractor in the PMA system*. This implies that the number of discovered tSCCs in the abstraction is a lower bound for the number of attractors in the corresponding PMA system. To interpret the results for the original system, local linearisation of non-linear vector field preserves topological equivalence implying preservation of hyperbolic attractors [10]. For complex attractors, we are not aware of any relevant mathematical results leaving it open for future research.

3.2   Case Studies

To demonstrate the applicability and benefits of our approach, it is applied to three biological models. Two of them are motifs in genetic regulatory networks and the third is the main part of the cell cycle control in mammalian cells. Note that all the models in this section are PMA approximated models of the original ODEs. Parameter sets for which the method is able to run, *allowed parameters*, consist of independent parameters and parameters not nested in PMA system.

Bi-stable repressilator. The first model to be presented is the smallest repressilator motif, studied in [6,14]. It includes two nodes which inhibit each other

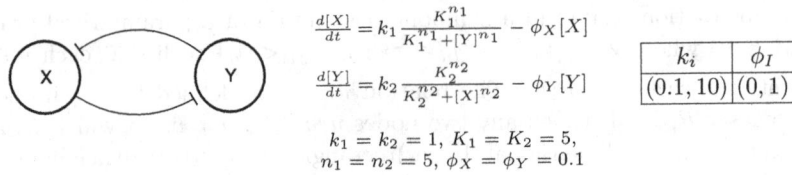

$$\frac{d[X]}{dt} = k_1 \frac{K_1^{n_1}}{K_1^{n_1} + [Y]^{n_1}} - \phi_X [X]$$

$$\frac{d[Y]}{dt} = k_2 \frac{K_2^{n_2}}{K_2^{n_2} + [X]^{n_2}} - \phi_Y [Y]$$

k_i	ϕ_I
$(0.1, 10)$	$(0, 1)$

$k_1 = k_2 = 1$, $K_1 = K_2 = 5$,
$n_1 = n_2 = 5$, $\phi_X = \phi_Y = 0.1$

Fig. 3. The bi-stable repressilator regulatory network (left) and its ODE model taken from [6] (middle). The parameters and their corresponding value intervals we have considered for all $i \in \{1, 2\}, I \in \{X, Y\}$ (right).

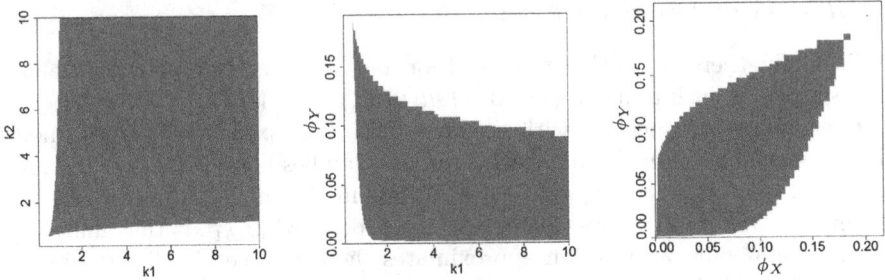

Fig. 4. The parameter space and the corresponding number of terminal components (one in white, two in green). The remaining parameter interval, which is not shown, exhibits one terminal component. Thanks to the symmetry of the model, there are only 3 pairs of allowed parameters (Color figure online).

(Fig. 3 left). In biology, this motif is very often present in gene regulatory networks, where X represents the product of *geneX* which inhibits the production of *geneY* and vice versa.

According to [21], there is a bistability in the model with parametrised ϕ_X. A bistability region has been discovered for $\phi_X \in (0.022, 0.119) \cup (0.120, 0.138)$ in [6]. Our algorithm has found a bistability region in $(0.014, 0.156)$ for parametrised ϕ_X. This extension of the parameter interval is caused by the presence of a non-trivial terminal component, instead of a *sink* [5].

Additionally, we have managed to analyse this model for all pairs of parameters allowed for the prototype implementation of the method (Fig. 4).

Tri-stable toggle switch. The tri-stable toggle switch is a model of a 3-variable repressilator in which each node inhibits not only one but both of its neighbours (Fig. 5 left). Just one of the two ingoing inhibitions is enough to repress any entity. Therefore the ODE model contains a multiplication of negative Hill functions in the entity regulation (Fig. 5 right).

We have analysed this model for all pairs of parameters allowed for the implementation. As predicted, the model shows tri-stability for specific parameter values (Fig. 6). Additionally, we have managed to analyse this model for a triple of parameters (ϕ_X, ϕ_Y, ϕ_Z) using a reduced state space.

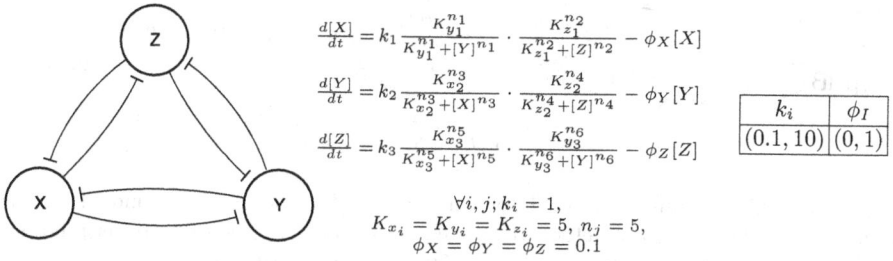

Fig. 5. The tri-stable toggle switch regulatory network (left) and its ODE model (middle). The parameter value intervals we have considered for all $i \in \{1, 2, 3\}, I \in \{X, Y, Z\}$ (right).

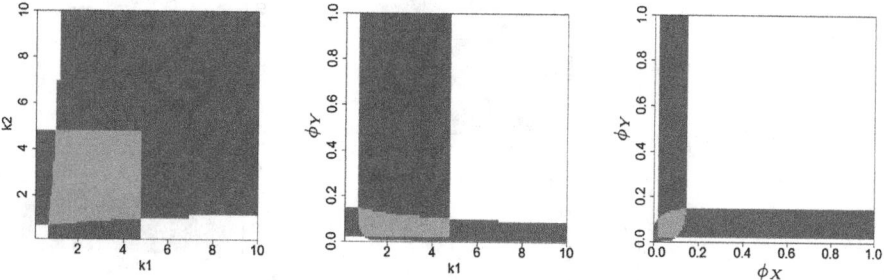

Fig. 6. The parameter space and the corresponding number of terminal components (one in white, two in green, three in blue). Thanks to the symmetry of the model, there are only 3 pairs of allowed parameters (Color figure online).

Regulation of the G_1/S Cell Cycle Transition. As the last case, we have investigated a well-known model representing the central module of the genetic regulatory network governing the G_1/S cell cycle transition in mammalian cells [24]. In particular, the model explains the mechanism behind the irreversible decision for cell division described by a two-gene regulatory network of interactions between the tumour suppressor protein pRB and the central transcription factor $E2F1$ (Fig. 7 left). In high concentration levels, $E2F1$ activates the G_1/S transition mechanism. In low concentration of $E2F1$, committing to S-phase is refused and that way the cell avoids DNA replication. For suitable parameter values, two distinct stable attractors exist. A numerical bifurcation analysis of $E2F1$ stable concentration depending on the degradation parameter of pRB (ϕ_{pRB}) has been provided in [24].

A bistability region has been discovered for $\phi_{pRB} \in [0.012, 0.0145]$ in [5]. Our algorithm has found a bistability region in $(0.010, 0.0146)$ for parametrised ϕ_{pRB}. This extension of parameter interval has the same reason as in the first case study—the presence of a non-trivial terminal component, instead of a *sink* [5].

Additionally, we have managed to analyse this model for all pairs of parameters allowed for the implementation (Fig. 8).

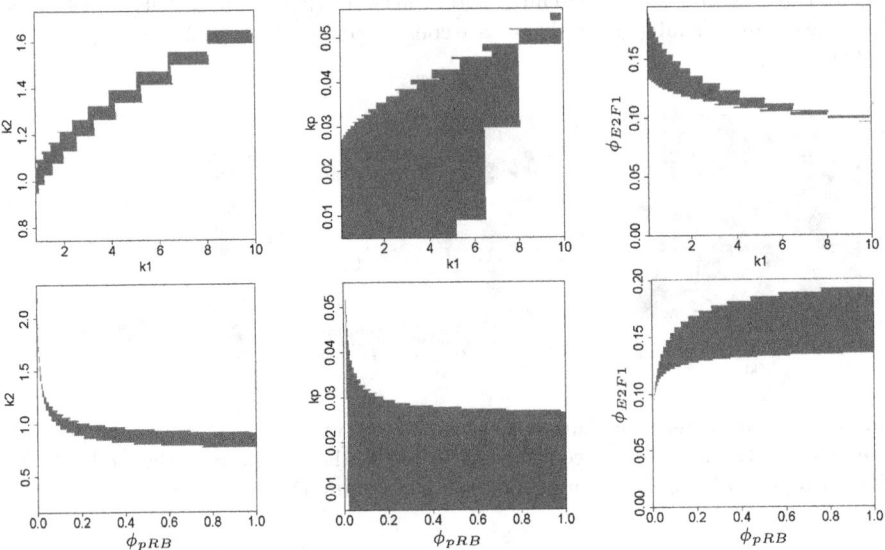

$$\frac{d[pRB]}{dt} = k_1 \frac{[E2F1]}{K_{m1}+[E2F1]} \frac{J_{11}}{J_{11}+[pRB]} - \phi_{pRB}[pRB]$$

$$\frac{d[E2F1]}{dt} = k_p + k_2 \frac{a^2+[E2F1]^2}{K_{m2}^2+[E2F1]^2} \frac{J_{12}}{J_{12}+[pRB]} - \phi_{E2F1}[E2F1]$$

$a = 0.04$, $k_1 = 1$, $k_2 = 1.6$, $k_p = 0.05$, $\phi_{E2F1} = 0.1$
$J_{11} = 0.5$, $J_{12} = 5$, $K_{m1} = 0.5$, $K_{m2} = 4$

Fig. 7. The G_1/S transition regulatory network (left) and its ODE model taken from [24] (right). The parameter value intervals we have considered are as follows: $k_1,(0.1, 10)$; $\phi_{pRB},(0, 1)$; $k_2,(0.16, 16)$; $k_p,(0.005, 0.5)$; $\phi_{E2F1},(0, 1)$.

Fig. 8. The parameter space and the corresponding number of terminal components (one in white, two in green). The remaining parameter interval, which is not shown, exhibits one terminal component (Color figure online).

4   Evaluation

We evaluate a prototype implementing the method from Sect. 2 in several aspects such as comparison with the naïve approach, scalability in model size, scalability in $|\mathbb{P}_{|\mathcal{M}|}|$, different algorithm types and the heuristics for the initial node selection. In this section, we use all biological models from Sect. 3 which are subject to approximation and abstraction described in that section. In addition, we employ a model of a four-stable switch—an extension of the tri-stable toggle switch with four genes where each of the four genes represses the others. It exhibits four different stable states. Moreover, there are the implementation details demanding deeper understanding used in this section which are described next. Note that all the time results in this section are in seconds and represent the average of four runs on a server with two eight-core processors (Intel Xeon X7560 2.26 GHz) and 448 GiB RAM.

Table 1. The second and third row represent the number of states and the number of parameter valuations for that particular model, respectively; B/R stands for the *Branch/Reach* algorithm.

Cores	Model1				Model2				Model3				Model4			
	720e3		320e3		1.5e6		750e3		125e3		75e3		390e3		280e3	
	7.1e6		2.8e6		5.5e6		1.8e6		160e6		57e6		255e6		124e6	
	B	R	B	R	B	R	B	R	B	R	B	R	B	R	B	R
2	842	377	651	176	798	886	326	340	1083	871	416	367	1516	1203	799	705
4	735	261	603	122	511	562	256	247	1000	683	391	301	1319	837	722	519
8	690	208	567	107	392	436	192	203	987	544	377	237	1271	723	680	473
12	706	237	595	120	471	477	219	222	1016	497	387	232	1299	718	702	464
16	719	237	590	119	459	467	216	216	1020	476	389	219	1281	704	706	456

Table 2. The second row represents $|\mathbb{P}_{|\mathcal{M}}|$, the number of parameter valuations for that particular model variant; B/R stands for the *Branch/Reach* algorithm.

Cores	Model1 (1250 states)				Model2 (1600 states)				Model3 (8e3 states)				Model4 (10e3 states)			
	1058e3		159e3		2.7e6		1.1e6		21e6		3.2e6		9.6e6		668e3	
	B	R	B	R	B	R	B	R	B	R	B	R	B	R	B	R
2	483	440	81	83	642	479	114	127	399	318	144	120	625	655	254	248
4	481	424	77	79	626	329	101	90	369	238	132	87	585	383	241	152
8	495	413	79	78	607	293	97	81	358	197	124	74	585	314	235	127
12	481	427	79	79	608	266	96	68	357	184	128	70	583	253	242	107
16	481	416	78	78	623	259	97	73	365	179	127	71	588	249	230	103

4.1 Implementation Details

In this section, we describe two algorithm variants: *Branch* and *Reach*. They differ in the form of the parallelism employed. The *Branch* algorithm runs two parallel workers each time the computation branches, as described in the pseudo-code. The *Reach* algorithm uses a parallel reachability procedure. Here, we describe some important implementation details of both algorithms:

Parameter representation. Due to the restrictions imposed on the model parameters in Sect. 3, we can represent each parameter set as a grid of disjoint hyper-rectangles. Each parameter set maintains its own grid which is refined or simplified as needed to maintain optimal resource usage.

Component counter. The mapping described in Sect. 2 is implemented as a list of disjoint parameter valuation sets. Intuitively, the set on position i contains all the parameter valuations for which i terminal components have been discovered so far.

Partition function. The *Reach* algorithm performs a partitioning of the state space in order to parallelise the reachability computation. To that end, we exploit the regular (rectangular) structure of our models and define a partitioning which splits the model into almost equally sized rectangular blocks. The number of blocks depends on the number of used cores.

Selection heuristics. We also compare three state selection heuristics. *None* is the naïve heuristics which selects the first available state in the set. The *CARD* heuristics tries to ensure that the symbolic parameter representation is well utilised during the computation. To that end, it selects the state with the highest parameter set cardinality as the initial state. Finally, the *CSTR* heuristics is designed to reduce the number of performed reachability computations by selecting states which are part of (or close to) the terminal components. Our observation is that a state is more likely to be a part of a terminal component if there are more transitions entering the state than leaving it. The *CSTR* heuristics therefore pre-computes this parametrised in/out ratio for all states and then uses it to select the best state. If two states agree on the in/out ratio, the parameter set cardinality is used to decide the winner, just as in the *CARD* heuristics.

4.2 Performance Evaluation

Comparison with the naïve approach. As the naïve approach we use Tarjan's SCC decomposition algorithm followed by the counting of terminal components; run once on G_p for each parameter valuation $p \in \mathbb{P}$. This approach was compared with the best results of our prototype for the same models. For the Bi-stable repressilator with 900 states and 866761 parameter valuations the naïve approach took 2520.46 s while our approach took 153.54 s. For the Tri-stable toggle switch with 1000 states and 436921 parameter valuations the naïve approach took 3815.66 s while our approach took 59.71 s.

Problem of the initial state. We analysed all heuristics on several models and for all cases using either *CSTR* or *CARD* was always a better option; sometimes even ten times faster. In some cases *CARD* was more efficient than *CSTR*.

Scalability in model size. These statistics were performed by both algorithm variants on 4 models each for 2 different sizes. Here, by size we mean the size of the state space together with $|\mathbb{P}_{|\mathcal{M}}|$. These cannot be separated because the size of $\mathbb{P}_{|\mathcal{M}}$ in this kind of models depends on the number of states due to the rectangular abstraction. In Table 1 you may observe that for the majority of models the *Reach* algorithm is the better option. For the models used we define abbreviations: *Model1* for the G_1/S switch, *Model2* for the bi-stable repressilator, *Model3* for the tri-stable switch and *Model4* for the four-stable switch.

Scalability in parameter space. These statistics were performed by both algorithm variants on the four previously mentioned models each for two differently sized parameter spaces with constant size of the state space. In Table 2 you may observe that for the majority of models the *Reach* algorithm is the better option.

5 Conclusion

The novel result of this paper is a parallel algorithm for the detection of terminal SCCs in parametrised graphs. The scalability of the algorithm has been analysed showing a significant speed-up w.r.t. the naïve approach using standard algorithms. We have shown that the algorithm can be sufficiently applied to detect attractors in dynamical systems. The case studies have shown the method can deal with two parameters in a reasonable time and even with three parameters in case of a smaller state space (for the tri-stable toggle switch model). The method provides a fully automated and parallel efficient alternative to traditional bifurcation analysis focused on multi-stability as in [24]. Note that the precision of the results is affected by settings of the approximation and abstraction procedures. Possible imprecisions can be observed as discontinuities in plotted results, see Fig. 4. This can be eliminated by manual fine-tuning of the approximation and abstraction. However, we have been primarily interested in the functionality of the algorithm here. Detailed study of the application aspects is left for future work.

References

1. Barnat, J., Chaloupka, J., Van De Pol, J.: Distributed algorithms for SCC decomposition. J. Logic Comput. **21**(1), 23–44 (2011)
2. Batt, G., Belta, C., Weiss, R.: Model checking genetic regulatory networks with parameter uncertainty. In: Bemporad, A., Bicchi, A., Buttazzo, G. (eds.) HSCC 2007. LNCS, vol. 4416, pp. 61–75. Springer, Heidelberg (2007). doi:10.1007/978-3-540-71493-4_8
3. Batt, G., Yordanov, B., Weiss, R., Belta, C.: Robustness analysis and tuning of synthetic gene networks. Bioinformatics **23**(18), 2415–2422 (2007)
4. Beneš, N., Brim, L., Demko, M., Pastva, S., Šafránek, D.: A model checking approach to discrete bifurcation analysis. In: Fitzgerald, J., Heitmeyer, C., Gnesi, S., Philippou, A. (eds.) FM 2016. LNCS, vol. 9995, pp. 85–101. Springer, Cham (2016). doi:10.1007/978-3-319-48989-6_6
5. Brim, L., Češka, M., Demko, M., Pastva, S., Šafránek, D.: Parameter synthesis by parallel coloured CTL model checking. In: Roux, O., Bourdon, J. (eds.) CMSB 2015. LNCS, vol. 9308, pp. 251–263. Springer, Cham (2015). doi:10.1007/978-3-319-23401-4_21
6. Brim, L., Demko, M., Pastva, S., Šafránek, D.: High-performance discrete bifurcation analysis for piecewise-affine dynamical systems. In: Abate, A., Šafránek, D. (eds.) HSB 2015. LNCS, vol. 9271, pp. 58–74. Springer, Cham (2015). doi:10.1007/978-3-319-26916-0_4
7. Chandy, K.M., Misra, J.: Distributed computation on graphs: shortest path algorithms. Commun. ACM **25**(11), 833–837 (1982)
8. Chatain, T., Haar, S., Jezequel, L., Paulevé, L., Schwoon, S.: Characterization of reachable attractors using petri net unfoldings. In: Mendes, P., Dada, J.O., Smallbone, K. (eds.) CMSB 2014. LNCS, vol. 8859, pp. 129–142. Springer, Cham (2014). doi:10.1007/978-3-319-12982-2_10
9. Choo, S.M., Cho, K.H.: An efficient algorithm for identifying primary phenotype attractors of a large-scale boolean network. BMC Syst. Biol. **10**(1), 95 (2016)
10. Coayla-Teran, E.A., Mohammed, S.E.A., Ruffino, P.R.C.: Hartman-grobman theorems along hyperbolic stationary trajectories. Discret. Contin. Dyn. Syst. **17**(2), 281–292 (2007)

11. Collins, P., Habets, L., van Schuppen, J., Černá, I., Fabriková, J., Šafránek, D.: Abstraction of biochemical reaction systems on polytopes. In: IFAC World Congress, pp. 14869–14875. IFAC (2011)
12. Demko, M., Beneš, N., Brim, L., Pastva, S., Šafránek, D.: High-performance symbolic parameter synthesis of biological models: a case study. In: Bartocci, E., Lio, P., Paoletti, N. (eds.) CMSB 2016. LNCS, vol. 9859, pp. 82–97. Springer, Cham (2016). doi:10.1007/978-3-319-45177-0_6
13. Dijkstra, E.W.: In reaction to Ernest Chang's "Deadlock Detection" (1979). http://www.cs.utexas.edu/users/EWD/ewd07xx/EWD702.PDF
14. Dilão, R.: The regulation of gene expression in eukaryotes: bistability and oscillations in repressilator models. J. Theor. Biol. **340**, 199–208 (2014)
15. Dudkowski, D., Jafari, S., Kapitaniak, T., Kuznetsov, N.V., Leonov, G.A., Prasad, A.: Hidden attractors in dynamical systems. Phys. Rep. **637**, 1–50 (2016)
16. Grosu, R., Batt, G., Fenton, F.H., Glimm, J., Le Guernic, C., Smolka, S.A., Bartocci, E.: From cardiac cells to genetic regulatory networks. In: Gopalakrishnan, G., Qadeer, S. (eds.) CAV 2011. LNCS, vol. 6806, pp. 396–411. Springer, Heidelberg (2011). doi:10.1007/978-3-642-22110-1_31
17. Guo, W., Yang, G., Wu, W., He, L., Sun, M.: A parallel attractor finding algorithm based on boolean satisfiability for genetic regulatory networks. PLOS ONE **9**(4), 1–10 (2014)
18. MacArthur, B.D., Ma'ayan, A., Lemischka, I.R.: Systems biology of stem cell fate and cellular reprogramming. Nat. Rev. Mol. Cell Biol. **10**(10), 672–681 (2009)
19. McLendon III, W., Hendrickson, B., Plimpton, S.J., Rauchwerger, L.: Finding strongly connected components in distributed graphs. J. Parallel Distrib. Comput. **65**(8), 901–910 (2005)
20. Milnor, J.: On the concept of attractor. Commun. Math. Phys. **99**(2), 177–195 (1985)
21. Müller, S., Hofbauer, J., Endler, L., Flamm, C., Widder, S., Schuster, P.: A generalized model of the repressilator. J. Math. Biol. **53**(6), 905–937 (2006)
22. Reif, J.H.: Depth-first search is inherently sequential. Inf. Process. Lett. **20**(5), 229–234 (1985). https://doi.org/10.1016/0020-0190(85)90024-9
23. Sullivan, D., Williams, R.: On the homology of attractors. Topology **15**(3), 259–262 (1976)
24. Swat, M., Kel, A., Herzel, H.: Bifurcation analysis of the regulatory modules of the mammalian G1/S transition. Bioinformatics **20**(10), 1506–1511 (2004)
25. Tarjan, R.E.: Depth-first search and linear graph algorithms. SIAM J. Comput. **1**(2), 146–160 (1972). https://doi.org/10.1137/0201010

Abduction Based Drug Target Discovery Using Boolean Control Network

Célia Biane and Franck Delaplace[✉]

IBISC, Univ Evry, Université Paris-Saclay, 91025 Evry, France
{celia.biane,franck.delaplace}@ibisc.univ-evry.fr

Abstract. A major challenge in cancer research is to determine the genetic mutations causing the cancerous phenotype of cells and conversely, the actions of drugs initiating programmed cell death in cancer cells. However, such a challenge is compounded by the complexity of the genotype-phenotype relationship and therefore, requires to relate the molecular effects of mutations and drugs to their consequences on cellular phenotypes. Discovering these complex relationships is at the root of new molecular drug targets discovery and cancer etiology investigation. In their elucidation, computational methods play a major role for the inference of the molecular causal actions from molecular and biological networks data analysis. In this article, we propose a theoretical framework where mutations and drug actions are seen as topological perturbations/actions on molecular networks inducing cell phenotype reprogramming. The framework is based on Boolean control networks where the topological network actions are modelled by control parameters. We present a new algorithm using abductive reasoning principles inferring the minimal causal topological actions leading to an expected behavior at stable state. The framework is validated on a model of network regulating the proliferation/apoptosis switch in breast cancer by automatically discovering driver genes and finding drug targets.

Keywords: Dynamical system reprogramming · Boolean control network · Abductive reasoning · Drug target prediction · Etiology of cancer

1 Introduction

In precision medicine, the discovery of causal genes and efficient drug targets is challenged by the complexity of the genotype-phenotype relationship. A key milestone in this challenge is the ability to understand how cell behaviour arises from the synergistic effect of local molecular interactions [32]. Accordingly, cells are envisioned as a web of macromolecular interactions constituting the "interactome" from which phenotype changes are explained by perturbations of molecular interactions [33]. At the molecular level, the phenotypic changes are assessed by the measure of the state of some molecules, called biomarkers, that are defined as observable and objective characteristics of biological processes. They are used

© Springer International Publishing AG 2017
J. Feret and H. Koeppl (Eds.): CMSB 2017, LNBI 10545, pp. 57–73, 2017.
DOI: 10.1007/978-3-319-67471-1_4

to assess the shift between normal and pathological conditions [31] and to predict the appropriate treatment [9]. Inferring, from the interactome, the molecular causes of phenotypic switches assessed by biomarkers will thus constitute the root for the development of efficient therapies, by predicting the actions at the molecular level directing cells from a diseased toward a healthy state.

In cancer, cells acquire phenotypes with characteristic cancerous hallmarks such as uncontrolled proliferative activity, apoptosis resistance and invasiveness [12]. These phenotypes are caused by multigenic mutations altering molecular interactions. Therefore, a preliminary issue concerns the definition of the effects of mutations on the interactome. In [38], the authors relate mutations to their network effect and introduce the notion of edgetic perturbations of molecular networks: nonsense mutation, out-of-frame insertion or deletion and defective splicing are interpreted as node or arc deletions whereas missense mutation and in-frame insertion or deletion can be modelled as node or arc addition. Moreover, in [7], the authors classify mutations according to the way they affect signalling networks and distinguish mutations that constitutively activate or inhibit enzymes and mutations that rewire the network interactions. The effect of mutations on molecular networks can thus be described as elementary topological actions of deletion or insertion of nodes and arcs. Symmetrically, targeted therapies switch cancer cells phenotype toward growth arrest and apoptosis. Their actions can also be interpreted as network rewiring [9]. A phenotypic switch following mutations or targeted therapies is therefore considered as the observable trait of a *dynamical system reprogramming* caused by *topological network actions* (TN-action).

The inference of TN-actions would provide major insights for etiological investigation of disease, molecular pathogenesis and drug targets prediction by assimilating them to the effects of causal gene mutations (*a.k.a*, drivers) or actions of drugs. In this endeavour, it is worth noticing that generate-and-test method checking the TN-actions exhaustively is often pointless. Indeed, assuming that an expected phenotypic switch results from the application of a specific gene action up to m amongst n genes, then the number of trials[1] equals $\sum_{k=1}^{m} \binom{n}{k}$. For example, the number of trials for targeting up to 10% on 100 genes exceeds 19 billions[2]. Hence, automatically inferring the TN-actions from observable effects is essential to meet this challenge. By considering biomarkers as the entry point of the inference, the issue thus refers to an inverse problem (*ie.*, causes discovery from effects) deducing the sufficient TN-actions from biomarker-based properties variation at stable states.

In this article, we introduce a theoretical framework for TN-action based system reprogramming formalized by Boolean control network. Based on this framework, we develop an algorithm inferring the causal TN-actions that reprogram a Boolean network, redirecting its dynamics to fulfil an expected property. The article is organised as follows: first, we define the Boolean control network framework (Sect. 2), then we present the inference of causal actions represented

---

[1] Corresponding to the number of parts of size 1 to m in a set with n elements.

[2] Exactly 19 415 908 147 835 trials.

by control parameters based on abduction principle (Sect. 3) and finally, we show its application in breast cancer (Sect. 4).

2 Boolean Control Network

In this section we first review the main theoretical elements used in this article, namely: propositional logic, Boolean network and then we introduce Boolean control network.

2.1 Propositional Logic

A propositional formula is inductively constructed from atoms composed of constants (False/0, True/1) and variables V, unary negation operator $\neg$, and binary logical operators (*e.g.*, $\wedge$/conjunction/AND, $\vee$/disjunction/OR). A *literal* is either an atom or its negation. Given a formula f, $V(f)$ denotes the set of variables occurring in f. For example, let f_α be the propositional formula representing the exclusive OR between atom x_1 and the negation of atom x_2, $f_\alpha = (x_1 \veebar \neg x_2)$, the variables are $V(f_\alpha) = \{x_1, x_2\}$ and the literals are x_1 and $\neg x_2$. Let $X' \subseteq X$ be a subset of variables $f_{\downarrow X'}$ is the restriction of a formula f to the literals involving the variables of X'.

A *cube* syntactically denotes a conjunction of literals and a *clause* a disjunction. In this article, cubes and clauses will be assimilated to literal sets when needed. A *disjunctive normal form* (DNF) of a formula is a disjunction of cubes (*ie.*, $\bigvee_i \bigwedge_{j_i} l_{j_i}$) whereas a *conjunctive normal form* (CNF) is a conjunction of clauses (*ie.*, $\bigwedge \bigvee_{j_i} l_{j_i}$). Any formula can be transformed in DNF or in CNF. For example, a DNF of f_α is $(x_1 \wedge x_2) \vee (\neg x_1 \wedge \neg x_2)$ and a CNF is $(\neg x_1 \vee x_2) \wedge (x_1 \vee \neg x_2)$.

Let an *interpretation* $I : V \to \{0, 1\}$ be a mapping assigning a truth value to each variable[3], a *model* of a formula f, $I \models f$, is an interpretation such that the formula is evaluated to True and a *satisfiable* formula admits a model at least. For example, f_α is satisfiable because the interpretations $I_1 = \{x_1 = 1, x_2 = 1\}$ and $I_2 = \{x_1 = 0, x_2 = 0\}$ are both models of f_α.

Formula f_1 *entails* formula f_2, denoted by $f_1 \models f_2$, if and only if any model of f_1 is also a model of f_2 (*ie.*, $f_1 \models f_2 \overset{\text{def}}{=} \forall I : I \models f_1 \implies I \models f_2$). Hence, the entailment defines a partial order on formulas.

A *minterm* C_I of an interpretation I is the unique cube such that $V(I) = V(C_I)$ fulfilling $I \models C_I$. For the example $C_1 = x_1 \wedge x_2$ and $C_2 = \neg x_1 \wedge \neg x_2$ are the minterms of I_1 and I_2 respectively. A cube C entailing a formula f is said an *implicant* of f and it is *prime* if it ceases to be one when deprived of any literal. Considering the example, C_1, C_2 are both prime implicants of f_α with I_1 and I_2 as model respectively, thus entailing f_α: $C_1 \models f_\alpha, C_2 \models f_\alpha$. Notice that by contrast to a minterm, an implicant does not necessary involve all the variables of the formula (*e.g.*, x_1 is an implicant of $(x_1 \vee x_2) \wedge (x_1 \vee x_3)$).

---

[3] A mapping will be described $x = v$ instead of $x \mapsto v$ for the sake of simplicity.

2.2   Boolean Network

A *Boolean network* is a discrete dynamical system operating on Boolean variables X that determines the *state* evolution of variables $x_i \in X$. It is defined as a system of Boolean equations of the form: $x_i = f_i(x_1, \ldots, x_n), 1 \leq i \leq n$ where each f_i is a propositional formula. A Boolean state of s is an interpretation of the variables (*ie.*, $s : X \to \mathbb{B}$) and S_X will denote the set of all states for variables of X.

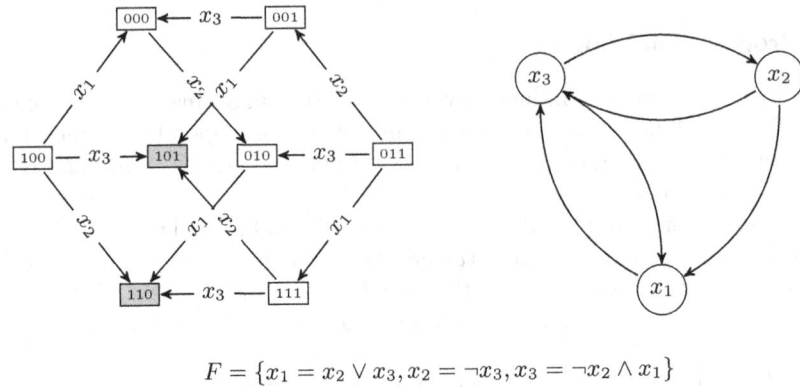

$$F = \{x_1 = x_2 \vee x_3, x_2 = \neg x_3, x_3 = \neg x_2 \wedge x_1\}$$

Fig. 1. Model of asynchronous dynamics and interaction graph.

The *model of dynamics* of a Boolean network describes all the trajectories of the system by a labelled transition system. For each transition the states of agents are updated with respect to a predefined updating policy. For the *asynchronous* updating used in the article, one agent only is updated per transition. Hence, the labelled transition system for the asynchronous updating is $\langle \longrightarrow, X, \mathbb{B}^n \rangle$ where the transition relation $\longrightarrow \subseteq S_X \times X \times S_X$ is labelled by the updated agent, $\xrightarrow{x_i}$ such that:

$$s_1 \xrightarrow{x_i} s_2 \stackrel{\text{def}}{=\!=} s_1 \neq s_2 \wedge s_2(x_i) = f_i(s_1) \wedge \forall x_j \in X \setminus \{x_i\} : s_2(x_j) = s_1(x_j).$$

We denote $\longrightarrow = \bigcup_{x_i \in X} \xrightarrow{x_i}$. A state s_2 is said *reachable* from state s_1 if and only if there exists a trajectory defined by the reflexive and transitive closure of the transition relation connecting s_1 to s_2, $s_1 \longrightarrow^* s_2$.

A state s is an *equilibrium* for $\longrightarrow$, if it can be infinitely reached once met, *ie.*, $\forall s' \in S_X : s \longrightarrow^* s' \implies s' \longrightarrow^* s$. An *attractor* is a set of equilibria that are mutually reachable and a *stable state* is an attractor of cardinality 1. In Fig. 1, the states 101 and 110 in grey are stable. Stable states remain identical whatever the updating policy as they comply to Definition 1:

$$\text{STBL}_F(s) \stackrel{\text{def}}{=\!=} \forall 1 \leq i \leq n : f_i(s) = s(x_i). \tag{1}$$

An *interaction graph* $\langle X, \longrightarrow \rangle$ portrays the causal interactions between variables of a Boolean network (*cf.*, Fig. 1). An interaction $x_i \dashrightarrow x_j$ exists if and only if x_i occurs as literal in a minimal DNF form of f_j, *ie.*,

$$x_i \dashrightarrow x_j \overset{\text{def}}{=\!=} x_i \in V(\text{DNF}(f_j)).$$

2.3 Boolean Control Network

Boolean Control Network (BCN) extends Boolean network by adding *control parameters* that are Boolean variables, $u_i \in U$ without equation definition. Hence, a BCN is defined as a function generating Boolean network parametrized by an interpretation of control parameters $\mu \in S_U$, called a *control input*: $F_u : S_U \rightarrow (S_X \rightarrow S_X)$. For example, an extension of the Boolean network in Fig. 1 to a BCN by adding four control parameters u_1, u_2, u_3, u_4 is:

$$F_{u_1, u_2, u_3, u_4} = \begin{cases} x_1 = (x_2 \wedge u_1) \vee x_3, \\ x_2 = \neg(x_3 \vee \neg u_2), \\ x_3 = ((\neg x_2 \wedge x_1) \vee \neg u_3) \wedge u_4 \end{cases} \tag{2}$$

The application of a control input μ to a Boolean control network F_μ therefore reprograms the dynamics. Figure 2 describes the dynamics resulting from the application[4] of two control inputs $\mu_1 = \{u_1 = 0, u_2 = 1, u_3 = 1, u_4 = 1\}$ and $\mu_2 = \{u_1 = 1, u_2 = 1, u_3 = 1, u_4 = 0\}$.

$$F_{\mu_1} = \begin{cases} x_1 = x_3, \\ x_2 = \neg x_3, \\ x_3 = \neg x_2 \wedge x_1 \end{cases} \qquad F_{\mu_2} = \begin{cases} x_1 = x_2 \vee x_3, \\ x_2 = \neg x_3, \\ x_3 = 1 \end{cases}$$

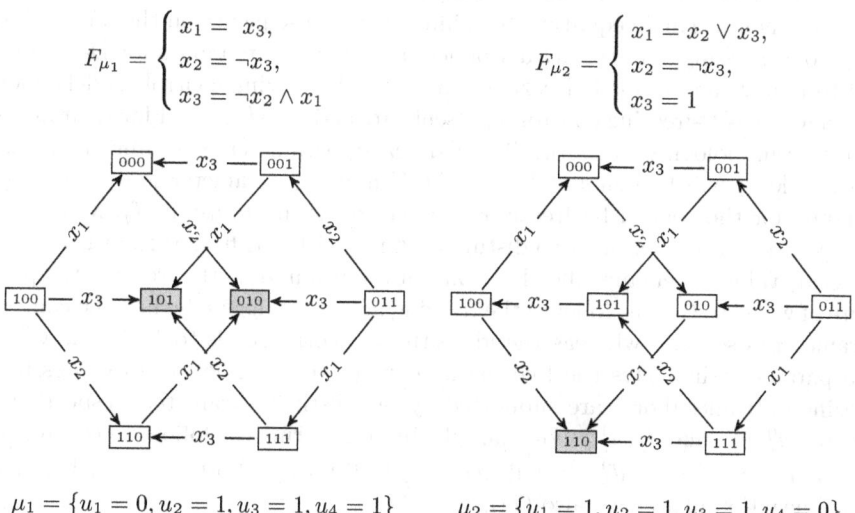

$$\mu_1 = \{u_1 = 0, u_2 = 1, u_3 = 1, u_4 = 1\} \qquad \mu_2 = \{u_1 = 1, u_2 = 1, u_3 = 1, u_4 = 0\}$$

Fig. 2. Modification of the dynamics by control inputs for the example of Fig. 1.

---

[4] The formulas resulting from the instantiation of the BCN by a control input are simplified.

Boolean control network provides a general framework for dynamical system reprogramming. Indeed, let F be an initial Boolean network reprogrammed into an other Boolean network G where the equations are modified, then the Boolean control network $F_u = (u \wedge F) \vee (\neg u \wedge G)$ behaves as F if $u = 1$ and as G if $u = 0$. The switch between F and its reprogramming G now depends on the value of u only. This encoding can be trivially extended to address a family of dynamical systems viewed as the different outcomes of the reprogramming by triggering each particular system from a particular valuation of several control parameters, e.g., $F_{u_1, u_2} = (u_1 \wedge u_2 \wedge F) \vee (\neg u_1 \wedge u_2 \wedge G_1) \vee (u_1 \wedge \neg u_2 \wedge G_2) \vee (\neg u_1 \wedge \neg u_2 \wedge G_3)$ with G_1, G_2, G_3 as reprogramming outcomes. However, the control will be practically specified in another way in order to represent the effective control operated in the real system (Sect. 2.4).

Finally, a Boolean control network can be associated to a *control constraint* $\Phi : U^m \to \mathbb{B}$ fixing the allowed control inputs.

2.4   Control-Freezing Category

Amongst the different possibilities to control a Boolean network, we focus on a particular category called *control-freezing* where the control action fixes (freezes) the variable states to a specific value. This category models the dynamical aftermaths on Boolean network of the TN-actions on the interaction graph. We define two categories of control actions: *Definition-freezing* (D-freezing) that controls the definition of a variable and *Use-freezing* (U-freezing) controlling the use of a variable in an equation defining another variable. Therefore, D-freezing directly assigns an invariant value to variables whereas U-freezing sets locally an invariant value for their use in an equation. The immediate consequence on the interaction graph of a freezing is to totally disconnect a node from its inputs for D-freezing and to remove an arc for U-freezing. Therefore, D-freezing control models node action whereas U-freezing control represents arc action (*cf.*, Sect. 4 for their interpretation in biological network). The D-freezing parameter governing the freeze of variable x_i will be denoted d_i and the U-freezing parameter is denoted $u_{i,j}$ standing for the control by freeze of the variable x_i in its use in f_j. Moreover, each control parameter has two distinct regimes: either it freezes the variable to a specific value or remains idle. The convention, inspired by the freezing temperature of water $0\,°C$, is as follows: the freezing action is triggered when the control parameter is set to 0 whereas the idle situation corresponds to 1. As the value of a parameter indicates the freezing activity (active or idle), the two possible freezing outcomes 0 or 1 are supported by two distinct parameters respectively denoted $d_i^0, u_{i,j}^0$ and $d_i^1, u_{i,j}^1$. For example, by considering the following controlled equation $x_1 = (\neg x_2) \wedge d_1^0$, d_i^0 will freeze x_1 to 0 if $d_1^0 = 0$ otherwise x_1 behaves as the negation of x_2 (See also (7)).

Control-Freezing Implementation to Boolean Network. The implementation of the freezing control on a Boolean network extends the formulas to obtain the expected control behaviour depending on the type of control parameters: D^0, D^1 or U^0, U^1.

D-Freezing Control Implementation. The D-freezing control of variable x_i consists in adding a D-freezing parameter to formula f_i such that setting $\mu(d_i^k) = 0, k \in \{0,1\}$ leads to freeze variable x_i to k and remains idle otherwise $(\mu(d_i^k) = 1)$. Formula f_i is completed according to this control behaviour:

$$x_i = f_i(x_1, \ldots, x_n) \wedge d_i^0 \qquad \text{for freezing to 0} \qquad (3)$$

$$x_i = f_i(x_1, \ldots, x_n) \vee \neg d_i^1 \qquad \text{for freezing to 1} \qquad (4)$$

D^0 and D^1 freezing parameters can be combined to trigger the freeze to different values. To avoid a contradictory freeze to 0 and 1 simultaneously, the constraint $\Phi = d_i^0 \vee d_i^1$ is added ensuring the mutual exclusion of the parameter activities.

U-Freezing Control Implementation. The U-freezing control application follows the same principles as the D-freezing control but applied on the occurrence of variables in the equations of other variables.

$$x_j = f_j(x_1, \ldots, x_i \wedge u_{i,j}^0, \ldots, x_n) \qquad \text{for freezing to 0} \qquad (5)$$

$$x_j = f_j(x_1, \ldots, x_i \vee \neg u_{i,j}^1, \ldots, x_n) \qquad \text{for freezing to 1} \qquad (6)$$

Both controls can be also combined with a constraint avoiding to trigger contradictory freezing controls simultaneously (*ie.*, $\Phi = u_{i,j}^0 \vee u_{i,j}^1$).

In Example (2), u_1 is assimilated to the U-freezing parameter of x_2 to 0 ($u_1 = u_{2,1}^0$) used in x_1 definition, u_2 can be interpreted as the U-freezing parameter of x_3 ($u_2 = u_{3,2}^1$), and u_3, u_4 are the D-freezing parameters of x_3 freezing the variable to 1 and 0 respectively ($u_3 = d_3^1, u_4 = d_3^0$). Consequently, the BCN (2) can be rewritten using the appropriate naming convention as:

$$F_{u_{2,1}^0, d_2^0, d_3^2, d_3^1} = \begin{cases} x_1 = \left(x_2 \wedge u_{2,1}^0\right) \vee x_3, \\ x_2 = \neg(x_3 \vee \neg u_{3,2}^1), \\ x_3 = \left((\neg x_2 \wedge x_1) \vee \neg d_3^1\right) \wedge d_3^0 \end{cases} \qquad (7)$$

The control activity is thus fully determined by the parameters assigned to 0 in a control input μ. The *set of active control parameters* collect these parameters to trace the control activity (*ie.*, $\{u_i \in U \mid \mu(u_i) = 0\}$). In the sequel U will represent the set of the freezing control parameters indifferently and $u_i \in U$ a generic freezing control parameter.

3   Control Parameters Inference

The issue is to formally characterize the basic patterns specifying the changes of the observable molecular traits resulting from biological system reprogramming. Such variations will be questioned at equilibrium conditions in a twofold way: either finding a particular property in some stable states, or finding a particular property in all of them. We thus define two modalities: the *possibility of meeting a property* in at least one stable state (PoP) and the *necessity of meeting a property* in all stable states (NoP). Let p be a Boolean function on states ($p : S_X \to \mathbb{B}$)

standing for a property, the PoP and NoP inference problems are defined as follows:

Find a control input μ fulfilling the constraints of Φ such that:

$$\exists s \in S_X : \text{STBL}_{F_\mu}(s) \wedge p(s). \qquad \text{(PoP)} \qquad (8)$$

$$\forall s \in S_X : \text{STBL}_{F_\mu}(s) \implies p(s). \qquad \text{(NoP)} \qquad (9)$$

Different control inputs may be suitable as solutions. For instance, gaining stable state 010 for Boolean network of Fig. 1 with parameters defined in (7) can be obtained with the following control inputs:

$$\left\{ u_{2,1}^0 = 0, u_{3,2}^1 = 1, d_3^1 = 1, d_3^0 = 1 \right\}$$
$$\left\{ u_{2,1}^0 = 0, u_{3,2}^1 = 1, d_3^1 = 1, d_3^0 = 0 \right\}$$
$$\left\{ u_{2,1}^0 = 0, u_{3,2}^1 = 1, d_3^1 = 0, d_3^0 = 0 \right\}$$

The plurality of solutions raises the question of their interpretation for identifying the root factors causing the expected effects. The causal factors are defined as the essential actions shifting the dynamics to the objective whereas the casual factors behave neutrally and do not interfere with the objective whatever their valuation. Focusing on the active parameters, only $u_{2,1}^0 = 0$ matters for shifting the dynamics to gain 010 (first solution) since it is shared by all solutions, and without this assignment the system reprogramming fail to reach the expected objective. The other parameters becoming active are casual because they can be set to 0 or 1 without deviating the dynamics to the result.

The set of causal control parameters forms a *core* K^* defined as a minimal active parameter set under the inclusion which is equivalent to the entailment order for cubes. Considering the example, the core $K^* = \{u_{2,1}^0\}$ is included in all other active parameter sets.

Several cores may be found for a given problem. For example, three different cores $\{d_3^1\}, \{u_{2,1}^0\}, \{u_{3,2}^1\}$ enable the loss of equilibrium 110. Hence, the inference algorithm aims at finding all the cores in regards to a reprogramming query formulated by the possibility or the necessity of meeting a property at steady-state.

3.1   Abduction Based Core Inference

Inferring a core corresponds to the determination of control parameters producing an expected effect. In logic finding causes from effects is an abduction problem. Abduction is a method of reasoning proposing hypotheses that provide the best explanation for observable facts in regards to knowledge of the problem constituting the theory [22,25,29]. In propositional logic, a cube C is an abductive explanation of a formula f formalizing the facts with respect to another formula Φ representing the theory if and only if: $C \wedge \Phi \models f$ and C is consistent with Φ (*ie.*, $\Phi \wedge C$ is satisfied). Finding a parsimonious hypothesis introduces the notion of minimal solution which is usually assimilated to a prime implicant. Within this framework, the possibility and the necessity of property (8, 9) are formulated as abduction problems in propositional logic (10, 11) by considering that p is a propositional formula. Lemma 1 demonstrates this equivalence.

Find a cube C_μ such that:

$$(C_s \wedge C_\mu) \wedge \phi \models (\text{STBL}_{F_u} \wedge p); \qquad \text{(PoP)} \qquad (10)$$
$$C_\mu \wedge \phi \models (\text{STBL}_{F_u} \implies p); \qquad \text{(NoP)} \qquad (11)$$

where C_s and C_μ are consistent with Φ, $V(C_\mu) = U, V(C_s) = X$ and the stability condition is defined as:

$$\text{STBL}_{F_u} \stackrel{\text{def}}{=} \bigwedge_{i=1}^{n} (x_i \iff f_i(x_1, \ldots, x_n, u_1, \ldots, u_m)).$$

In Example (7), the components of the problem for gaining state 010 (Fig. 2, μ_1) are:

$$
\begin{aligned}
\text{STBL}_{F_u} = \quad & x_1 \iff \left(x_2 \wedge u_{2,1}^0\right) \vee x_3 & \text{Stability condition} \\
\wedge \; & x_2 \iff \neg(x_3 \vee \neg u_{3,2}^1) \\
\wedge \; & x_3 \iff \left((\neg x_2 \wedge x_1) \vee \neg d_3^1\right) \wedge d_3^0 \\
\Phi = \; & d_3^0 \vee d_3^1 & \text{Exclusive activity of } d_3 \\
p = \; & \neg x_1 \wedge x_2 \wedge \neg x_3 & \text{Minterm of } s = 010
\end{aligned}
$$

For the loss of stable state 101 (Fig. 2, μ_2), only the property differs, now defined as: $p = \neg(x_1 \wedge \neg x_2 \wedge x_3)$ corresponding to the negation of the minterm of 101.

Lemma 1. *(10) and (11) define the PoP (8) and NoP (9) problems as abductive problems in propositional logic. (See the extended version for the proof [2].)*

3.2 Core Inference Algorithm

For a formula f, the core inference consists in finding a satisfiable implicant C^* fulfilling $C^* \models f$ that minimizes the number of negative control parameters $(\neg u_i)$ with respect to the inclusion. The resulting core K^* is trivially deduced by collecting the negative control parameters of C^*. Computing a core is an NP-Hard problem[5]. In this section, we present an algorithm adapted from the method developed for prime implicants computation in [28] and based on 0-1 Integer Linear Programming (0-1 ILP). A 0-1 ILP problem is formulated as:

$$\text{Minimize} \sum_{j=1}^{h} m_j \cdot y_j, \text{subject to} \sum_{j=1}^{h} W_{i,j} \cdot y_j \leq v_i, \text{for } 1 \leq i \leq r, y \in \{0,1\}^h.$$

where y is the unknown vector, and m, v vectors, W matrix are the parameters of the problem.

---

[5] By reduction of the minimum hitting set problem.

The method, called ILP-CORE, operates on a formula f in CNF and computes the set of all the cores $\mathcal{K}^*$. The method is based on the translation of the constraints related to core definition into 0-1 ILP constraints such that a solution y is a binary representation of an implicant C^*. The algorithm is outlined in Algorithm 1 and the main steps are fully described in the proof of Theorem 2.

Function ILP-CORE$(f: \textit{CNF formula })$

$(\min m.y^T, Wy \leq v) = $ Describe constraints on core as 0-1 ILP problem ;
`// ` $C^* \models f$ `minimizing the number of negative control parameters.`
$\mathcal{K}^* = \emptyset$;
repeat

    $y = $ Solve $(\min m.y^T, Wy \leq v)$ with a 0-1 ILP solver ;
    if *a solution y is found* **then**

        $K^* = $ Collect the negative control parameters from y;
        $\mathcal{K}^* = \mathcal{K}^* \cup \{K^*\}$;
        Exclude all solutions $K, K^* \subseteq K$ by adding constraints to $Wy \leq v$;

    end

until *No solution y is found*;

return $\mathcal{K}^*$ `// the set of all cores`

end

Algorithm 1. Outline of the ILP-CORE algorithm.

Theorem 2. *The* ILP-CORE *algorithm finds all and only the cores. (See the extended version [2] for the proof.)*

To properly specify the PoP and NoP resolutions, the method is called with different formulas specifying the query. Applied to PoP (10), the complete formula is passed as parameter since literals of C^* contain control parameters as well as variables identifying the state. For NoP (11), as C^* must contain control parameters only, each clause is then restricted to control parameters by removing the literals involving state variables (*ie.,* $x_i \in X$). The constraints on control parameters Φ are already in CNF form by definition (Sect. 2.4).

$$\text{ILP-CORE}(\text{CNF}(\text{STBL}_{F_u} \wedge p) \wedge \Phi) \qquad \text{(PoP)}$$
$$\text{ILP-CORE}(\text{CNF}(\text{STBL}_{F_u} \implies p)_{\downarrow U} \wedge \Phi) \ \text{(NoP)}$$

3.3   Related Works

BCN was recently introduced in systems biology to provide the theoretical foundations and computational methods for investigating cell fate reprogramming and therapeutic target discovery. In [17] the authors apply a stuck-at fault model to simulate drug intervention in an acyclic growth factors pathway by a generate-and-test method. Stuck-at fault model mimics the defects on combinatorial logic circuit which were assimilated here to malignant mutations. Based on this model, authors identify drug actions for single mutations by correcting all possible single faults. This framework was improved by [19] using a Max-SAT based method dedicated to acyclic networks in order to directly compute the control parameter values and final states. Inferring the drug targets on a network is also developed by [23] using algebraic techniques (Gröbner basis) in order to modify the system dynamics for creating or avoiding particular stable states. In [37], the authors propose a heuristic method with the same goal but focused on the control of key-nodes stabilizing "motifs" identifying sub-networks. Finally, we have introduced the principle of the abductive inference of cores for drug target discovery in [3] which is significantly extended here, in particular with the formalization and the generalization of the TN-actions as control freezing, and with a more efficient method for the core inference.

Our approach follows a similar orientation of these works by using BCN for modelling disease and drug actions. By comparison, the target discovery is modelled in an original way as an abductive problem. The resulting framework supports any kind of networks including cycles with actions applied on both nodes and arcs and find multiple targets qualifying the parsimonious TN-actions (cores) reprogramming the system. The proposed algorithm infers the causes of expected properties met at stable states and we formalize their query in a general setting using propositional formulas with the Necessity and Possibility modalities.

4   Application to Breast Cancer

This section shows the application of TN-actions inference for the study of breast cancer. Mainly, cancer cells differ from normal cells by their uncontrolled proliferation and apoptotic evasion. Accordingly, targeted drugs aim at inducing apoptosis or stop the proliferation of cancer cells [12]. We therefore developed a model (Sect. 4.1) focusing on the regulation of division and apoptosis. We infer the causal TN-actions leading to a loss or gain of apoptosis (Sect. 4.2) and then analyse the results (Sect. 4.3).

4.1   Aptoptosis/Cell Division Boolean Network

The model focuses on the regulation of cell division and apoptosis by the EGFR signalling pathway and a BRCA1/TP53 DNA damage response module. These genes have been identified as central in the process of tumor formation in breast cancer [16,24]. The model incorporates the positive and negative interactions between nuclear TP53 and MDM2 described by [6], the main messengers of the PI3K/AKT and MAPK signalling following EGFR activation described by [35] and adds BRCA1 and PARP1 regulation of DNA damage. These pathways are gathered into a unique Boolean network through the lens of their role in the regulation of the G1/S transition and the triggering of apoptosis in case of

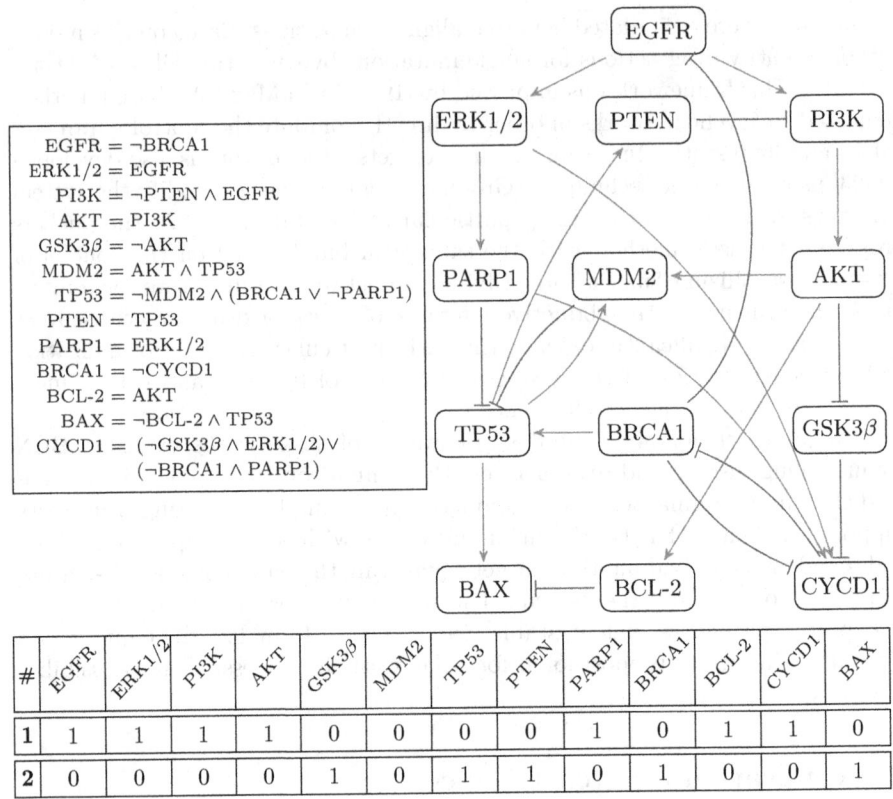

The Boolean network equations (left):

EGFR = ¬BRCA1
ERK1/2 = EGFR
PI3K = ¬PTEN ∧ EGFR
AKT = PI3K
GSK3β = ¬AKT
MDM2 = AKT ∧ TP53
TP53 = ¬MDM2 ∧ (BRCA1 ∨ ¬PARP1)
PTEN = TP53
PARP1 = ERK1/2
BRCA1 = ¬CYCD1
BCL-2 = AKT
BAX = ¬BCL-2 ∧ TP53
CYCD1 = (¬GSK3β ∧ ERK1/2)∨
        (¬BRCA1 ∧ PARP1)

#	EGFR	ERK1/2	PI3K	AKT	GSK3β	MDM2	TP53	PTEN	PARP1	BRCA1	BCL-2	CYCD1	BAX
1	1	1	1	1	0	0	0	0	1	0	1	1	0
2	0	0	0	0	1	0	1	1	0	1	0	0	1

Fig. 3. Boolean network (left) with its regulatory graph (right) representing the activatory (green) and inhibitory (red) interactions, and stable states (below). (Color figure online)

DNA damage. The corresponding Boolean network[6], constructed from published litterature and signalling pathways databases (KEGG [15] and Signor [26]), is shown in Fig. 3 and the molecular mechanism for each interaction is detailed and referenced in the extended version [2]). The Boolean dynamics is bistable characterizing two cellular functions in normal cells: either (1) the cell enters division by activation of the G1/S transition and inhibition of apoptosis, or (2) it enters in apoptosis and arrest the cell cycle.

4.2   Inference Query

As network reprogramming effects biomarker profile changes, it is required to (1) identify the biomarkers discriminating phenotypes and (2) define the reprogramming queries based on these biomarkers for causal genes and drug actions inference.

---

[6] For the sake of simplicity, the names of genes (by convention written in upper case letters) can also denominate the proteins they encode.

Since the proliferative activity of cells depends on the balance between division and apoptosis, we selected CYCLIN D1 and BAX as biomarkers as they are the key effector of the G1/S transition of cell division and initiation of apoptosis [1,11]. The pair (CYCLIN D1, BAX) distinguishes four phenotypes: apoptosis, division, quiescence (apoptosis balanced by division) and dormancy (neither apoptosis nor division) [30] through to the following signatures: $(0,1)$ for apoptosis, $(1,0)$ for division, $(0,0)$ for quiescence and $(1,1)$ for dormancy.

Since cancer cells are characterized by their inability to trigger apoptosis, the reprogramming query for the inference of causal genes corresponds to the loss of apoptosis. Conversely, as drugs induce apoptosis in cancer cells, the reprogramming query for the inference of drug actions corresponds to the gain of apoptosis. Apoptosis is formalized as a property by the minterm of $(0,1)$ signature: $p = \neg\text{CYCD1} \wedge \text{BAX}$. The loss of apoptosis thus corresponds to the necessity of $\neg p$ since the apoptosis must not occur in any stable state. To recover this marking, the query can be either the necessity or the possibility of p. We have tested both and the solutions providing stable states are the same.

Finally, the genetic events are modelled by control parameters as follows: the loss of expression of a gene following loss-of-function mutations or other genetic events such as gene deletion corresponds to D^0-freezing; gene over-expression following gain-of-function mutations or other genetic events such as gene amplification are represented by D^1-freezing; and the loss of interactions between two molecules is interpreted as U^0-freezing. The Boolean network (Fig. 3) is automatically completed with control parameters by following the rules set out in Sect. 2.4. Notice that U^1-freezing does not seem interpretable in terms of biological events and not used here.

4.3   Analysis of the Results

We inferred the actions from combination of D^0/D^1-freezing on all variables (molecules) except markers and the U^0-freezing on all interactions separately to compare them. The computed TN-actions are shown in Table 1. The TN-actions for the gain of apoptosis have been inferred from the model with BRCA1-deficiency (BRCA1 $= 0$).

Applied to the loss of apoptosis with D-freezing, the method retrieves the main driver genes identified in breast cancer namely BRCA1, TP53, PI3K and EGFR [5,14]. Moreover, it segregates tumor suppressor genes (*ie.*, frequently affected by gain-of-function mutations in cancers) from oncogenes (*ie.*, frequently affected by loss-of-function mutations in cancers) [8,21]: D^0-frozen genes all correspond to tumour suppressors and D^1-frozen genes to oncogenes. For the gain of apoptosis after application of BRCA1 deficiency, the single D-freezing inferred actions recover the necessity of blocking PARP1, the synthetic lethal partner of BRCA1. The pair BRCA1/PARP1 are called synthetic lethal partners because the use of PARP inhibitors in patients with BRCA1-deficiency prevents any possibility of DNA-repair resulting in permanent DNA damage inducing apoptosis of the cancer cell [10,20]. Finding such partnerships is critical for anticancer

Table 1. Freezing actions causing the gain or loss of apoptosis.

- Health → Cancer: necessary loss of apoptosis -

	Single D-freezing	Single U^0-freezing	
NODE ACTION	BRCA1 = 0	TP53 —▷ BAX	ARC ACTION
	TP53 = 0	*Double U^0-freezing*	
	PI3K = 1	BRCA1 —▷ EGFR, TP53 —▷ PTEN	
	AKT = 1	BRCA1 —▷ EGFR, BRCA1 —▷ CYCD1	
	BCL-2 = 1	BRCA1 —▷ EGFR, BRCA1 —▷ TP53	
	MDM2 = 1	BRCA1 —▷ EGFR, PTEN —▷ PI3K	
	Double D-freezing	BRCA1 —▷ EGFR, GSK3β —▷ CYCD1	
	GSK3β = 0, ERK1/2 = 1		
	PTEN = 0, EGFR = 1		
	GSK3β = 0, EGFR = 1		

- BRCA1 mutation (Cancer) → Cell death: possible gain of apoptosis -

	Single D-freezing	Single U^0-freezing	
NODE ACTION	BRCA1 = 1	ERK1/2 —▷ PARP1	ARC ACTION
	PARP1 = 0	EGFR —▷ ERK1/2	
	ERK1/2 = 0	*Double U^0-freezing*	
	EGFR = 0	PARP1 —▷ CYCD1, PARP1 —▷ TP53	

treatment [13] but since the cancer target differs from the drug target, they are hard to recover experimentally and computationally.

The algorithm also predicts double D-freezing actions for the necessary loss of apoptosis which suggest that overexpression of EGFR alone would not be sufficient to provoke a cancerous phenotype and must be combined with either loss of PTEN or GSK3β. The validation of such result is less obvious than the former and is based on the concomittent overexpression of EGFR and loss of PTEN/GSK3β. Work in [18] confirms the existence of a co-occurence of EGFR over-expression and loss of PTEN in 20% of the tumors of the studied population. Moreover, authors also show that PTEN loss is associated to resistance to EGFR inhibitors. Similarly in erlotinib resistant model cell lines [4] it has been observed that GSK3β was upregulated. Thus, these works suggest the existence of the predicted cooperation between these genes.

It is also predicted that EGFR inhibition would be synthetic lethal with BRCA1 mutations. This is supported by the observation that the proliferation properties of BRCA-deficient cells are sensitive to EGFR inhibition by erlotinib [5]. We found no published work suggesting that ERK1/2 inhibition in such cells would be synthetic lethal.

In summary, in the studied model the method accurately predicts cancerous genes and drug targets and segregate oncogenes from tumor suppressors.

The inference also recovers cooperative gene mutations and synthetic lethal partnerships. The double freezing results provide some insights on the necessary cooperative combination of perturbations that are difficult to assess experimentally [27, 36]. Moreover by inferring cores, the method separate causal genes to casual ones (passengers) and determine frequent drivers as well as rare ones which is more difficult to obtain by statistical analysis that prioritize genes from the frequency of their occurrence [34]. Usually, drivers are classified in subtypes where a specific drug target is associated for each subtype. In the proposed approach the drug target may be directly inferred from the application of the TN-actions corresponding to drivers on the initial boolean network. Finally, arc inference (U^0-freezing) refines the results on nodes (D^0-freezing) and, to the best of our knowledge, the resulting predictions are not experimentally confirmed.

5 Conclusion

In this article, we have proposed a modelling framework discovering the reprogramming actions of a dynamical system using BCN and designed a new inference method based on abduction that identifies the minimal causes reprogramming the network. A library called PROTAXION was developed in Mathematica to support the application on concrete cases. It has been validated on a breast cancer model and has shown that the method can retrieve driver genes and drug targets.

A perspective of this work is to include the notion of resistance in the inference. Two sorts of resistances were established: the primary arising prior to a classical treatment and the secondary which is an adaptive negative response to a treatment. As the method infers all the causes responsible for a biomarker profile shift, the primary resistance is interpreted in our framework as the variation of the input Boolean network of a patient in comparison to a generic one in which the drug targets were deduced. In this context, we need to specialize the network to a patient. The issue for the secondary resistance is more complex and necessitates to predict the further alterations of the network once a TN-action is applied. The prediction of secondary resistance requires to extend the BCN model by including the notion of temporal sequence of control inputs instead of a single control input.

References

1. Baldin, V., Lukas, J., Marcote, M.J., Pagano, M., Draetta, G.: Cyclin D1 is a nuclear protein required for cell cycle progression in G1. Genes Dev. **7**(5), 812–821 (1993)
2. Biane, C., Delaplace, F.: Abduction based drug target discovery using boolean control network. In: HAL Archive (2017). https://hal.archives-ouvertes.fr/hal-01522072
3. Biane, C., Delaplace, F., Melliti, T.: Abductive network action inference for targeted therapy. In: Static Analysis and Systems Biology (2016)

4. Botting, G.M., Rastogi, I., Chhabra, G., Nlend, M., Puri, N.: Mechanism of resistance and novel targets mediating resistance to EGFR and c-Met tyrosine kinase inhibitors in non-small cell lung cancer. PloS one **10**(8), e0136155 (2015)

5. Burga, L.N., Hai, H., Juvekar, A., Tung, N.M., Troyan, S.L., Hofstatter, E.W., Wulf, G.M.: Loss of BRCA1 leads to an increase in epidermal growth factor receptor expression in mammary epithelial cells, and epidermal growth factor receptor inhibition prevents estrogen receptor-negative cancers in BRCA1-mutant mice. Breast Cancer Res. **13**(2), R30 (2011)

6. Ciliberto, A., Novák, B., Tyson, J.J.: Steady states and oscillations in the p53/Mdm2 network. Cell Cycle **4**(3), 488–493 (2005)

7. Creixell, P., Schoof, E.M., Simpson, C.D., Longden, J., Miller, C.J., Lou, H.J., Perryman, L., Cox, T.R., Zivanovic, N., Palmeri, A., Wesolowska-Andersen, A., Helmer-Citterich, M., Ferkinghoff-Borg, J., Itamochi, H., Bodenmiller, B., Erler, J.T., Turk, B.E., Linding, R.: Kinome-wide decoding of network-attacking mutations rewiring cancer signaling. Cell **163**(1), 202–217 (2015)

8. Croce, C.M.: Oncogenes and cancer. New Engl. J. Med. **358**(5), 502–511 (2008). PMID: 18234754

9. Csermely, P., Korcsmàros, T., Kiss, H.J.M., London, G., Nussinov, R.: Structure and dynamics of molecular networks: a novel paradigm of drug discovery: a comprehensive review. Pharmacol. Therapeutics **138**(3), 333–408 (2013)

10. Farmer, H., McCabe, N., Lord, C.J., Tutt, A.N.J., Johnson, D.A., Richardson, T.B., Santarosa, M., Dillon, K.J., Hickson, I., Knights, C., et al.: Targeting the DNA repair defect in BRCA mutant cells as a therapeutic strategy. Nature **434**(7035), 917–921 (2005)

11. Gupta, S.: Molecular signaling in death receptor and mitochondrial pathways of apoptosis (review). Int. J. Oncol. **22**(1), 15–20 (2003)

12. Hanahan, D., Weinberg, R.A.: Hallmarks of cancer: the next generation. Cell **144**(5), 646–674 (2011)

13. Kaelin, W.G.: The concept of synthetic lethality in the context of anticancer therapy. Nat. Rev. Cancer **5**(9), 689–98 (2005)

14. Kandoth, C., McLellan, M.D., Vandin, F., Ye, K., Niu, B., Charles, L., Xie, M., Zhang, Q., McMichael, J.F., Wyczalkowski, M.A., et al.: Mutational landscape and significance across 12 major cancer types. Nature **502**(7471), 333–339 (2013)

15. Kanehisa, M., Furumichi, M., Tanabe, M., Sato, Y., Morishima, K.: KEGG: new perspectives on genomes, pathways, diseases and drugs. Nucl. Acids Res. **45**(D1), D353–D361 (2017)

16. Kolch, W., Halasz, M., Granovskaya, M., Kholodenko, B.N.: The Dynamic Control of Signal Transduction Networks in Cancer Cells. Nature Publishing Group (2015)

17. Layek, R., Datta, A., Bittner, M.: ER Dougherty: cancer therapy design based on pathway logic. Bioinformatics **27**(4), 548–555 (2011)

18. Lee, J.Y., Hong, M., Kim, S.T., Park, S.H., Kang, W.K., Kim, K.-M., Lee, J.: The impact of concomitant genomic alterations on treatment outcome for trastuzumab therapy in HER2-positive gastric cancer. Sci. Rep. **5**, 9289 (2015)

19. Lin, P.-C.K., Khatri, S.P.: Application of Max-SAT-based ATPG to optimal cancer therapy design. BMC Genomics **13**(Suppl 6), S5 (2012)

20. Livraghi, L., Garber, J.E.: PARP inhibitors in the management of breast cancer: current data and future prospects. BMC Med. **13**(1), 1 (2015)

21. Lodish, H., Zipursky, S.L.: Molecular cell biology. Biochem. Mol. Biol. Educ. **29**, 126–133 (2001)

22. Marquis, P.: Extending abduction from propositional to first-order logic. In: Jorrand, P., Kelemen, J. (eds.) FAIR 1991. LNCS, vol. 535, pp. 141–155. Springer, Heidelberg (1991). doi:10.1007/3-540-54507-7_12

23. Murrugarra, D., Veliz-Cuba, A., Aguilar, B., Laubenbacher, R.: Identification of control targets in boolean molecular network models via computational algebra. BMC Syst. Biol. **10**(1), 94 (2016)

24. Narod, S.A., Foulkes, W.D.: BRCA1 and BRCA2: 1994 and beyond. Nat. Rev. — Cancer **4**(9), 665–676 (2004)

25. Peirce, C.S.: On the natural classification of arguments. Proc. Am. Acad. Arts Sci. **7**, 261–287 (1867)

26. Perfetto, L., Briganti, L., Calderone, A., Perpetuini, A.C., Iannuccelli, M., Langone, F., Licata, L., Marinkovic, M., Mattioni, A., Pavlidou, T., Peluso, D., Petrilli, L.L., Pirro, S., Posca, D., Santonico, E., Silvestri, A., Spada, F., Castagnoli, L., Cesareni, G.: SIGNOR: a database of causal relationships between biological entities. Nucl. Acids Res. **44**(D1), D548–D554 (2016)

27. Phillips, P.C.: Epistasis - the essential role of gene interactions in the structure and evolution of genetic systems. Nat. Rev. Genetics **9**(11), 855–867 (2008)

28. Pizzuti, C.: Computing prime implicants by integer programming. In: Proceedings Eighth IEEE International Conference on Tools with Artificial Intelligence, pp. 332–336. IEEE Computer Society Press (1996)

29. Quine, W.V.: On cores and prime implicants of truth functions. Am. Math. Mon. **66**(9), 755–760 (1959)

30. Spiliotaki, M., Mavroudis, D., Kapranou, K., Markomanolaki, H., Kallergi, G., Koinis, F., Kalbakis, K., Georgoulias, V., Agelaki, S.: Evaluation of proliferation and apoptosis markers in circulating tumor cells of women with early breast cancer who are candidates for tumor dormancy. Breast Cancer Res. **16**(6), 485 (2014)

31. Strimbu, K., Tavel, J.A.: What are biomarkers? Curr. Opin. HIV AIDS **5**(6), 463–466 (2011)

32. Vidal, M.: A unifying view of 21st century systems biology. FEBS Lett. **583**(24), 3891–3894 (2009)

33. Vidal, M., Cusick, M.E., Barabási, A.-L.: Interactome networks and human disease. Cell **144**(6), 986–998 (2011)

34. Vogelstein, B., Papadopoulos, N., Velculescu, V.E., Zhou, S., Diaz, L.A., Kinzler, K.W.: Cancer genome landscapes. Science **339**(6127), 1546–1558 (2013)

35. Von der Heyde, S., Bender, C., Henjes, F., Sonntag, J., Korf, U., Beissbarth, T.: Boolean ErbB network reconstructions and perturbation simulations reveal individual drug response in different breast cancer cell lines. BMC Syst. Biol. **8**(1), 75 (2014)

36. Wang, X., Fu, A.Q., McNerney, M.E., White, K.P.: Widespread genetic epistasis among cancer genes. Nat. Commun. **5**, 4828 (2014)

37. Zanudo, J.G.T., Albert, R.: Cell fate reprogramming by control of intracellular network dynamics. PLoS Comput. Biol. **11**(4), e1004193 (2015)

38. Zhong, Q., Simonis, N., Li, Q.-R., Charloteaux, B., Heuze, F., Klitgord, N., Tam, S., Haiyuan, Y., Venkatesan, K., Mou, D., Swearingen, V., Yildirim, M.A., Yan, H., Dricot, A., Szeto, D., Lin, C., Hao, T., Fan, C., Milstein, S., Dupuy, D., Brasseur, R., Hill, D.E., Cusick, M.E., Vidal, M.: Edgetic perturbation models of human inherited disorders. Mol. Syst. Biol. **5**(321), 321 (2009)

Probably Approximately Correct Learning of Regulatory Networks from Time-Series Data

Arthur Carcano[1], François Fages[2(✉)], and Sylvain Soliman[2]

[1] Ecole Normale Supérieure, Paris, France
arthur.carcano@ens.fr
[2] Inria, University Paris-Saclay, Lifeware Group, Palaiseau, France
{Francois.Fages,Sylvain.Soliman}@inria.fr

Abstract. Automating the process of model building from experimental data is a very desirable goal to palliate the lack of modellers for many applications. However, despite the spectacular progress of machine learning techniques in data analytics, classification, clustering and prediction making, learning dynamical models from data time-series is still challenging. In this paper we investigate the use of the Probably Approximately Correct (PAC) learning framework of Leslie Valiant as a method for the automated discovery of influence models of biochemical processes from Boolean and stochastic traces. We show that Thomas' Boolean influence systems can be naturally represented by k-CNF formulae, and learned from time-series data with a number of Boolean activation samples per species quasi-linear in the precision of the learned model, and that positive Boolean influence systems can be represented by monotone DNF formulae and learned actively with both activation samples and oracle calls. We consider Boolean traces and Boolean abstractions of stochastic simulation traces, and study the space-time tradeoff there is between the diversity of initial states and the length of the time horizon, and its impact on the error bounds provided by the PAC learning algorithms. We evaluate the performance of this approach on a model of T-lymphocyte differentiation, with and without prior knowledge, and discuss its merits as well as its limitations with respect to realistic experiments.

1 Introduction

Modelling biological systems is still an art which is currently limited in its applications by the number of available modellers. Automating the process of model building is thus a very desirable goal to attack new applications, develop patient-tailored therapeutics, and also design experiments that can now be largely automated with a gain in both the quantification and the reliability of the observations, at both the single cell and population levels.

Machine learning is revolutionising the statistical methods in biological data analytics, data classification and clustering, and prediction making. However, learning dynamical models from data time-series is still challenging. A recent survey on probabilistic programming [14] highlighted the difficulties associated

© Springer International Publishing AG 2017
J. Feret and H. Koeppl (Eds.): CMSB 2017, LNBI 10545, pp. 74–90, 2017.
DOI: 10.1007/978-3-319-67471-1_5

with modelling time, and concluded that existing frameworks are not sufficient in their treatment of dynamical systems. There has been early work on the use of machine learning techniques, such as inductive logic programming [19] combined with active learning in the vision of the "robot scientist" [4], to infer gene functions, metabolic pathway descriptions [1,2] or gene influence systems [3], or to revise a reaction model with respect to CTL properties [5]. Since a few years, progress in this field is measured on public benchmarks of the "Dream Challenge" competition [15,18]. Logic Programming, and especially *Answer Set Programming* (ASP), provide efficient tools such as CLASP [11] to implement learning algorithms for Boolean models. They have been applied in [12] to the detection of inconsistencies in large biological networks, and have been subsequentially applied to the inference of gene networks from gene expression data and to the design of discriminant experiments [26]. Furthermore, ASP has been combined with CTL model-checking in [20] to learn mammalian signalling networks from time series data, and identify erroneous time-points in the data.

Active learning extends machine learning with the possibility to call oracles, e.g. make experiments, and budgeted learning adds costs to the calls to the oracle. The original motivation for the budgeted learning protocol came from medical applications in which the outcome of a treatment, drug trial, or control group is known, and the results of running medical tests are each available for a price [8]. In this context, multi-armed bandit methods [7] currently provide the best strategies. In [16], a bandit-based active learning algorithm is proposed for experiment design in dynamical system identification.

In this paper, we consider the framework of Probably Approximately Correct (PAC) Learning which was introduced by Leslie Valiant in his seminal paper on a theory of the learnable [24]. Valiant questioned what can be learned from a computational viewpoint, and introduced the concept of PAC learning, together with a general-purpose polynomial-time learning protocol. Beyond the algorithms that one can derive with this methodology, Valiant's theory of the learnable has profound implications on the nature of biological and cognitive processes, of collective and individual behaviors, and on the study of their evolution [25].

Here we present PAC learning as a possible basis to develop a method for the automated discovery of influence models of biochemical processes from time-series data. To the best of our knowledge, the application of PAC learning to dynamical models of biochemical systems has not been reported before. We show that Thomas' gene regulatory networks [22,23] can be naturally represented by Boolean formulae in conjunctive normal forms with a bounded number of litterals (i.e. k-CNF formulae), and can be learned from Boolean traces with a number of Boolean transition samples per species quasi-linear in the precision of the learned model, using Valiant's PAC learning algorithm for k-CNF formulae. We also show that Boolean influence systems with their positive Boolean semantics discussed in [9] can be naturally represented by monotone DNF formulae, and learned actively from a set of positive samples with calls to an oracle.

For the sake of evaluation, we consider Boolean traces and Boolean abstractions of stochastic simulation traces, and study the space-time tradeoff there is between the diversity of initial states and the length of the time horizon, and its impact on the error bounds provided by PAC learning algorithms. In the following, we first illustrate our results[1] with a toy example, the Lotka-Volterra prey-predator system as running example, and then on a Thomas regulatory network of the differentiation of the T-helper lymphocytes from [17,21], composed of 32 influences and 12 variables. We evaluate the performance of PAC learning on this model, with and without prior knowledge, and discuss its merits as well as its limitations with respect to realistic experiments.

2   Preliminaries on PAC Learning

2.1   PAC Learning Protocol

Let n be the dimension of the model to learn, and let us consider a finite set of Boolean variables $x_1, \ldots, x_n$, A vector is an assignment of the n variables to $\mathbb{B} = \{0,1\}$; A Boolean function $G : \mathbb{B}^n \rightarrow \mathbb{B}$; assigns a Boolean value to each vector. The idea behind the PAC learning protocol is to discover a Boolean function[2], G, which approximates a hidden function F, while restricting oneself to the two following operations:

- SAMPLE(): returns a positive example, i.e. a vector v such that $F(v) = 1$. The output of SAMPLE() is assumed to follow a given probability distribution $D(v)$, which is used to measure the approximation of the result.
- ORACLE(v): returns the value of $F(v)$ for any input vector v.

Definition 1 ([24]). *A class $\mathcal{M}$ of Boolean functions is said to be* learnable *if there exists an algorithm $\mathcal{A}$ with some precision parameter $h \in \mathbb{N}$ such that:*

- *$\mathcal{A}$ runs in polynomial time both in n and h;*
- *for any function F in $\mathcal{M}$, and any distribution D on the positive examples, $\mathcal{A}$ deduces with probability higher than $1 - h^{-1}$ an approximation G of F such that*
  - *$G(v) = 1$ implies $F(v) = 1$ (no false positive)*
  - $$\sum_{v \ s.t. \ F(v) = 1 \wedge G(v) = 0} D(v) < h^{-1} \ \textit{(low probability of false negatives)}$$

Note that it is possible to use two different parameters h_1 and h_2 for the *probability of false negatives* and the *quality of the approximation*, but here, we used $h_1 = h_2 = h$ for the sake of simplicity.

---

[1] For the sake of reproducibility, the code used in this article is available at http://lifeware.inria.fr/wiki/software/#CMSB17.

[2] More generally, the PAC learning protocol can discover partial vectors, but for the applications discussed in the current article it is enough to only consider total vectors.

2.2   PAC Learning Algorithms

Valiant showed the learnability of some important classes of functions in this framework, in particular for Boolean formulae in conjunctive normal forms with at most k literals per conjunct (k-CNF), and for monotone (i.e. negation free, positive literals only) Boolean formulae in disjunctive normal form (DNF).

The computational complexity of the PAC learning algorithms for these classes of functions is expressed in terms of a function $L(h, S)$, defined as the smallest integer i such that in i independent Bernoulli trials, each with probability at least h^{-1} of success, the probability of having fewer than S successes is less than h^{-1}. Interestingly, this function is *quasi-linear in h and S*, more precisely for all integers $S \geq 1$ and reals $h > 1$, we have $L(h, S) \leq 2h(S + \log_e h)$ [24].

Theorem 1 ([24]). *For any k, the class of k-CNF formulae on n variables is learnable with an algorithm that uses $L(h, (2n)^{k+1})$ positive examples and no call to the oracle.*

The proof is constructive and relies on Algorithm 1 below. In this algorithm, the initialization of the learned function g to the false constraint expressed as the conjunction of all possible *clauses* (i.e. disjunctions of litterals) leads to the learning of a minimal generalization of the positive examples with no false positive and low probability of false negatives.

Algorithm 1. PAC-learning of k-CNF formulae.

---

1. initialise g to the conjunction of all the $(2n)^k$ possible clauses of at most k literals,
2. do $L(h, (2n)^{k+1})$ times
    (a) $v := \text{SAMPLE}()$
    (b) delete all the clauses in g that do not contain a literal true in v
3. output: g

In our implementation of the PAC-learning algorithm for k-CNF formulae, we shall make use of the lattice structure of k-clauses ordered by implication. Interestingly, this data structure allows for

- $O(1)$ access to any k-clause;
- and for a clause c, $O(1)$ access to the smallest clauses implied by c and to the biggest clauses that imply c.

The class of monotone DNF formulae is also learnable. Let the *degree* of a Boolean formula be the largest number of prime implicants (i.e., minimal formulae covering one of the product-terms of the Boolean formula expressed as a sum of products) in an equivalent rewriting of the formula as a non-redundant sum of prime-implicants.

Theorem 2 ([24]). *The class of monotone DNF formulae on n variables is also learnable with an algorithm that uses $L(h, d)$ examples and dn calls to the oracle, where d is the* degree *of the function to learn.*

The proof relies on Algorithm 2. As previously, the algorithm guarantees that a minimal generalization is learned from both the samples and the oracle. The polynomial computational complexity follows from the fact that each monomial m is a prime implicant of f by construction, and that it is constructed by at most n calls to the oracle.

Algorithm 2. PAC-learning of monotone DNF formulae.

---

1. initialise g with false (constant zero),
2. do $L(h, d)$ times
   (a) $v :=$ SAMPLE()
   (b) if $v \Rightarrow g$ exit
   (c) for $i := 1$ *to* n
       i. if x_i is determined in v
          A. $v^* := v[x_i \leftarrow *]$
          B. if ORACLE(v^*) then
             – $v := v^*$
             – $m := \bigwedge_{v \Rightarrow x_j} x_j \wedge \bigwedge_{v \Rightarrow \neg x_k} \neg x_k$
             – $g := g \vee m$
3. output: g

---

3   Influence Models of Molecular Cell Processes

In this section, we present the formalism of influence systems used to model regulatory networks in cell molecular biology. We assume again a finite set of molecular species $\{x_1, \ldots, x_n\}$ and consider Boolean states that represent the activation or presence of each molecular species of the system, i.e. vectors in $\mathbb{B}^n$ that specify whether or not the ith species is present, or the ith gene activated.

3.1   Influence Systems with Forces

Influence systems with forces have been introduced in [9] to generalize the widely used logical models of regulatory networks *à la* Thomas [22], in order to provide them with a hierarchy of semantics including quantitative differential and stochastic semantics, similarly to reaction systems [10].

Definition 2 ([9]). *An* influence system I *is a set of quintuples* (P, I, t, σ, f) *called* influences, *noted in the examples below in Biocham v4[3] syntax,*
f for P/I -> t *if* $\sigma = +$, *and* f for P/I -< t *if* $\sigma = -$, *where*

---
[3] http://lifeware.inria.fr/biocham4.

- *P is a multiset on S, called* positive sources *of the influence,*
- *I a multiset of* negative sources,
- *t ∈ S is the* target,
- *σ ∈ {+, −} is the* sign *of the influence, accordingly called either* positive *or* negative influence,
- *and f : $\mathbb{R}_+{}^n \rightarrow \mathbb{R}_+$ is a function[4] called the* force *of the influence.*

The positive sources are distinguished from the negative sources of an influence (positive or negative), in order to annotate the fact that in the differential semantics, the source increases or decreases the force of the influence, and in the Boolean semantics with negation whether the source, or the negation of the source, is a condition for a change in the target.

Example 1. The classical birth-death model of Lotka–Volterra can be represented by the following influence system between a proliferating prey A and a predator B:

```
k1 * A * B for A, B -< A.
k1 * A * B for A, B -> B.
k2 * A for A -> A.
k3 * B for B -< B.
```

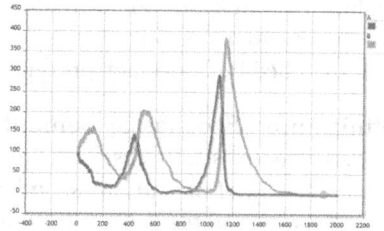

Time

The influence forces can be used for differential or stochastic simulation as above. This example contains both positive and negative influences but no influence inhibitor, i.e. no negative source in the influences: $(\{A, B\}, \emptyset, A, -, k1 * A * B)$, $(\{A, B\}, \emptyset, B, +, k1 * A * B)$, $(\{A\}, \emptyset, A, +, k2 * A)$ and $(\{B\}, \emptyset, B, -, k3 * B)$. For an example of influence with inhibitor, one can consider the specific inhibition of the proliferation rate of A by some variable C (which is distinguished from a general negative influence of C on A) by writing C as an inhibitor of the positive influence of A on A: `k2 * A/(1 + C) for A/C -> A.`

Definition 3 (Boolean Semantics). *The Boolean semantics (resp. positive Boolean semantics) of an influence system $\{(P_i, I_i, t_i, \sigma_i, f_i)\}_{1 \leq i \leq n}$ over a set S of n variables, is the Boolean transition system $\longrightarrow$ defined over Boolean state vectors in $\mathbb{B}^n$ by $x \longrightarrow x'$ if there exists an influence $(P_i, I_i, t_i, \sigma_i, f_i)$ such that $x \models \bigwedge_{p \in P_i} p \bigwedge_{n \in I_i} \neg n$ (resp. $x \models \bigwedge_{p \in P_i} p$) and $x' = x \, \sigma_i \, t_i$.*

where adding (resp. subtracting) t amounts to making the corresponding coordinate true (resp. false).

Equivalently, the Boolean semantics of an influence system over n species, $x_1, \ldots, x_n$, can be represented by n activation and n deactivation Boolean functions, which determine the possible transitions from each Boolean state:

---

[4] More precisely, in a well-formed influence system, f is assumed to be partially differentiable; $x_i \in P$ if and only if $\sigma = +$ (resp. $-$) and $\partial f / \partial x_i(x) > 0$ (resp. < 0) for some value $x \in \mathbb{R}_+^n$; and $x_i \in I$ if and only if $\sigma = +$ (resp. $-$) and $\partial f / \partial x_i(x) < 0$ (resp. > 0) for some value $x \in \mathbb{R}_+^n$.

Definition 4 (Boolean Activation Functions). *The* Boolean activation functions $x_k{}^+$, $x_k{}^-$: $\{0,1\}^n \rightarrow \{0,1\}$, $1 \leq k \leq n$, *of an influence system* $\mathcal{M}$ *are*

$$x_k{}^+ = \bigvee_{(P,I,x_k,+,f)\in\mathcal{M}} \bigwedge_{p\in P_i} p \bigwedge_{n\in I_i} \neg n \qquad\qquad x_k{}^- = \bigvee_{(P,I,x_k,-,f)\in\mathcal{M}} \bigwedge_{p\in P_i} p \bigwedge_{n\in I_i} \neg n$$

The positive activation functions are defined without negation by ignoring the inhibitors.

Conversely any system of Boolean activation functions can be represented by an influence system by putting the activation functions in DNF, and associating an influence to each conjunct.

Note that the positive Boolean semantics simply ignores the negative sources of an influence. This is motivated by the abstraction and approximation relationships that link the Boolean semantics to the stochastic semantics and to the differential semantics, for which the presence of an inhibitor decreases the force of an influence but does not prevent it to apply [9].

Definition 5 (Stochastic Semantics). *The stochastic semantics (resp. positive stochastic semantics) of an influence system* $\{(P_i, I_i, t_i, \sigma_i, f_i)\}_{1\leq i\leq n}$ *over a set* S *of* n *variables, relies on the transition system* $\longrightarrow$ *defined over discrete states, i.e. vectors in* $\mathbb{N}^n$, *by* $\forall(P_i, I_i, t_i, \sigma_i, f_i), \boldsymbol{x} \longrightarrow \boldsymbol{x}'$ *with propensity* f_i *if* $\boldsymbol{x} \geq P_i, \boldsymbol{x} < I_i$ *(resp. no condition on* I_i*) and* $\boldsymbol{x}' = \boldsymbol{x}\,\sigma_i\,t_i$. *Transition probabilities between discrete states are obtained through normalization of the propensities of all enabled transitions, and the time of next transition is given by an exponential distribution* [13].

We call a positive influence system, an influence system without inhibitors or interpreted under the positive semantics.

3.2   Monotone DNF Representation of Positive Influence Systems

Definition 4 shows how to represent an influence system by $2 * n$ activation functions in DNF, and *positive* influence systems by *monotone* DNF activation functions.

Example 2. The activation functions of the Lotka–Volterra influence system of Example 1 are monotonic DNF formulae with only one conjunct since in this example there is only one signed influence per variable:

$$A^+ = (A) \qquad\qquad B^+ = (A \wedge B)$$
$$A^- = (A \wedge B) \qquad\qquad B^- = (B)$$

3.3 k-CNF Representation of General Influence Systems

Monotone DNF formulae cannot encode the Boolean dynamics of influence systems with negation, which tests the absence of inhibitors, i.e., negative literals. This is possible using a k-CNF representation of the activation functions, provided that there are at most k species that can play a given "role". For instance, in a hypothetic activation function in CNF $(a \vee b \vee c) \wedge (d \vee e) \wedge \neg f$, each clause can be interpreted as a role, and each role can be played by a limited number of species, at most k.

Example 3. The activation functions of the prey-predator model with inhibition of Example 1 cannot be represented by monotone formulae. They can however be represented by the following 1-CNF formulae ($k = 1$ since there is only one positive and one negative influence for each target):

$$A^+ = (A) \wedge (\neg C) \qquad\qquad A^- = (A) \wedge (B)$$
$$B^+ = (A) \wedge (B) \qquad\qquad B^- = (B)$$

Example 4. In Sect. 5, we shall study a model of T lymphocyte differentiation which contains 2-CNF activation functions, for instance

$$\text{IFNg}^+ = (\text{STAT4} \vee \text{TBet}) \qquad \text{IFNg}^- = (\neg \text{STAT4}) \wedge (\neg \text{TBet})$$

3.4 k-CNF Models of Thomas Functional Influence Systems

Definition 6 ([22]). *A Thomas network on a finite set of genes $\{x_1, \ldots, x_n\}$ is defined by n Boolean functions $\{f_1, \ldots, f_n\}$ which give for each gene its possible next state, given the current state.*

The difference with the previous general influence systems is that the activation and deactivation functions are exclusive and defined by one single function. As shown in [9], non-terminal self-loops cannot be represented in Thomas functional influence systems. Given a general influence system with activation functions $x_i{}^+$ and $x_i{}^-$, one can associate a Thomas network with attractor function[5]

$$f_i(v) = \begin{cases} 1 \text{ if } \begin{cases} v_i = 0 \text{ and } x_i{}^+(v) = 1 \\ v_i = 1 \text{ and } x_i{}^-(v) = 0 \end{cases} \\ 0 \text{ if } \begin{cases} v_i = 0 \text{ and } x_i{}^+(v) = 0 \\ v_i = 1 \text{ and } x_i{}^-(v) = 1 \end{cases} \end{cases}$$

k-CNF formulae can again be used to represent Thomas gene regulatory network functions with some reasonable restrictions on their connectivity. In particular, it is worth noticing that in Thomas networks of degree bounded by k, each gene has at most k regulators, each gene activation function f_i thus depends of at most k variables and can consequently be represented by a k-CNF formula.

---

[5] Note that this function ignores the cases where $v_i = 0$ and $x_i{}^-(v) = 0$, or $v_i = 1$ and $x_i{}^+(v) = 1$ which may create loops in non-terminal states in general influence systems.

Example 5. The above translation applied to Example 1 gives $f_A = A \wedge \neg B, f_B = 0$. Note that the form of f_B means that the only possible state change for B is from 1 to 0.

Example 6. The T-lymphocyte model studied in Sect. 5 is originally a Thomas' network, where we have, for instance: $f_{\text{IFNg}} = (\text{STAT4} \vee \text{TBet})$

4   PAC Learning from Traces

4.1   Diverse Initial States Versus Long Time Horizon

In practice, one cannot assume to have full access to the hidden Boolean function as required by SAMPLE and ORACLE, but rather to data time-series, or traces, produced from biological experiments. For the scope of this paper, we consider simulation traces obtained from a hidden model which we wish to discover. Two types of traces are considered: Boolean and stochastic simulation traces. In both cases, the mapping to the concepts of PAC-learning is easy: a SAMPLE for the activation function x^+ (resp. deactivaction function x^-) is a state s_i such that $x_i < x_{i+1}$ (resp. $x_i > x_{i+1}$). See Fig. 1 for an example.

$$\cdots \rightarrow \begin{pmatrix} 0 \\ 1 \\ 0 \\ 1 \end{pmatrix} \rightarrow \begin{pmatrix} \mathbf{1} \\ 1 \\ 0 \\ 1 \end{pmatrix} \rightarrow \begin{pmatrix} 1 \\ 1 \\ 0 \\ \mathbf{0} \end{pmatrix} \rightarrow \cdots$$
$$\quad\quad (a) \quad\quad\quad (b) \quad\quad\quad (c)$$

Fig. 1. Illustration of a Boolean trace with three steps. Between a and b, the first gene has been activated, and between b and c, the last one has been deactivated.

One striking feature of PAC learning is to associate a guarantee on the quality h of each learnt Boolean function depending on the number of samples used, namely $L(h, (2n)^{k+1})$, where n is the number of genes/molecules observed, and k is the maximum number of literals per conjunct. In practice, k seems to be limited to 3 or 2, and the number $2n$ of different possible literals in a clause, can also be reduced through prior knowledge (e.g. in Sect. 5.3).

It is worth noticing that the global guarantee on the learnt model is the minimum of all precision bounds h. In order to perfectly recover a hidden model, it is thus necessary to have sufficiently diverse samples. For this reason, one should expect to get better performance with large sets of short traces obtained from a uniformly distributed set of initial states, rather than with a small set of long traces which introduce a bias in the distribution of the transition samples (e.g. when looping in an attractor). The important point is that PAC learning algorithms do provide bounds on the error according to this space-time trade-off.

On the other hand, the ORACLE procedure needs to evaluate the (de)activation function on a given vector v, that is, it needs to be able to set the system in a state abstracted by v and say whether or not a given gene can be (de)activated from this state. In practice, this cannot be achieved without approximation. The intuitive solution would be to set the system in the desired state and see whether or not the gene is (de)activated. However, different atomic steps are possible from a given state and we have no guarantee that the one we are interested in will happen in a given finite number of runs. These considerations militate for studying an extension of the PAC-learning framework with an oracle that would be only probabilistic.

4.2   PAC Learning from Boolean Traces

A first experiment was to produce Boolean (de)activation traces by simulation of a given influence model, and use them to learn the hidden model. Figure 2 reports our results obtained with 25 Boolean traces of short length equal to 2 (i.e. when trading time for space) on Example 1, where to increase readability we used long names for the species. It is worth noticing that in this particular model, the positive infuences cannot be learned from (de)activation traces, since they contain their target as positive source and thus do not correspond to an activation function. Indeed, the activation functions in the Lokta–Volterra models report the apparition on extinction of the species' population as a whole and not of individuals of it. The results in this tradeoff are perfect in the sense that the negative influences are correctly inferred.

```
Prey, Predator -< Prey.
Prey, Predator -> Predator.
Prey -> Prey.
Predator -< Predator.
```

```
Predator -< Predator
Predator, Prey -< Prey
```

Fig. 2. The Lokta-Volterra prey vs. predator influence model of Example 1 with long names (left panel) and the (most likely) influence model PAC-learned on 25 simulations of length 2 (right panel) from random initial states.

```
Predator+ : False
Predator- : Predator /\ !Prey
Prey+ : False
Prey- : Predator /\ Prey
```

```
Predator / Prey -< Predator
Predator, Prey -< Prey
```

Fig. 3. Most likely PAC-learned activation functions (left pane, where !A stands for $\neg A$), and corresponding influence model (rigth panel obtained by CNF-DNF conversion), on a *single random Boolean trace of length 50* from the standard initial state with prey and predator present.

On the other hand, PAC learning from a single Boolean trace obtained from the standard initial state where both the prey and the predator are present (i.e. trading space for time), most likely leads to the influence model shown in Fig. 3. For the prey to go extinct, there must be both a prey in the first place and a predator to eat it. This is correct. For the predator to disappear, it is necessary that there is a predator in the first place and that there is no prey. The first part of this conjunction is true, but the second is false: predators may disappear even if there are preys left. However, this case is unlikely, the most likely case is that the predator will go extinct only once there are no more preys left for it to eat. As can be seen even on this very simple example, the "approximately" in PAC has a precise meaning. Yet, as explained in Definition 1, the quantification of this approximation relies on the knowledge of the distributions of the samples. In the present case, the probability of a positive example v of (de)activation function $x\pm$ to be sampled is strongly and intuitively correlated to both the probability that the system reaches state v and the probability of the actual (de)activation of gene x from state v.

4.3   PAC Learning from Stochastic Traces

Let us now consider sets of stochastic traces. They can be produced from an influence system with forces, using Gillespie's algorithm (Definition 5), assuming here mass-action kinetics with rate 1 for all influences. The initial states are random, but with equal probability to be 0 or > 0 in order to facilitate the observation of the inhibitions in the influences. The states in $\mathbb{N}^n$ can be

```
biocham: pac_learning('library:examples/lotka_volterra/LVi.bc
      ', 50, 1).
% Maxmimum K used:  minimum number of samples for h=1: 18

% 14 samples (max h ~ 0.7777777777777778)
Predator -< Predator

% 7 samples (max h ~ 0.3888888888888889)
Predator,Prey -> Predator

% 1 samples (max h ~ 0.05555555555555555)
Predator,Prey -< Prey

% 21 samples (max h ~ 1.1666666666666667)
Prey -> Prey
```

Listing 1: Biocham running the k-CNF PAC learning algorithm on the Lotka–Volterra influence model from stochastic simulation traces of length 1, obtained from 50 random initial states. Among those 50 initial states, 7 had both prey and predator absent, leading to no sample.

abstracted to Boolean samples by the usual $\{0, > 0\}$ abstraction for the states, and the increasing/decreasing abstraction for choosing samples for the activation/deactivation functions. Using the same $\{0, > 0\}$ abstraction to detect samples would again forbid to learn autocatalytic influences like `Prey -> Prey` for the same reason as in the Boolean case.

Interestingly, Listing 1 shows that here again, even with a low number of samples, and therefore a very low precision bound h, one can find the full model with less than 50 simulations of length 1, all starting from random initial states.

5   Evaluation on a Model of T-Helper Lymphocytes Differentiation

5.1   Boolean Thomas Network

In this section we evaluate the performance of the k-CNF PAC learning algorithm on an influence system of 12 variables and 32 influences that models the differentiation of the T-helper lymphocytes. This model, presented in [21] is actually a Boolean simplification of the original multi-level model of [17]. It studies the regulatory network of stimuli leading to differentiation between Th-1 and Th-2 lymphocytes from an original CD4+ T helper (Th-0). The model has three different stable states corresponding to Th-0 (naive lymphocyte), Th-1 and Th-2 when IL12 is off, and two others when IL12 is on (the Th-0 one is lost). Figure 4 shows the influence graph of the model. The influence model is given in Listing 2.

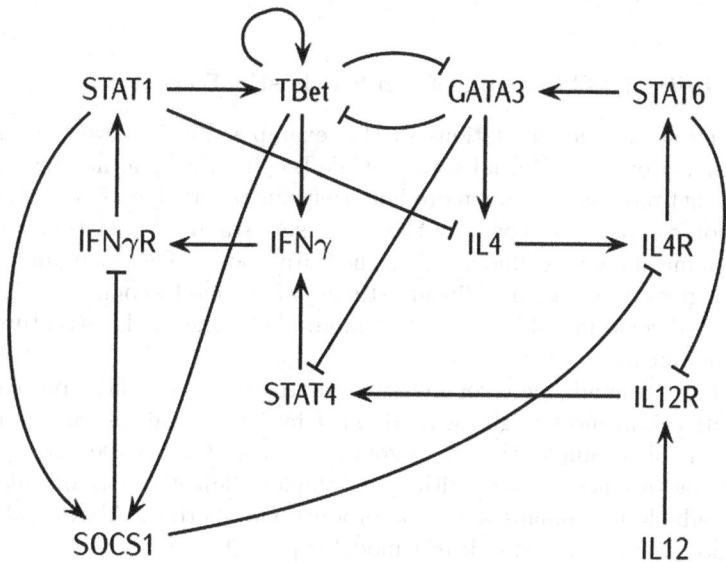

Fig. 4. Figure 4 of [21] displaying the Th-lymphocyte differentiation model.

```
STAT4 , TBet -> IFNg.            / IFNgR -< STAT1.
/ STAT4 -< IFNg.                 IL4R -> STAT6.
/ TBet -< IFNg.                  / IL4R -< STAT6.
GATA3 / STAT1 -> IL4.            IL12R / GATA3 -> STAT4.
/ GATA3 -< IL4.                  / IL12R -< STAT4.
STAT1 -< IL4.                    GATA3 -< STAT4.
IFNg / SOCS1 -> IFNgR.           STAT1 -> SOCS1.
/ IFNg -< IFNgR.                 TBet -> SOCS1.
SOCS1 -< IFNgR.                  / STAT1 , TBet -< SOCS1.
IL4 / SOCS1 -> IL4R.             STAT6 / TBet -> GATA3.
/ IL4 -< IL4R.                   STAT1 / GATA3 -> TBet.
SOCS1 -< IL4R.                   TBet / GATA3 -> TBet.
IL12 / STAT6 -> IL12R.           GATA3 -< TBet.
/ IL12 -< IL12R.                 / STAT1 , TBet -< TBet.
STAT6 -< IL12R.                  / STAT6 -< GATA3.
IFNgR -> STAT1.                  TBet -< GATA3.
```

Listing 2: Influence system for the lymphocyte differentiation of Example 5.

All learning experiments described below run on a 3 GHz Linux desktop in less than 3 s. However, the CNF (activation functions) to DNF (influence model) conversions could be very slow, reaching more than 4 min in the worst cases (e.g. with a single simulation of 10^6 steps). Note also that since IL12 is an input, in all experiments the PAC learning algorithm only finds *false* as Boolean function for its activation or deactivation. We thus removed it from the results below for readability.

5.2   *Ab initio* PAC Learning from Stochastic Traces

When using stochastic simulations in this example, Fig. 5 shows that a simple randomization of the initial states (while keeping the total number of samples constant) provides a much more homogeneous repartition of activation and deactivation samples (as shown by the decreasing standard deviation), and, as expected, a much higher confidence h in the learnt model. The minimum number of samples gives in fact a quasi-linear estimate of the model confidence h. Obviously, more diverse initial states reveal more about the model structure than longer experiments.

On the other hand, the error measured as the number of false positive and false negative influences (right scale divided by 10), reveals a non monotonic behavior: in this example, there is a zone with sets of 10 to 100 traces, where PAC learning produces models with very complex (de)activation formulae that are not readable by humans and that produce many errors. Above 500 traces from random initial states the learnt model is perfect.

The guarantee on the accuracy of the learnt model comes directly from Valiant's work with the approximation bounds. Note however that Valiant's results stand only if we actually have at least L samples for each of the

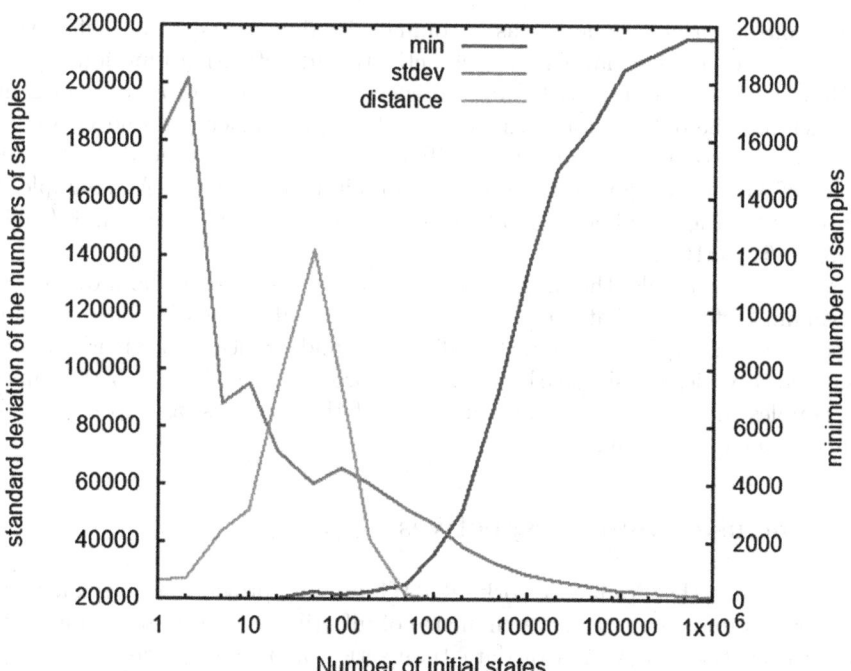

Fig. 5. Minimum numbers of (de)activation samples with standard deviations, and model errors (i.e. false positive and negative influences) obtained for the 24 boolean functions of the Th-lymphocyte example, as a function of the number of initial states (with total number of samples kept constant by adjusting the time horizon).

(de)activation functions, where L is Valiant's bound. A first naive approach might be to simply let the trace run for $2nL$ steps, or a constant factor of it. Nevertheless, the repartition of samples for each function can be pretty non-uniform. Interestingly, the minimal number of samples gives us the lowest $L(h, S)$ and thus quasi-linearly the lowest guarantee h.

Our simulation results show that when using PAC-learning to find the structure of a regulatory model, an approach based on mutants (knock-offs, over-expression, etc.) is much more informative than an approach based on (repeated or longer) similar observations. Note that this is in line with the pratice of integrative analyses such as [6] and its more than 130 mutants.

5.3   PAC Learning with Prior Knowledge on the Influence Graph

Furhtermore, to improve the guarantee h and the corresponding accuracy of the learnt models, especially for bigger models, it is necessary to look again at what constrains h. We have $samples = 2h(S + \log h)$, where S is the number of possible k-CNF clauses, bounded by $(2n)^{k+1}$.

The previous section explored the diversity of the samples, another option is to reduce S for a given n. This can be done by formalizing possible/known interactions as prior knowledge, as is common in Machine Learning, effectively restricting the possible clauses for each activation/deactivation function.

Here we want the user to be able to specify, for each gene x, a set of gene V_x which are the only ones on which x^+ and x^- may depend. If one views the influences as a graph, this is akin to specifying a set of possible (undirected) edges outside of which the algorithm cannot build its influence system. An example of such hints for the lymphocyte model are given at http://lifeware.inria.fr/wiki/software/#CMSB17.

In such an example, the number of possible clauses becomes bounded by 3^3 (maximum 3 effectors that are either a positive literal, a negative literal or not in the clause at all) instead of 26^4 (Valiant's bound). Since for a given number of samples h varies quasi-linearly in S, the improvement is drastic (50000 times less samples for the same h). The accuracy of the model is, as expected with such guarantee, improving a lot.

6 Conclusion and Perspectives

We have shown that Valiant's work on PAC learning provides an elegant trail, with error bounds, to attack the challenge of inferring the structure of influence models from the observation of data time series, and more precisely to automatically discover possible regulatory networks of a biochemical process, given sufficiently precise observations of its executions.

The Boolean dynamics of biochemical influence systems, including Thomas regulatory networks, can be represented by k-CNF formulae without loss of generality, and k-CNF PAC learning algorithm can be used to infer the structure of the network, and bound the errors made according to the distribution of the state transition samples and the space-time tradeoff in the traces. When dimension increases, we have shown on an example of T-lymphocyte differentiation from the literature that the k-CNF PAC learning algorithm can also leverage available prior knowledge on the system to deliver precise results with a reasonable amount of data.

The Boolean dynamics of positive influence systems can also be straightforwardly represented by monotone DNF activation and deactivation functions, and monotone DNF PAC learning algorithm applied with an interesting recourse to oracles which are particularly relevant in the perspective of online active learning and experimental design. More work is needed however to make comparisons on common benchmarks with other approaches already investigated in this context, such as Answer Set Programming (ASP) and budgeted learning, and to investigate the applicability of these methods to real experiments taking into account the noise in the observations.

Acknowledgements. This work is partly supported by the ANR project Hyclock.

References

1. Angelopoulos, N., Muggleton, S.H.: Machine learning metabolic pathway descriptions using a probabilistic relational representation. Electron. Trans. Artif. Intell. **7**(9), 1–11 (2002). also in Proceedings of Machine Intelligence
2. Angelopoulos, N., Muggleton, S.H.: Slps for probabilistic pathways: Modeling and parameter estimation. Technical Report TR 2002/12. Department of Computing, Imperial College, London, UK (2002)
3. Bernot, G., Comet, J.P., Richard, A., Guespin, J.: A fruitful application of formal methods to biological regulatory networks: Extending Thomas' asynchronous logical approach with temporal logic. J. Theor. Biol. **229**(3), 339–347 (2004)
4. Bryant, C.H., Muggleton, S.H., Oliver, S.G., Kell, D.B., Reiser, P.G.K., King, R.D.: Combining inductive logic programming, active learning and robotics to discover the function of genes. Electron. Trans. Artif. Intell. **6**(12), 1–36 (2001)
5. Calzone, L., Chabrier-Rivier, N., Fages, F., Soliman, S.: Machine learning biochemical networks from temporal logic properties. In: Priami, C., Plotkin, G. (eds.) Transactions on Computational Systems Biology VI. LNCS, vol. 4220, pp. 68–94. Springer, Heidelberg (2006). doi:10.1007/11880646_4
6. Chen, K.C., Calzone, L., Csikász-Nagy, A., Cross, F.R., Györffy, B., Val, J., Novàk, B., Tyson, J.J.: Integrative analysis of cell cycle control in budding yeast. Mol. Biol. Cell **15**(8), 3841–3862 (2004)
7. Deng, K., Bourke, C., Scott, S.D., Sunderman, J., Zheng, Y.: Bandit-based algorithms for budgeted learning. In: ICDM (2007)
8. Deng, K., Zheng, Y., Bourke, C., Scott, S., Masciale, J.: New algorithms for budgeted learning. Mach. Learn. **90**, 59–90 (2013)
9. Fages, F., Martinez, T., Rosenblueth, D.A., Soliman, S.: Influence systems vs Reaction systems. In: Bartocci, E., Lio, P., Paoletti, N. (eds.) CMSB 2016. LNCS, vol. 9859, pp. 98–115. Springer, Cham (2016). doi:10.1007/978-3-319-45177-0_7
10. Fages, F., Soliman, S.: Abstract interpretation and types for systems biology. Theor. Comput. Sci. **403**(1), 52–70 (2008)
11. Gebser, M., Kaufmann, B., Neumann, A., Schaub, T.: *clasp*: A conflict-driven answer set solver. In: Baral, C., Brewka, G., Schlipf, J. (eds.) LPNMR 2007. LNCS (LNAI), vol. 4483, pp. 260–265. Springer, Heidelberg (2007). doi:10.1007/978-3-540-72200-7_23
12. Gebser, M., Schaub, T., Thiele, S., Usadel, B., Veber, P.: Detecting inconsistencies in large biological networks with answer set programming. In: Garcia de la Banda, M., Pontelli, E. (eds.) ICLP 2008. LNCS, vol. 5366, pp. 130–144. Springer, Heidelberg (2008). doi:10.1007/978-3-540-89982-2_19
13. Gillespie, D.T.: Exact stochastic simulation of coupled chemical reactions. J. Phys. Chemis. **81**(25), 2340–2361 (1977)
14. Gordon, A.D., Henzinger, T.A., Nori, A.V., Rajamani, S.K.: Probabilistic programming. In: Proceedings of the on Future of Software Engineering, FOSE 2014, pp. 167–181, NY, USA. ACM, New York (2014)
15. Hill, S.M., et al.: Inferring causal molecular networks: empirical assessment through a community-based effort. Nat. Method. **1**(4), 310–318 (2016)
16. Llamosi, A., Mezine, A., dÁlché-Buc, F., Letort, V., Sebag, M.: Experimental design in dynamical system identification: a bandit-based active learning approach. In: Calders, T., Esposito, F., Hüllermeier, E., Meo, R. (eds.) ECML PKDD 2014. LNCS, vol. 8725, pp. 306–321. Springer, Heidelberg (2014). doi:10.1007/978-3-662-44851-9_20

17. Mendoza, L.: A network model for the control of the differentiation process in Th cells. Biosystems **84**(2), 101–114 (2006)
18. Meyer, P., Cokelaer, T., Chandran, D., Kim, K.H., Loh, P.R., Tucker, G., Lipson, M., Berger, B., Kreutz, C., Raue, A., Steiert, B., Timmer, J., Bilal, E., Sauro, H.M., Stolovitzky, G., Saez-Rodriguez, J.: Network topology and parameter estimation: from experimental design methods to gene regulatory network kinetics using a community based approach. BMC Syst. Biol. **8**(1), 1–18 (2014)
19. Muggleton, S.H.: Inverse entailment and progol. New Gener. Comput. **13**, 245–286 (1995)
20. Ostrowski, M., Paulevé, L., Schaub, T., Siegel, A., Guziolowski, C.: Boolean network identification from perturbation time series data combining dynamics abstraction and logic programming. Biosystems **149**, 139–153 (2016)
21. Remy, E., Ruet, P., Mendoza, L., Thieffry, D., Chaouiya, C.: From logical regulatory graphs to standard petri nets: dynamical roles and functionality of feedback circuits. In: Priami, C., Ingólfsdóttir, A., Mishra, B., Riis Nielson, H. (eds.) Transactions on Computational Systems Biology VII. LNCS, vol. 4230, pp. 56–72. Springer, Heidelberg (2006). doi:10.1007/11905455_3
22. Thomas, R.: Boolean formalisation of genetic control circuits. J. Theor. Biol. **42**, 565–583 (1973)
23. Thomas, R.: Regulatory networks seen as asynchronous automata : a logical description. J. Theor. Biol. **153**, 1–23 (1991)
24. Valiant, L.: A theory of the learnable. Commun. ACM **27**(11), 1134–1142 (1984)
25. Valiant, L.: Probably Approximately Correct. Basic Books (2013)
26. Videla, S., Konokotina, I., Alexopoulos, L.G., Saez-Rodriguez, J., Schaub, T., Siegel, A., Guziolowski, C.: Designing experiments to discriminate families of logic models. Front. Bioeng. Biotechnol. **3**, 131 (2015)

Identifying Functional Families of Trajectories in Biological Pathways by Soft Clustering: Application to TGF-β Signaling

Jean Coquet[1,2]([✉]), Nathalie Theret[1,2], Vincent Legagneux[2], and Olivier Dameron[1]

[1] Université de Rennes 1 - IRISA/INRIA, UMR6074,
263 avenue du Général Leclerc, 35042 Rennes, Cedex, France
`jean.coquet@inria.fr`
[2] INSERM U1085 IRSET, Université de Rennes 1,
2 avenue Pr Léon Bernard, 35043 Rennes, Cedex, France

Abstract. The study of complex biological processes requires to forgo simplified models for extensive ones. Yet, these models' size and complexity place them beyond understanding. Their analysis requires new methods for identifying general patterns. The Transforming Growth Factor TGF-β is a multifunctional cytokine that regulates mammalian cell development, differentiation, and homeostasis. Depending on the context, it can play the antagonistic roles of growth inhibitor or of tumor promoter. Its context-dependent pleiotropic nature is associated with complex signaling pathways. The most comprehensive model of TGF-β-dependent signaling is composed of 15,934 chains of reactions (trajectories) linking TGF-β to at least one of its 159 target genes. Identifying functional patterns in such a network requires new automated methods.

This article presents a framework for identifying groups of similar trajectories composed of the same molecules using an exhaustive and without prior assumptions approach. First, the trajectories were clustered using the Relevant Set Correlation model, a shared nearest-neighbors clustering method. Five groups of trajectories were identified. Second, for each cluster the over-represented molecules were determined by scoring the frequency of each molecule implicated in trajectories. Third, Gene set enrichment analysis on the clusters of trajectories revealed some specific TGF-β-dependent biological processes, with different clusters associated to the antagonists roles of TGF-β. This confirms that our approach yields biologically-relevant results. We developed a web interface that facilitates graph visualization and analysis.

Our clustering-based method is suitable for identifying families of functionally-similar trajectories in the TGF-β signaling network. It can be generalized to explore any large-scale biological pathways.

Keywords: TGF-β · Signaling pathways · Discrete dynamic model · Soft clustering · RSC model

© Springer International Publishing AG 2017
J. Feret and H. Koeppl (Eds.): CMSB 2017, LNBI 10545, pp. 91–107, 2017.
DOI: 10.1007/978-3-319-67471-1_6

1   Introduction

Living cells use molecular signaling networks to adapt their phenotype to the microenvironment modifications. In order to decipher the dynamic of signaling pathways, mathematical models have been developed using different strategies [4,10]. Differential equation-based models are limited to small networks due to the explosion in the number of variables in complex networks and the lack of known quantitative values for the parameters [1]. Qualitative modeling approaches based on events discretization have been successfully applied to large networks. In qualitative models, signaling networks are represented as a graph where each node (genes or proteins) is represented by a finite-state variable and edges describe interactions between biomolecules as rules [17]. Such models proved to be suitable for describing the qualitative nature of biological information whithin large and complex signaling pathways [19].

Signaling by the polypeptide Transforming Growth Factor TGF-β is one of the most intriguing signaling networks that govern complex multifunctional profiles. TGF-β was first described as a potent growth inhibitor for a wide variety of cells. It affects apoptosis and differentiation thereby controlling tissue homeostasis [7]. At the opposite, upregulation and activation of TGF-β has been linked to various diseases, including fibrosis and cancer through promotion of cell proliferation and invasion [24]. The pleiotropic effects of TGF-β are associated to the diversity of signaling pathways that depend on the biological context [13]. TGF-β binding to the receptor complex induces the phosphorylation of intracellular substrates, R-Smad proteins which hetero-dimerize with Smad4. The Smad complexes move into the nucleus where they regulate the transcription of TGF-β-target genes. Alternatively, non-Smad pathways are activated by ligand-occupied receptor to modulate downstream cellular responses [14]. These non-Smad pathways include mitogen-activated protein kinase (MAPK) such as p38 and Jun N-terminal kinase (JNK) pathways, Rho-like GTPase signaling pathways, and phosphatidylinositol-3-kinase/protein kinase B (PKB/AKT) pathways. Combinations of Smad and non-Smad pathways contribute to the high heterogeneity of cell responses to TGF-β. Additionally, many molecules from these pathways are involved in other signaling pathways activated by other microenvironment inputs, which leads to complex crosstalks [12].

Numerical approaches using differential models have been developed to describe the behavior of TGF-β canonical pathway involving Smad proteins [27]. Because of the numerous components and the lack of quantitative data, the non canonical pathways have never been included in these TGF-β models. To solve this problem, Andrieux et al. recently developed a qualitative discrete formalism compatible with large-scale discrete models [2]. The Cadbiom language is a state-transition formalism based on a simplified version of guarded transition [16]. It allows a fine-grained description of the system's dynamic behavior by introducing temporal parameters to manage competition and cooperation between parts of the models (http://cadbiom.genouest.org). Based on the Cadbiom formalism, Andrieux et al. integrated the 137 signaling pathways from the Pathway Interaction Database (PID) [20] and derived an exhaustive TGF-β signaling network

that includes canonical and non-Smad pathways [2]. Using this model they identified 15,934 signaling trajectories regulating 145 TGF-β target genes and found
specific signatures for activating TGF-β-dependent genes.

Characterizing these 15,934 signaling trajectories remains a challenging task.
They are mainly composed of signaling molecules whose modularity and combination are the base of cell response plasticity and adaptability [9,15,21]. We
developed a methodological approach to identify families of trajectories with
functional biological signature based on their signaling molecules content. The
major difficulty were the inner complexity of the networks, and the fact that
some molecules may be involved in multiple families, as suggested by TGF-β's
context-dependent roles. To address these challenges, we used an unsupervised
soft-clustering method to compare signaling trajectories according to their molecular composition. The clusters correspond to families of trajectories, and can
share common molecules. Our analysis does not rely on a priori knowledge on
the number of clusters nor on the membership of a molecule to a cluster. Based
on this approach, we identified five groups of signaling trajectories. Importantly
we further show that these five groups are associated with specific biological
functions thereby demonstrating the relevance of soft clustering to decipher cell
signaling networks.

2   Materials and Methods

Cellular signaling pathways are chains of biochemical reactions. Typically, they
encompass the interaction of signaling molecules such as growth factors with
receptors at the cell surface, the transmission of signal through signaling cascades involving many molecules such as kinases and finally the molecular networks involved in regulation of target gene transcription within the nucleus. In
order to decipher the complexity of signaling TGF-β-dependent networks and
for characterizing these trajectories, we focus on the proteins involved in the
reactions (reactants, products and catalyzers). Note that a gene can encode for
a protein implicated elsewhere in the pathway, so proteins and genes form non-
disjoint sets.

The trajectories are first submitted to a pre-processing step to generate a non
redundant set of signaling trajectories. The second step groups similar trajectories using soft clustering. The third step characterizes the specificity of groups
of trajectories by determining the over-represented proteins and their biological
function using semantic annotations.

2.1   Available Data and Pre-processing

The original data-set contained the 15,934 signaling trajectories involved in the
regulation of 145 TGF-β-dependent genes as previously described in [2]. A signaling trajectory is defined as a set of molecules required for activation of TGF-β-
dependent genes (Fig. 1A). Each original trajectory T_k was composed of TGF-β,
signaling molecules and a single target gene (Fig. 1B). There were 321 signaling

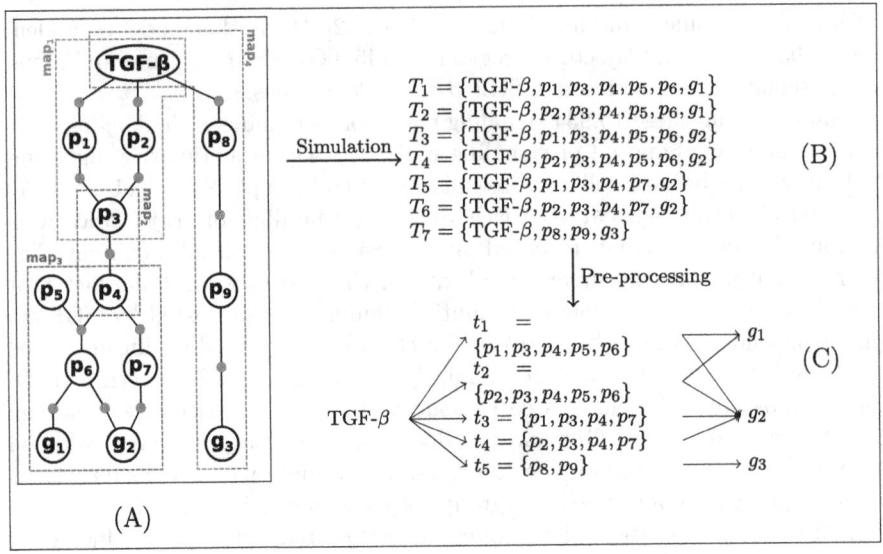

Fig. 1. Example of the generation of trajectories from PID maps and their pre-processing. (A) The signaling network made of 4 maps and is composed of proteins, TGF-β and genes. (B) Trajectories are defined by a set of proteins containing TGF-β, signaling proteins (p_i) and target genes (g_i). (C) Pre-processed trajectories are restricted to signaling proteins. After pre-processing, the trajectories T_1 and T_3 are represented by the trajectory t_1; T_2 and T_4 are represented by t_2; T_5 is represented by t_3; T_6 is represented by t_4; T_7 is represented by t_5.

molecules (identified by their uniprot ID) involved in at least one of the 15,934 signaling trajectories. To compare the trajectories based on their molecule composition, we first discarded TGF-β which was belonging to all the trajectories. Next we observed that several trajectories were composed of the same signaling molecules but differed only by the target genes. We decided to discard the target genes from the trajectories, and to represent separately the associations between trajectories and target genes (Fig. 1C). The motivation was (i) to avoid the artificial duplication of trajectories, and (ii) to have a model that represents explicitly the fact that a single chain of reactions can influence several genes. In the remainder of the article, the pre-processed trajectories are noted t_k and their set is noted S.

2.2 Clustering Method

We used the Relevant Set Correlation (RSC) model to identify clusters of trajectories [6]. This model uses as input a function $Q(t)$ that returns for every trajectory $t \in S$ a list of all the other trajectories in S sorted by their decreasing correlation with t.

$Q(t)$ function for ranking the trajectories by decreasing correlation to t. A trajectory $t_i \in S$ is represented by a binary vector v_i whose dimension is equal to the number of all proteins. The coordinate value of "1" indicates that the trajectory contains the protein, and the coordinate value of "0" indicates that the trajectory does not (see Table 1).

Table 1. Example of binary matrix representing the protein composition of trajectories. If a protein p_j is present in a trajectory t_i then the cell (i,j) is "1" else "0".

	p_1	p_2	p_3	p_4	p_5	p_6	p_7	p_8	p_9
t_1	1	0	1	1	1	1	0	0	0
t_2	0	1	1	1	1	1	0	0	0
t_3	1	0	1	1	0	0	1	0	0
t_4	0	1	1	1	0	0	1	0	0
t_5	0	0	0	0	0	0	0	1	1

Based on the binary vectors, we apply the Pearson correlation formula and construct a similarity matrix (see Table 2):

$$r(t_i, t_j) = \frac{\sum_{k=1}^{n}(t_{i,k} - \overline{t_i})(t_{j,k} - \overline{t_j})}{\sqrt{\sum_{k=1}^{n}(t_{i,k} - \overline{t_i})^2 \sum_{k=1}^{n}(t_{j,k} - \overline{t_j})^2}} \qquad (1)$$

where $(t_{i,1}, t_{i,2}, ..., t_{i,n})$ and $(t_{j,1}, t_{j,2}, ..., t_{j,n})$ are the vectors of trajectories t_i and t_j with $\overline{t_i}$ and $\overline{t_j}$ their respective average.

Table 2. Example of correlation matrix of trajectories t_i obtained from the trajectories' composition of Table 1. If two trajectories t_i, t_j have exactly the same proteins the value of the cell (i,j) is 1.0. If the trajectories do not share any proteins the value is 0.0.

	t_1	t_2	t_3	t_4	t_5
t_1	1.000	0.550	0.350	−0.100	−0.598
t_2	0.550	1.000	−0.100	0.350	−0.598
t_3	0.350	−0.100	1.000	0.550	−0.478
t_4	−0.100	0.350	0.550	1.000	−0.478
t_5	−0.598	−0.598	−0.478	−0.478	1.000

For each trajectory $t_k \in S$, the Pearson correlation gives a partial ordering $<t_i>_{i=1}^{|S|}$ of trajectories where $i < j$ implies that $r(t_k, t_i) \geq r(t_k, t_j)$ (see Table 3). If two trajectories have the same correlation score, they are sorted alphabetically. We define the $Q(t)$ function as follows:

$$Q(t_k) = <t_i>_{i=1}^{|S|} \qquad \forall (i,j) \in [1, |S|]^2, i < j \Rightarrow r(t_k, t_i) \geq r(t_k, t_j) \qquad (2)$$

Table 3. Example of partial ordering of all trajectories for every trajectory t_i. All trajectories are sorted for each trajectory t_k in function of their Pearson correlation score.

Q	1	2	3	4	5
t_1	t_1	t_2	t_3	t_4	t_5
t_2	t_2	t_1	t_4	t_3	t_5
t_3	t_3	t_4	t_1	t_2	t_5
t_4	t_4	t_3	t_2	t_1	t_5
t_5	t_5	t_3	t_4	t_1	t_2

Heuristic Algorithm for Clustering the Trajectories. The GreedyRSC method is an heuristic algorithm to apply the RSC model [6]. It performs a soft clustering, where the clusters may overlap and do not necessarily cover the entire data set. In addition to the $Q(t)$ function, it requires four parameters:

- x_1: Minimum size of cluster
- x_2: Maximum size of cluster
- x_3: Maximum interset significance score between two clusters.
- x_4: Minimum significance score.

Houle [6] defines the significance score by the function $Z_1(A)$ and the inter-set significance score by the function $Z_1(A, B)$ where A and B are two clusters.

The minimum size x_1 of pattern means that all clusters would be composed of at least x_1 trajectories. To respect this constraint, we have to choose the minimum significance score $x_4 = \sqrt{x_1(|S| - 1)}$ where $|S|$ is the number of trajectories. We can prove the computation of the minimum significance score as follows:

Let A be a cluster (set of trajectory),

$$|A| \geq x_1 \geq 0$$
$$SR_1(A)\sqrt{|A|(|S| - 1)} \geq SR_1(A)\sqrt{x_1(|S| - 1)}$$
$$Z_1(A) \geq SR_1(A)\sqrt{x_1(|S| - 1)}$$

where $SR_1(A)$ is the intra-set correlation measure. A value of 1 indicates total identity among the trajectories of A, whereas a value approaching 0 indicates total difference. Because $0 \leq SR_1(A) \leq 1$, we need a minimum significance score x_4 equal to $\sqrt{x_1(|S| - 1)}$ to ensure that all clusters have a minimum of x_1 trajectories.

For studying the RSC clustering robustness, we performed 64 ($= 4 \times 4 \times 4$) analyses with four different values covering a wide range for the variables x_1, x_2 and x_3:

- $x_1 = [2, 5, 10, 50]$
- $x_2 = [1500, 2000, 3000, 6000]$
- $x_3 = [0.1, 0.5, 1.0, 2.0]$

Because RSC is a non-deterministic clustering method, we performed five replicates of each of the 64 clustering analyses.

Next hierarchic clustering based on Jaccard index permitted to compare the different clusters obtained by the 320 clustering. The clusters were classified in several groups and we extracted the intersection for each group. We named "core i" the intersection to the "group i", for example i.e. the set of trajectories that belong to all the clusters of "group i".

2.3 Identification of the Over-Represented Proteins in Each Core

Trajectories clustering was performed using correlation score based on the presence and the absence of proteins. The core of each group can be characterized by a set of over-represented proteins, i.e. the proteins that appear more often in the trajectories of the core than we would expect if we had selected the same number of trajectories randomly (Fig. 2).

We can compute the protein level of representation for each cluster with a zScore of protein frequency:

$$Z_A(p) = \frac{N_A(p) - F_S(p)|A|}{\sqrt{F_S(p)|A|(1 - F_S(p))}} \tag{3}$$

where p is a protein and A is a cluster of trajectories, $N_A(p)$ is the number of trajectories in A involving p, $F_S(p)$ is the frequency of p in all trajectories S and $|A|$ is the size of cluster.

The zScore allows to normalize the frequency of proteins in the cluster of trajectories compared to all trajectories. For each core, we computed the zScore of all the proteins. We then identified a list of over-represented proteins with a high zScore.

Based on the scores of over-representation of proteins in trajectories, we next searched for the biological significance of the protein signatures that characterized the three cores. The Gene Set Enrichment Analysis (GSEA) is a method which permits to identify significantly enriched classes of genes or proteins in a large set of genes or proteins, that are associated with specific biological functions. The analyses were performed using the GSEA tool developed by the Broad Institute [22]. The lists of proteins and their respective score frequency were used as input and biological processes from Gene Ontology database were selected as *gene sets database*. The outputs were the "biological processes" terms significantly enriched in the submitted lists of proteins from each core when compared with the other cores.

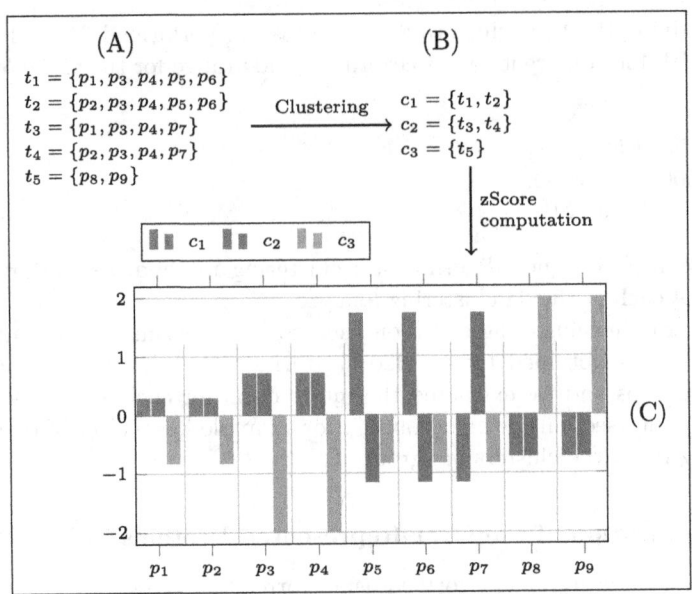

Fig. 2. Example of calculation for determining over-represented proteins between three cores of trajectories. (A) t_1, t_2, t_3, t_4 and t_5 are five trajectories containing proteins p_1, p_2, p_3, p_4, p_5, p_6, p_7, p_8 and p_9. (B) the clustering method identifies three cores c_1, c_2 and c_3. (C) distribution of representation level of proteins in c_1, c_2 and c_3 cores. For example, p_1 and p_2 are slightly over-represented in the cores c_1 and c_2 but not over-represented in c_3, contrary to p_9. The core c_3 can be characterized by p_8 and p_9.

3   Results

3.1   TGF-β Signaling Trajectories Are Highly Connected

In order to identify functional families of signaling trajectories based on the comparison of their signaling molecules (proteins) content, we performed a pre-processing step as described in material and method. Discarding TGF-β and the target genes from the 15,934 trajectories led to 6017 trajectories composed of 321 different proteins.

As illustrated in Fig. 3, the number of proteins per trajectory varied from 1 to 50, with more than 90% of trajectories containing at least 10 proteins. Analyses of the distribution of each protein in all trajectories showed a great heterogeneity. More than 70 proteins were present in at least 500 trajectories, and 6 proteins were present in more than 3000 trajectories (FOS, JUN, ATF2, MAP2K4, ELK1, JAK2). Conversely 75 proteins appeared in fewer than 10 trajectories. Together these results showed that many proteins are shared by many trajectories suggesting high degree of connectivity of TGF-β-dependent signaling pathways.

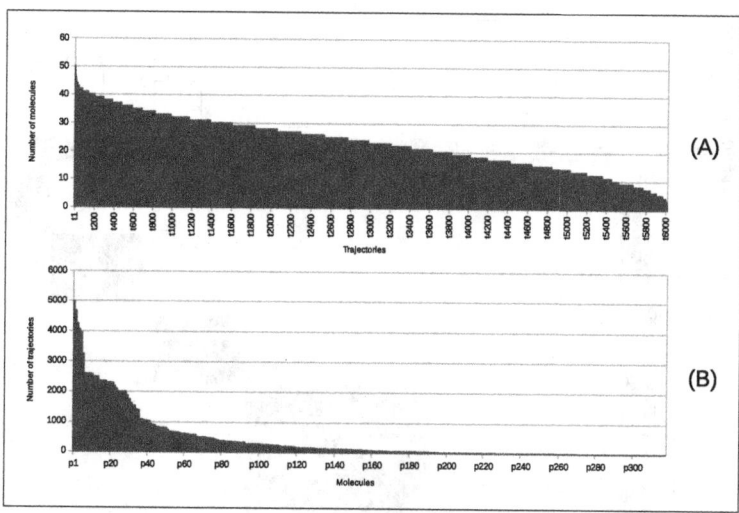

Fig. 3. Distribution of (A) the number of molecules for each trajectory and (B) the number of trajectories involving each molecule. These results showed that most proteins are shared by many trajectories suggesting high degree of connectivity of TGF-β-dependent signaling pathways.

3.2   Relevant Set Correlation Method Identifies Five Families of Trajectory Clusters

Using a greedy strategy and a large variety of parameters, we performed 320 clusterings over the 6017 trajectories. Each clustering generated 3, 4 or 5 clusters leading to 1139 different clusters of trajectories. In order to compare their similarity, we calculated the Jaccard index based on the number of shared trajectories between two given clusters. Using a hierarchical classification of this similarity between clusters, we identified five groups of clusters (Fig. 4).

To characterize the five groups of clusters, we analyzed the number of clusters associated with each group, the number of trajectories associated with these clusters (average cluster size) and the redundancy between clusters (union and intersection). As described in Table 4, the groups 1 and 2 were characterized by clusters generated from 320 and 319 clusterings respectively, suggesting a robust classification of trajectories. The three other groups 3, 4 and 5 contained clusters generated from 160 clusterings suggesting higher sensitivity to parameters. The average cluster size expressed as the average number of trajectories contained in clusters varied from 202 in group 4 to 2170 in group 1. The core of a group is the intersection of the clusters of a group. It is the set of the trajectories that belong to all the clusters of the group, so it allows to focus on the most stable trajectories of the group. The cores of groups 1 and 2 contained 1485 (57%) and 1458 (67%) trajectories respectively, while the core size of groups 3, 4 and 5 were either identical or very similar to the union of clusters. To further characterize these cores, we determined the number of proteins implicated in the trajectories

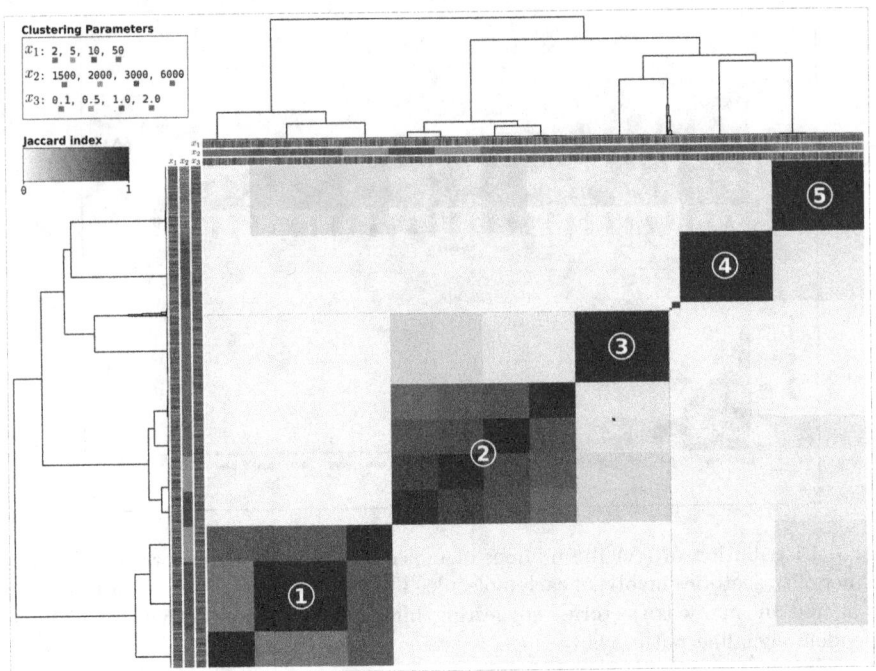

Fig. 4. Hierarchical classification of the clusters generated by the 320 clusterings using varying parameters (x_1, x_2 and x_3) according to their similarities (Jaccard index). The parameter values are indicated by four different colors. Each cluster results from a clustering characterized by a combination of the three parameters. The five groups of clusters identified are numbered from 1 to 5 and the intensity of blue color indicates the Jaccard index between two clusters. (Color figure online)

and the number of target genes activated by these signaling trajectories. While the total number of proteins involved in trajectories from each group was almost similar, the number of target genes was highly variable. The trajectories from the most important core 1 (1485) were characterized by 114 proteins but only 3 target genes suggesting complex combinations of signaling for these genes. At the opposite the trajectories from core 4 that contained only 202 trajectories were characterized by 156 proteins that activate 19 genes.

3.3 Cores Are Characterized by Specific Over-Represented Protein Signatures Associated with Biological Processes

In order to characterize the protein signature of each core, we investigated the level of representation of proteins within all the trajectories from each core. For that purpose, we calculated the zScore of protein frequency in each clusters. The list of protein zScores for each core was provided as supplementary tables[1].

---

[1] http://www.irisa.fr/dyliss/public/tgfbVisualization/supplementaryData.

Table 4. Statistics of clusters.

	Group 1	Group 2	Group 3	Group 4	Group 5
Number of clusters	320	319	160	160	160
Average cluster size (Number of trajectories)	2170.0	1905.58	899.62	202.0	877.12
Union of clusters (Number of trajectories)	2590	2289	904	202	888
Core size = Intersection of clusters (Number of trajectories)	1485	1458	894	202	870
Number of proteins for each core	114	188	110	156	151
Number of target genes for each core	3	68	58	19	16

As shown in Fig. 5, the zScore distribution of the 321 proteins from trajectories of each core was highly heterogeneous. Interestingly, the zScore distribution from core 1 was inversely correlated with that of core 2 suggesting different biological functions associated with trajectories. Together these observations suggested that each core of trajectories was characterized by specific protein signatures. During the course of the analysis of the zScore values, we showed that the probability to randomly find a protein in a group of trajectories with a zScore higher than 4.0 is less than 0.006%. As a consequence, we decided to select the proteins with a zScore superior to 4.0 to refine the protein signatures of the five cores of trajectories.

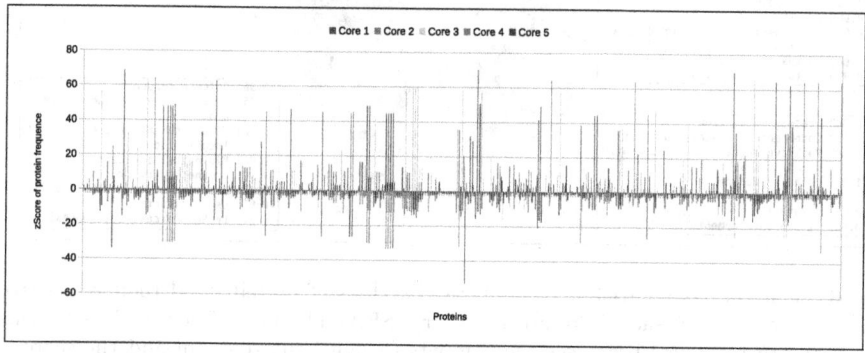

Fig. 5. Distribution of zScore values of the frequences of 321 proteins in the trajectories from cores of the five cluster groups. The zScore distribution of the 321 proteins from trajectories of each core is highly heterogeneous. These observation suggested that each core of trajectories was characterized by specific protein signatures

Based on the scores of proteins over-representation in trajectories, we next searched for the biological significance of the protein signatures that characterized the five cores. Gene Set Enrichment Analysis (GSEA) is a method for identifying significantly the elements of a set that appear more often in the set that one would expect if the set had been randomly assembled. It is typically used for determining which specific biological functions are specific of a set of genes or proteins. The analyses were performed using the GSEA tool developed by the Broad Institute [22]. The lists of proteins and their respective score frequency were used as input for GSEA analysis and the outputs are the lists of enriched biological processes (see supplementary tables, Footnote 1). As shown in Fig. 6, each core was characterized by specific set of biological functions since 57%, 90%, 80%, 81% and 88% of GO-terms were specific of core 1, core 2, core 3, core 4, and core 5, respectively. In order to identify the representative terms, we used Revigo [23] that reduces the list of GO terms on the basis of semantic similarity measures. Consequently trajectories from core 1 and core 2 were mainly associated with antigen receptor-mediated signaling and serine-threonine kinase activity, respectively (Fig. 6). The functional annotation of cores 3 and 4 were more heterogeneous while core 5 clustered signaling trajectories that are clearly involved in immune response. An important conclusion from these results is that even if signaling trajectories share many proteins, our analysis revealed groups of trajectories that correspond to different functional families.

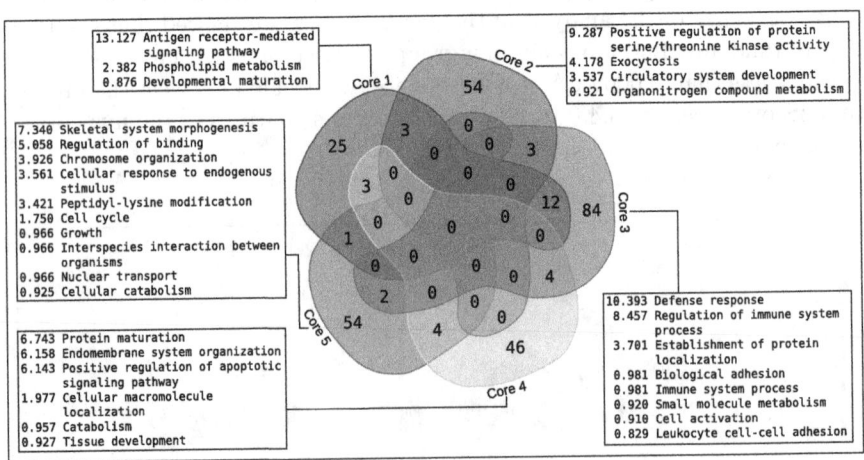

Fig. 6. Gene ontology enrichment analysis. The lists of proteins and their respective score frequency from each Core are used for GSEA. The lists of enriched GO terms associated to biological processes are compared using Venn diagram and the score is uniqueness score of the GO-term calculated by REVIGO tool.

Together our data demonstrate that our approach for clustering signaling trajectories based on their protein content is powerful to discriminate TGF-β-influenced networks. To illustrate the complexity of TGF-β-dependent signaling

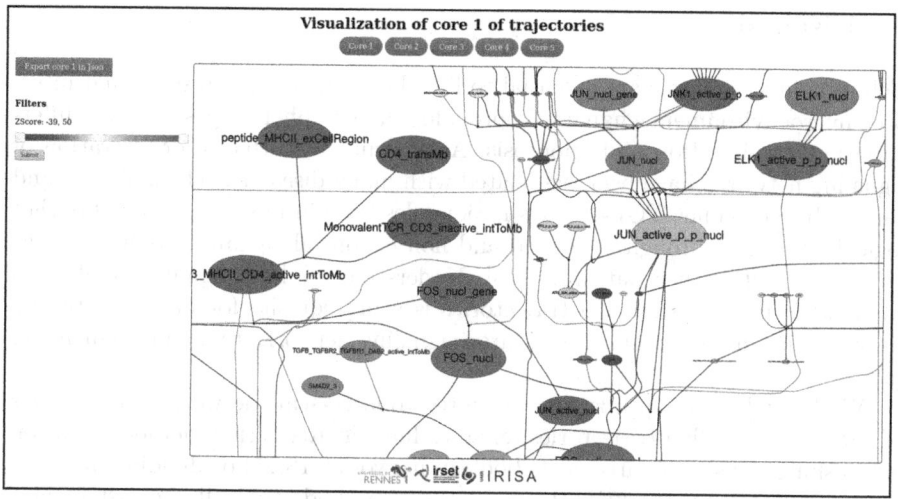

Fig. 7. Screenshot of the Web visualization of core 1. A node is a bio-molecule, the node size corresponds to the number of trajectories involving this bio-molecule and the node color correspond to the representation of the bio-molecule in the core (blue the bio-molecule is under-represented and red it is over-represented). (Color figure online)

pathways, we compiled the trajectories from each core and the resulting networks were illustrated in Fig. 7.

3.4   Web Visualization of TGF-β-Influenced Networks

To facilitate the exploration of the signaling trajectories clustered in each core, we developed a web interface:

    http://www.irisa.fr/dyliss/public/tgfbVisualization/

The interface is based on the Cytoscape JavaScript library (http://js. cytoscape.org). Nodes are proteins and their size is correlated to the number of trajectories involving this protein. Node color indicates the occurrence of the protein in the trajectories from a core. The occurrence is based on the zScore of protein frequency (blue for zScore < 0 and red for zScore > 0) and selection of the level of occurrence permits to filter information. The black circle nodes illustrate biological reactions (association, dissociation, phosphorylation,degradation, migration etc.) as described in [2]. The black edges link proteins to the input or the output of a reaction, green edges link the protein that regulates positively the reaction and red edges link the protein that regulate negatively the reaction (Fig. 7). Exploration of the graphs is facilitated by manually repositioning nodes and edge. The graph can be exported in JSON format.

4   Discussion

Cell signaling networks are essential to life. They allow cells to sense and interpret microenvironment changes to provide adapted phenotypes such as differentiation, proliferation and apoptosis. As a result, disturbance or alteration of signaling networks have been associated with many diseases such as fibrosis and cancer. In particular, TGF-β plays major roles both in physiological and pathological processes through canonical and non canonical signaling pathways that cross-react with other pathways [13]. Understanding how signaling molecules combine to provide signaling trajectories is a prerequisite for future therapeutic strategies, however analyses of large signaling networks remain a challenging task.

While qualitative approaches are suited to large-scale networks, the analysis of numerous signaling trajectories remains difficult. Reduction methods focus on diminishing the size of large-scale boolean networks [18,25] or dividing methods in several sub-networks [26]. However, these methods typically consist in performing the reduction before the analysis, whereas for TGF-β we focused on an exhaustive analysis of the signaling network.

In addition to exhaustivity, the originality of our approach lies in analyzing the signaling trajectories according to their protein composition rather than the genes they influence. Our approach was motivated by the fact that signaling pathways share a large number of "modular domains" in various combinations [11]. These combinations support the functional diversity of signaling pathways.

These modular domains provide the underlying structure of the signaling trajectories. Our goal was to identify groups of similar trajectories. When considering two trajectories, the more modules they share, the more similar they are. There are many clustering methods (for example hierarchical, K-means, distribution-based, density-based) [8]. As we mentioned previously, a modular domain can be involved in multiple combinations, so their study required soft-clustering methods which allows clusters to overlap and share some elements. We selected shared nearest-neighbours (SNN) clustering, which have successfully been applied to handle the heterogeneity and large-scale of trajectories [5]. The Relevant Set Correlation method is further appropriate in that there is no need to define the neighborhood size. Likewise, our approach does not rely on a priori assumption on the number of clusters.

Relevant Set Correlation proved to be a robust clustering method for our dataset. All 64 combinations of parameter values generated clusters that systematically belonged to group 1 and group 2 and one of groups 3, 4 and 5. Half the simulations produced clusters that belonged to groups 3, 4 or 5. In Fig. 4, the analysis of the influence of the parameter values for groups 3, 4 and 5 showed that x_1 and x_3 had no influence on the groups, whereas pairs of values of x_2 were associated to different groups: the two lowest with group 5, the two highest with group 4 and a combination of the highest and the lowest with group 3. Surprisingly, the two intermediate values of x_2 (2000 and 3000) were markers of groups 4 and 5, for which they were associated with their closest extreme value, whereas the lowest and highest values of x_2 were associated to group 3. This indicates that RSC produced either groups 3 and 5 for the low values of the range of the clusters' maximum size (x_2), or groups 3 and 4 for the high values. At this point, further analysis is required for determining either which of the low or high values are the more adapted to our dataset, or if groups 3, 4 and 5 are all biologically-relevant and we are facing a limitation of RSC. Overall, our study with the various combinations of parameter values showed that (1) because it is non-deterministic, performing multiple runs with the same parameter values is useful, (2) RSC is a robust clustering method for our dataset, (3) groups 1 and 2 were independent from the parameter values whereas groups 3, 4 and 5 were not, and (4) low values of clusters' maximum size produced clusters in groups 3 and 5, whereas high values produced clusters in groups 3 and 4. According to this observation, the over-represented proteins in trajectories from core 1 and 2 clearly discriminate the canonical pathways associated with TGF-β receptor-dependent cell response during injury and development (core 1) and the non canonical pathways involving all other kinase-dependent signaling (core 2), respectively. Together these two cores of clusters illustrated the so-called "Jekyll and Hyde" aspects of TGF-β in cancer [3].

Although it does not rely on a priori knowledge, our approach may be dependent on annotation bias. Since biological knowledge is by nature incomplete, some well studied signaling processes may be described in details in databases, whereas some lesser studied ones would be incompletely described, or with a coarser granularity (usually both). This would then result in a higher frequency of the well studied modules and give a misleading impression of being more important. It should be noted that this is an intrinsic bias of the data we rely on, and not of our analysis method. This bias should be taken into account by the experts when analyzing the results.

5   Conclusion

We proposed an exhaustive and without prior assumption soft-clustering-based method for identifying families of functionally-similar trajectories in signaling network. Among 15,934 trajectories involved in TGF-β signaling, our approach identified five groups of trajectories based on their molecular composition. The functional characterization of these groups revealed that each group is involved in different roles of TGF-β, which confirmed that our approach yields biologically-relevant results. The approach can be generalized to explore any large-scale biological pathways.

References

1. Aldridge, B.B., Burke, J.M., Lauffenburger, D.A., Sorger, P.K.: Physicochemical modelling of cell signalling pathways. Nat. Cell Biol. 8(11), 1195–1203 (2006)
2. Andrieux, G., Le Borgne, M., Théret, N.: An integrative modeling framework reveals plasticity of TGF-β signaling. BMC Syst. Biol. 8(1), 1 (2014)

3. Bierie, B., Moses, H.L.: Tumour microenvironment: TGFβ: the molecular Jekyll and Hyde of cancer. Nat. Rev. Cancer **6**(7), 506–520 (2006)
4. ElKalaawy, N., Wassal, A.: Methodologies for the modeling and simulation of biochemical networks, illustrated for signal transduction pathways: a primer. Biosystems **129**, 1–18 (2015)
5. Hamzaoui, A., Joly, A., Boujemaa, N.: Multi-source shared nearest neighbours for multi-modal image clustering. Multimedia Tools Appl. **51**(2), 479–503 (2011)
6. Houle, M.E.: The relevant-set correlation model for data clustering. Stat. Anal. Data Min. **1**(3), 157–176 (2008)
7. Ikushima, H., Miyazono, K.: Biology of transforming growth factor-β signaling. Curr. Pharm. Biotechnol. **12**(12), 2099–2107 (2011)
8. Joshi, A., Kaur, R.: A review: comparative study of various clustering techniques in data mining. Int. J. Adv. Res. Comput. Sci. Softw. Eng. **3**(3) (2013)
9. Kashtan, N., Alon, U.: Spontaneous evolution of modularity and network motifs. Proc. Natl. Acad. Sci. U.S.A. **102**(39), 13773–13778 (2005)
10. Kestler, H.A., Wawra, C., Kracher, B., Kühl, M.: Network modeling of signal transduction: establishing the global view. BioEssays **30**(11–12), 1110–1125 (2008)
11. Lim, W.A.: Designing customized cell signalling circuits. Nat. Rev. Mol. Cell Biol. **11**(6), 393–403 (2010)
12. Luo, K.: Signaling cross talk between TGF-β/Smad and other signaling pathways. Cold Spring Harbor Perspect. Biol. **9**(1), a022137 (2017)
13. Massagué, J.: TGFβ signalling in context. Nat. Rev. Mol. Cell Biol. **13**(10), 616–630 (2012)
14. Mu, Y., Gudey, S.K., Landström, M.: Non-smad signaling pathways. Cell Tissue Res. **347**(1), 11–20 (2012)
15. Peisajovich, S.G., Garbarino, J.E., Wei, P., Lim, W.A.: Rapid diversification of cell signaling phenotypes by modular domain recombination. Science **328**(5976), 368–372 (2010)
16. Rauzy, A.: Guarded transition systems: a new states/events formalism for reliability studies. Proc. Inst. Mech. Eng. Part O: J. Risk Reliab. **222**(4), 495–505 (2008)
17. Saadatpour, A., Albert, R.: Discrete dynamic modeling of signal transduction networks. In: Liu, X., Betterton, M.D. (eds.) Computational Modeling of Signaling Networks, pp. 255–272. Humana Press, Totowa (2012)
18. Saadatpour, A., Albert, R., Reluga, T.C.: A reduction method for boolean network models proven to conserve attractors. SIAM J. Appl. Dyn. Syst. **12**(4), 1997–2011 (2013)
19. Samaga, R., Klamt, S.: Modeling approaches for qualitative and semi-quantitative analysis of cellular signaling networks. Cell Commun. Signal. **11**(1), 1 (2013)
20. Schaefer, C.F., Anthony, K., Krupa, S., Buchoff, J., Day, M., Hannay, T., Buetow, K.H.: PID: the pathway interaction database. Nucleic Acids Res. **37**(suppl 1), D674–D679 (2009)
21. Scott, J.D., Pawson, T.: Cell signaling in space and time: where proteins come together and when they're apart. Science **326**(5957), 1220–1224 (2009)
22. Subramanian, A., Tamayo, P., Mootha, V.K., Mukherjee, S., Ebert, B.L., Gillette, M.A., Paulovich, A., Pomeroy, S.L., Golub, T.R., Lander, E.S., et al.: Gene set enrichment analysis: a knowledge-based approach for interpreting genome-wide expression profiles. Proc. Natl. Acad. Sci. **102**(43), 15545–15550 (2005)
23. Supek, F., Bošnjak, M., Škunca, N., Šmuc, T.: Revigo summarizes and visualizes long lists of gene ontology terms. PLoS ONE **6**(7), e21800 (2011)

24. Tian, M., Neil, J.R., Schiemann, W.P.: Transforming growth factor-β and the hallmarks of cancer. Cell. Signal. **23**(6), 951–962 (2011)
25. Zañudo, J.G., Albert, R.: An effective network reduction approach to find the dynamical repertoire of discrete dynamic networks. Chaos Interdisc. J. Nonlinear Sci. **23**(2), 025111 (2013)
26. Zhao, Y., Kim, J., Filippone, M.: Aggregation algorithm towards large-scale boolean network analysis. IEEE Trans. Autom. Control **58**(8), 1976–1985 (2013)
27. Zi, Z., Chapnick, D.A., Liu, X.: Dynamics of TGF-β/Smad signaling. FEBS Lett. **586**(14), 1921–1928 (2012)

Strong Turing Completeness of Continuous Chemical Reaction Networks and Compilation of Mixed Analog-Digital Programs

François Fages[1]([✉]), Guillaume Le Guludec[1,2], Olivier Bournez[3], and Amaury Pouly[4]

[1] Inria, Université Paris-Saclay, EP Lifeware, Palaiseau, France
`francois.fages@inria.fr`
[2] Sup Telecom, Paris, France
[3] LIX, CNRS, Ecole Polytechnique, Palaiseau, France
[4] Max Planck Institute for Computer Science, Saarbrücken, Germany

Abstract. When seeking to understand how computation is carried out in the cell to maintain itself in its environment, process signals and make decisions, the continuous nature of protein interaction processes forces us to consider also analog computation models and mixed analog-digital computation programs. However, recent results in the theory of analog computability and complexity establish fundamental links with classical programming. In this paper, we derive from these results the strong (uniform computability) Turing completeness of chemical reaction networks over a finite set of molecular species under the differential semantics, solving a long standing open problem. Furthermore we derive from the proof a compiler of mathematical functions into elementary chemical reactions. We illustrate the reaction code generated by our compiler on trigonometric functions, and on various sigmoid functions which can serve as markers of presence or absence for implementing program control instructions in the cell and imperative programs. Then we start comparing our compiler-generated circuits to the natural circuit of the MAPK signaling network, which plays the role of an analog-digital converter in the cell with a Hill type sigmoid input/output functions.

1 Introduction

"The varied titles of Turing's published work disguise its unity of purpose. The central problem with which he started, and to which he constantly returned, is the extent and the limitations of mechanistic explanations of nature.", Max Newman.

The Church-Turing thesis states that there is only one notion of effective computation over discrete structures (integers, words, ...), and in fact all mechanistic computation models devised up to now (Church's λ-calculus, Post's rewriting systems, random access machines, programming languages, ...) have

© Springer International Publishing AG 2017
J. Feret and H. Koeppl (Eds.): CMSB 2017, LNBI 10545, pp. 108–127, 2017.
DOI: 10.1007/978-3-319-67471-1_7

always been shown to be encodable in Turing machines. The more recent physical Church-Turing thesis goes beyond the original thesis by stating that all physically computable functions are Turing-computable.

In this view, it is theoretically possible to give a computational meaning to information processing in the cell in terms of algorithms and programs. However, while one lesson of Computer Science is that digital computation scales up to very large circuits and programs, contrarily to analog computation, one has to face the paradox that in a cell, even if one can observe an all-or-nothing activation of genes, one cannot deny the importance of the continuous gradual activations of protein complexes, of the time it takes, of the absence of clock signals, i.e. the importance of analog computation in the cell [20,41,43].

Classical computability and complexity theories mainly focus on computation over discrete domains, i.e. words or integers. When dealing with reals or functions, several approaches can be considered. In computational analysis, the notion of computation over the real numbers is defined in terms of approximation in arbitrary but finite precision:

Definition 1 ([48]). *A real number $r \in \mathbb{R}$ is computable (resp. in polynomial time) in the sense of computational analysis if there exists an effective approximation program of r in arbitrary precision, i.e. a Turing machine which takes as input a precision $p \in \mathbb{N}$ and outputs a rational number $r_p \in \mathbb{Q}$ s.t. $|r - r_p| \leq 2^{-p}$ (resp. in a time polynomial in p).*

Clearly, every real number can be represented as an infinite string representing a converging Cauchy sequence as above, and a computable real is one whose representation is computable. In this setting, a computable real number can thus be seen as a program which takes as input an accuracy, and returns as output an approximation of the real number by a rational number at the requested precision. A computable function is then a program that maps any (computable or not[1]) approximation of a real x to an approximation of $f(x)$.

Definition 2 ([48]). *A function $f : \mathbb{R} \to \mathbb{R}$ is computable if there exists a Turing machine with oracle which computes an approximation of $f(x)$ given x as oracle. It is computable in polynomial time if this is done in a time polynomial in p and m for $x \in [-2^m, 2^m]$.*

In this paper, we consider these notions to give a mathematical meaning to the notion of biochemical computation with continuous concentrations. In this view, the language of biochemical reactions is seen as a programming language for computing with non negative real valued concentrations, i.e. over $\mathbb{R}_+$. We consider elementary reactions, i.e. reactions with at most two reactants and with mass-action-law kinetics. It is well known that the other classical biochemical rate functions, such as Michaelis-Menten, Hill kinetics, are derived by reduction of elementary reaction systems with mass-action law kinetics, using for instance quasi-steady state or quasi equilibrium approximations [44].

---

[1] Restricting the definition to computable arguments might seem quite natural but is not the classical definition of computable analysis, see the Appendix of [48].

We first show the Turing completeness in the strong sense of uniform computability, of elementary biochemical reactions without polymerisation under the differential semantics on a finite universe of molecular species. This solves an open problem explicitly mentioned in [17]. where it was shown that a Turing machine could be simulated by a chemical reaction network with a small probability of error. Although not surprising, this result is in sharp contrast to the discrete semantics of reaction systems which are not Turing complete without either the tolerance of a small probability error [17], or the addition of other mechanisms such as the unbounded dynamic creation of membranes [2,9,38], or the presence of polymerization reactions on an infinite universe of polymers [10] or DNA stacks [40].

Furthermore, following [6] we generalize the purely analog characterization of the complexity class PTIME to positive binary reaction systems which stabilize on one component with a trajectory length bounded by a polynomial of the input and the precision.

Then we derive from the proof of these results a compiler of behavioural specifications[2] into elementary reaction systems, without prejudging of their biochemical implementation, by enzymatic reactions [37], DNA [13] or RNA for instance.

We illustrate this approach with the compilation of trigonometric functions, such as the cosine function, as either functions of time or of an input variable. Then, we study different sigmoid functions which can serve as markers of presence or absence for implementing program control instructions and compiling imperative programs.

Then we start comparing our compiler-generated circuits to natural circuits, with the example of the MAPK signaling network, which plays the role of an analog-digital converter in the cell with a Hill type sigmoid input/output function [28].

2   Computational Functions and Computational Complexity over the Reals

The *General Purpose Analog Computer* (GPAC) of Shannon [46] is a model of computation based on circuits built from analog blocks. A set of variables or entries $x, y, z, \ldots$ including time t are considered and four types of blocks (constants, sums, products, and Stieltjes integral of one variable with respect to another variable - by default the time variable when it is not indicated) are connected (with possibly feedback connexions) in order to generate a system whose dynamic is considered as "generating" functions. Shannon's original presentation suffers from several problems, including the fact that some circuits may or may not have a solution. This problem was solved in [26] which gives a satisfactory

---

[2] For the sake of reproducibility, all the examples described in this paper are directly executable online in Biocham v4 (http://lifeware.inria.fr/biocham4) notebooks available at http://lifeware.inria.fr/wiki/software/#CMSB17.

definition of GPAC-generable functions in terms of the solution to polynomial initial value problems in polynomial differential equations (PIVP):

Definition 3 *[26]. A function* $f : \mathbb{R}_+ \to \mathbb{R}$ *is GPAC-generable[3] if it is one component of the* $y(t)$ *solution of some ordinary differential equation* $y'(t) = p(y(t))$ *for a polynomial vector* $p \in \mathbb{R}^n[\mathbb{R}^n]$ *and initial values* $y(0) \in \mathbb{R}^n.$

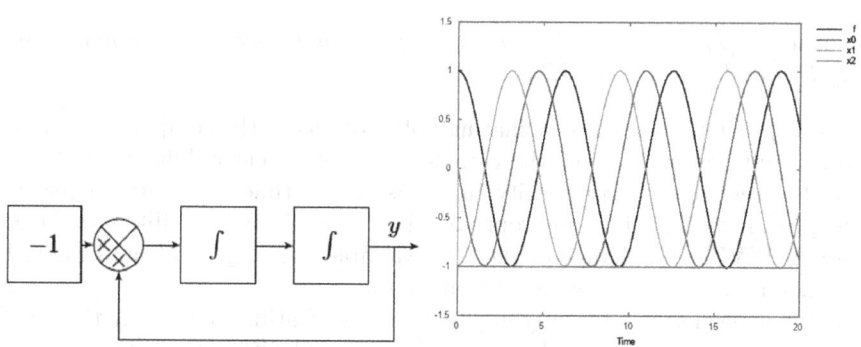

Fig. 1. GPAC circuit for generating the cosine function as a function of time, and numerical simulation trace.

For example, the GPAC (y = integral integral -1 * y) shown in Fig. 1 is constructed with two integral blocks and a multiplication by -1 which gives $y''(t) = -y(t)$. This circuit when initialized with $y(0) = 1$ *generates* the cosine function, $cos(t)$, as a function of time. The class of GPAC-generable functions enjoys a number of properties, such as stability by addition, multiplication and composition, and also contains elementary functions such as trigonometric functions, exponential functions, logarithms, etc. This notion of generality has for some time been considered synonymous with analog-computability, which made the GPAC a computation model less expressive than computational analysis as some functions such as Rieman's Zeta function or Euler's Gamma functions are known not to be differentially algebraic [46].

However, it is possible to define a notion of GPAC-computability which is both natural in terms of PIVP and equivalent to computational analysis. The idea is to proceed by approximation of the result for any entry on one component of the system, as follows:

Definition 4 *[4]. A function* $f : \mathbb{R} \to \mathbb{R}$ *is GPAC-computable if there are polynomial vectors* $p \in \mathbb{R}^n[\mathbb{R}^n]$, *a polynomial* $q \in \mathbb{R}^n[\mathbb{R}]$ *such that for all* x *there exists some (necessarily unique) function* $y : \mathbb{R} \to \mathbb{R}^n$ *such that*

$$y(0) = q(x), \ y'(t) = p(y(t))$$

and $|y_1(t) - f(x)| \leq y_2(t)$, *with* $y_2(t) \geq 0$ *decreasing and* $\lim_{t \to \infty} y_2(t) = 0.$

---

[3] This definition can be generalized to functions of several variables over different domains [7].

In other words, the computation of f with the argument x consists in putting the system in a polynomially dependent state of x, then letting the system evolve according to the dynamics described by p. The result of the computation is obtained in one component of the system, say the first, with arbitrary precision given by some other component of the system, say the second, is decreasing[4] to 0.

Then the following theorem perfectly reconciles the notions of digital (i.e. by Turing machines) and analog (i.e. by PIVP) computability:

Theorem 1 *[4, 5].* *A function is computable in the sense of computational analysis if and only if it is GPAC-computable.*

While previous result is conciliating both notions at the computability level, such a result was missing at the complexity level. A clear difficulty is that a naive definition of the complexity in terms of the time necessary to reach a given precision can not be appropriate, since it is always possible to contract time in a PIVP by a change of the time variable, e.g. $t_{\mathrm{fast}} = e^t$, and multiply the differential equations by an arbitrary term.

This has been solved recently in [39] by demonstrating that taking the *length of the trajectory* as measure of computational complexity, i.e. a combination of time and space (amplitude), which takes into account the cost of computing for instance $t_{\mathrm{fast}} = e^t$, yields a valid notion of time complexity, equivalent to classical time complexity. In particular, a purely analog characterization of the complexity class PTIME has been given in [6]. Let $\|y\|$ refers to the infinite norm of y (i.e. the maximum absolute value of its components).

Definition 5 *[6].* *A function $f : \mathbb{R} \to \mathbb{R}$ is said to be Ω -computable in length, where $\Omega : \mathbb{R}_+^2 \to \mathbb{R}$, if there are polynomial vectors $p \in \mathbb{R}^n[\mathbb{R}^n]$, a polynomial $q \in \mathbb{R}^n[\mathbb{R}]$ such that for all x there exists some (necessarily unique) function $y : \mathbb{R} \to \mathbb{R}^n$ satisfying for all $t \in \mathbb{R}_+$:*

- $y(0) = q(x)$ and $y'(t) = p(y(t))$ with $\|y'(t)\| \geq 1$ (holds if t is one variable),
- for any μ, if $\int_0^t \|y'(\tau)\| \, d\tau \geq \Omega(|x|, \mu)$ then $|y_1(t) - f(x)| \leq e^{-\mu}$.

Theorem 2 *[6].* *The Ω-computable functions in length, where Ω is a polynomial, are exactly the functions computable in polynomial time in the sense of the computational analysis.*

Taking unrestricted Ω leads back to the previous notion of computable functions in the sense of computational analysis.

Theorem 1 implies in particular that polynomial differential equations (PIVP) are universal. This is in a strong sense, compared to notions of universality used in articles such as [27, 34] where it is basically shown that boolean circuits can be realized, yielding a *non-uniform* notion of computability: for each

---

[4] The decreasing assumption is here to yield a simple way to decide when the result on the first component is correct with the required precision: given some precision ϵ, just wait until the second component is less than ϵ.

input there exists an ODE system computing the result. Here this is a *uniform computability* result: a given polynomial differential equation is able to simulate a Turing machine on all inputs, independently of the size of the input.

In order to be more concrete on the encoding of Turing machines, let us rephrase [6]. One can fix a finite alphabet $\Gamma = \{0, .., k - 2\}$ and encode a word $w = w_1 w_2 \ldots w_{|w|}$ by the couple $\psi(w) = \left(\sum_{i=1}^{|w|} w_i k^{-i}, |w| \right)$. There is nothing special about this encoding, other encodings may be used, however, two crucial properties are necessary: (i) $\psi(w)$ must provide a way to recover the word without ambiguity, (ii) $\|\psi(w)\|$ is $O(|w|)$. In particular, over the alphabet $\Gamma = \{0, 1\}$, the use of base 3 (instead of base 2) simplifies the decoding.

Now consider any decision problem (language) $\mathcal{L} \subset \Gamma^*$. If $\mathcal{L}$ is decidable, then there is a Turing machine that decides it. Then [6] provides (effectively from the Turing machine) some polynomial vectors $p \in \mathbb{R}^n[\mathbb{R}^n]$ and a polynomial $q \in \mathbb{R}^n[\mathbb{R}]$ such that for all $w \in \Gamma^*$ there is a (unique) $y : \mathbb{R}_+ \to \mathbb{R}^d$ such that for all $t \in \mathbb{R}_+$:

1. $y(0) = q(\psi(w))$ and $y'(t) = p(y(t))$ with $\|y'(t)\| \geq 1$,
2. if $|y_1(t)| \geq 1$ for some t then $|y_1(u)| \geq 1$ for all $u \geq t$ (the decision is stable)
3. if $w \in \mathcal{L}$ (resp. $\notin \mathcal{L}$) then there is some t with $y_1(t) \geq 1$ (resp. ≤ -1)

Furthermore, if $\mathcal{L}$ is decided in polynomial time (i.e. is in class PTIME) then there is some polynomial Ω (that can be obtained effectively from the polynomial bound for $\mathcal{L}$ and from the Turing machine) such that this happens in polynomial length: condition 3. is replaced by

3. if $w \in \mathcal{L}$ (resp. $\notin \mathcal{L}$) and $\int_0^t \|y'(\tau)\| \, d\tau \geq \Omega(|w|)$ then $y_1(t) \geq 1$ (resp. ≤ -1)

In other words, [6] is considering a notion of termination given by the fact that some variable becomes of absolute value greater than 1: if the value is greater than 1 (respectively: less than -1) this corresponds to acceptance (resp. rejection). Other criteria for acceptance could be considered as seen from the proofs of [6]. The fact that the acceptance region is at some distance from the rejectance region (a value between -1 and 1 means the absence of decision) is here only to avoid representation problems if one wants to simulate the involved equations.

Notice that [6] was leaving open the issue whether the involved polynomial in the polynomial ordinary differential equations could have non-rational coefficients (notice that the constructions were however using only computable coefficients, but possibly irrational). It has been proved recently that only rational coefficients are needed [3].

The notion of uniform computability is the strong notion of Turing universality involved in the rest of this paper.

3   Turing Completeness of Elementary Chemical Reaction Networks

The previous results provide a solid foundation for studying biochemical analog computation. However, a biochemical reaction system is a *positive* dynamical system living in the cone $\mathbb{R}^n_+$, where the state is defined by the positive concentration values of the molecular species[5]. Furthermore, we wish to restrict ourselves to elementary reaction systems, governed by the mass-action-law kinetics and where each reaction has at most two reactants.

Let $\mathcal{M}$ be a *finite set* of n molecular species $\{y_1, \ldots, y_n\}$.

Definition 6 [21]. *A reaction is a triple* (R, P, f), *where* $R : \mathcal{M} \to \mathbb{N}$ *is a multiset of reactants,* $P : \mathcal{M} \to \mathbb{N}$ *is a multiset of products and* $f : \mathbb{R}^n_+ \to \mathbb{R}_+$, *called the* rate function, *is a partially differentiable function verifying* $R(y_i) > 0$ *iff* $\frac{\partial f}{\partial y_i}(y) > 0$ *for some* $y \in \mathbb{R}^n_+$.
A reaction system *is a finite set of reactions.*

A mass-action-law reaction *is a reaction in which the rate function* f *is a monomial of the form* $k * \Pi_{y \in \mathcal{M}} y^{R(y)}$ *where* k *is called the* rate constant.

An elementary reaction *is a mass-action-law reaction with at most two reactants.*

For the sake of both readability and reproducibility, the examples will be noted in the sequel in Biocham syntax, where a reaction (R, P, f) is written `f for R => P`, or just `R => P` if the rate function is a mass action kinetics with rate constant is equal to 1; the multisets are written with linear expressions and _ stands for the empty multiset. Furthermore, a reaction with catalysts `f for R+C => C+P` is abbreviated as `f for R = [C] => P`.

Definition 7. *The differential semantics of a reaction system* $\{(R_i, P_i, f_i)\}_{i \in I}$ *is the ODE system*

$$\{y' = \Sigma_{i \in I}(R_i(y) - Pi(y)) * f_i\}_{y \in \mathcal{M}}.$$

The dynamics given by the law of mass action leads to a polynomial ODE system of the form $y'(t) = p(y(t))$ with $p(y)_i = \sum_j (P_j(y_i) - R_j(y_i)) * k_j * \Pi^n_{i=1} y_i^{R_j(y_i)}$. There are thus additional constraints, compared to general PIVPs: the components y_i must always be positive, and the monomials of p_i whose coefficient is negative must have a non-zero y_i exponent. These constraints are necessary conditions for the existence of a set of biochemically realizable reactions that react according to the dynamics $y' = p(y)$. Note however that we shall not discuss here the choice of their possible implementations by particular biochemical devices, such as DNA polymers [40], DNA double strands [33] or enzymatic reactions [19,37] as this is beyond the scope of this paper.

Interestingly, the previous computability and complexity results can be generalized to elementary biochemical reaction systems. First, the restriction to positive systems can be shown complete, by encoding each component y_i by the difference between two positive components y_i^+ and y_i^-, which can be normalized by a mutual annihilation reaction, $y_i^+ + y_i^- \Rightarrow$ _, so that one variable is null. It is worth noting that this encoding has been used in [36] for implementing linear I/O systems.

---

[5] Note that we do not impose that concentration values are small values, less than 1 for instance. We consider arbitrary large concentration and molecule numbers [25].

Definition 8. *A function $f : \mathbb{R}_+ \to \mathbb{R}_+$ is chemically-computable if there exist a mass-action-law reaction system $\{(R_i, P_i, f_i)\}_{i \in I}$ over some molecular species $\{y_1, \ldots, y_n\}$, and a polynomial $q \in \mathbb{R}_+{}^n[\mathbb{R}_+]$ defining the initial concentration values, such that f is GPAC-computed by q and its (polynomial) differential semantics $p \in \mathbb{R}_+{}^n[\mathbb{R}_+{}^n]$.*

A function $f : \mathbb{R}_+ \to \mathbb{R}$ is chemically-computable if there exists a chemically computable function $f^+ : \mathbb{R}_+ \to \mathbb{R}_+{}^2$ (by straightfortward generalization of Definition 4 to multiple computations) over $\{y_1^+, \ldots, y_n^+, y_1^-, \ldots, y_n^-\}$ such that $f = f_1^+ - f_2^-$.

In this definition, to compute $f(x)$, one has thus to design a reaction system over a finite set of molecular species, initialized to some values defined by a vector of polynomials $q(x)$ (e.g. following [8,12]), which guarantees that the result is obtained in the concentration of one distinguished molecular species, with a precision indicated by another distinguished molecular species (see Definition 4). Note however that in practice, in the examples of the following sections, the precision parameter will be left.

How to design such a reaction system is shown by the proofs of the following results.

Theorem 3. *Any GPAC-computable function can be computed by a mass-action-law reaction system under the differential semantics preserving the polynomial length complexity.*

Proof. Let us consider a GPAC-computable function by a polynomial differential equation $p \in \mathbb{R}^n[\mathbb{R}^n]$. Each variable $y_i \in \mathbb{R}$ can be encoded by a couple of variables $(y_i^+, y_i^-) \in \mathbb{R}_+^2$ such that at any time, $y_i = y_i^+ - y_i^-$.

Let $\hat{p}_i(y_1^+, y_1^-, \ldots, y_n^+, y_n^-) = p_i[y = y^+ - y^-]$, we write $\hat{p}_i = \hat{p}_i^+ - \hat{p}_i^-$, where the monomials of $\hat{p}_i^+$ and $\hat{p}_i^-$ have positive coefficients. A positive system is then defined by:

$$\forall i \leq n, \quad \begin{cases} y_i^{+\,'} = \hat{p}_i^+ - f_i y_i^+ y_i^- \\ y_i^{-\,'} = \hat{p}_i^- - f_i y_i^+ y_i^- \\ y_i^+(0) = \max(0,\ y_i(0)) \\ y_i^-(0) = \max(0,\ -y_i(0)) \end{cases}$$

where the f_i's are polynomials with positive coefficients such that $f_i \geq \max(\hat{p}_i^+, \hat{p}_i^-)$, for instance $f_i = \hat{p}_i^+ + \hat{p}_i^-$. The terms $-f_i y_i^+ y_i^-$ can be implemented by annihilation reactions

$$f_i \text{ for } y_i^+ + y_i^- \xrightarrow{\ y^+, y^-\ } \_$$

which ensure that one of the $y_i^\pm$ always remains small.

Note that we have: $y_i^{+\,'} \leq \hat{p}_i^+(1 - y_i^+ y_i^-)$ and $y_i^{-\,'} \leq \hat{p}_i^-(1 - y_i^+ y_i^-)$, so that $(y_i^+ y_i^-)' \leq q \cdot (1 - y_i^+ y_i^-)$ where q is a polynomial with positive coefficients. Since at $t = 0$ we have $y_i^+ y_i^- = 0$, we deduce by a Gronwall inequality that we always have $y_i^+ y_i^- \leq 1$. Therefore, $|y_i^\pm| \leq |y_i| + 1$, and $|y^\pm| \leq |y| + n$. Consequently, if the original system is increased in space by a polynomial in the size of the input and the time, then this is still the case for the positive system obtained by the preceding construction. Furthermore, each monoid of the form $\lambda y_1^{\alpha_1} \ldots y_m^{\alpha_m}, \lambda > 0$ appearing in the right term of an equality of the form $y = p$ can be implemented by a reaction of the form

$$\alpha_1 y_1 + \ldots + \alpha_m y_m \xrightarrow{\lambda} y + \alpha_1 y_1 + \ldots + \alpha_m y_m. \qquad \square$$

Second, one can remark that we can also restrict ourselves to elementary reactions, since every PIVP is equivalent to a quadratic PIVP.

Theorem 4 *[11]. Any solution of a PIVP is the solution of a PIVP of degree at most two.*

Proof. The proof consists in introducing variables for each monomial as follows

$$v_{i_1,\ldots,i_n} = y_1^{i_1} y_2^{i_2}, \ldots, y_n^{i_n}.$$

We have $y_1 = v_{1,0,\ldots,0}$ and so on. The substitution of these variables in the differential equations of y_i' gives equations of the first degree in the variables $v_{i_1,\ldots,i_n}$. The differential equations for variables that are not y_i are of the form

$$v_{i_1,\ldots,i_n}' = \sum_{k=0}^{n} i_k * v_{i_1,\ldots,i_k-1,\ldots,i_n} * y_k'$$

i.e. a polynomial of degree two since the y_k' differentials are linear combinations of the variables $v_{i_1,\ldots,i_n}$. $\qquad \square$

These results show that elementary biochemical reaction systems under the differential semantics have the expressive power of PIVPs. By Theorems 1, 3 and 4, we get

Theorem 5. *Elementary reaction systems on finite universes of molecules are Turing-complete under the differential semantics.*

It is worth noticing that this result differs from previous results on the universality of continuous chemical reaction networks or neural networks which were based on a *non-uniform* notion of computability [27,34]. Here we obtain a *uniform computability* result: a given reaction system on a finite set of molecular species is able to simulate a Turing machine on all inputs, independently of the size of the input. This result can be considered as solving the open problem mentioned explicitly in Sect. 8 of [17].

Furthermore, our translation of PIVPs to positive quadratic PIVPs preserves the polynomial time complexity defined in PIVPs as the trajectory length up to some precision. The translation of Theorem 2 together with Theorems 3 and 4 give.

Theorem 6. *A function over the reals is computable (resp. in polynomial time) if and only if it is computable by an elementary reaction system using only synthesis reactions with at most two catalysts of the form*

$$- => z \ or \ \_ = [x] => z \ or \ \_ = [x+y] => z$$

and degradation reactions by annihilation of the form

$$x\_p + x\_m => \_$$

(resp. with trajectories of polynomial length).

Proof. In the proof of Theorem 3, we have shown that one consequence of the annihilation reactions with fast kinetics is to make x_p and x_m not larger than $|x|+1$ for all x, and thereby ensure the preservation of the polynomial complexity. This inequality also shows that annihiliation reactions are useful to ensure the convergence of the result components.

One can remark in this proof that the encoding of real valued variables by two signed variables allows us to replace substractions by additions in the ODEs just by sorting the monomials according to their sign. Furthermore, the proof of Theorem 4 rewrites the terms with terms of degree at most 2 without changing their sign. As a consequence, all the terms of the ODE are monomials of the forms k, $k * x$, $k * x * y$ or $-f * x_p * x_m$ which can be encoded with synthesis reactions with at most two catalysts, and annihiliation reactions.   □

The possible implementations of the particular synthesis and degradation reactions used in Theorem 6 are beyond the scope of this paper. Let us just remark that a formal synthesis reaction as _ =[x] => z does not need to be a real synthesis reaction with DNA or RNA, but can be implemented with proteins, for instance by a phosphorylation reaction by kinase x, i.e. of the form iz = [x] => z where iz assumed to be in excess is the (inactive) dephosphorylated form of z. Similarly, the annihilation reaction z_p + z_m => _ might be thought as representing in reality, among many other possibilities, a complexation reaction which produces an inactive (stable) complex.

4 Biochemical Compilation of Analog Functions

4.1 Compilation of GPAC-Generable Functions

The proof of Theorem 3 shows how a PIVP can be implemented with biochemical reactions by doubling the number of variables for the positive and negative parts, and by implementing each monomial of the differential equations by a catalytic reaction of synthesis or degradation according to its sign. Similarly, the proof of Theorem 4 shows how to restrict code generation to elementary reactions of at most two reactants, by increasing the number of variables (i.e. molecular species), that is to say by sacrificing the dimension of the system to the minimization of the degrees.

These are the principles of our biochemical compiler which translates a mathematical function defined by a PIVP into a system of elementary reactions. For implementation reasons however, our compiler departs from the previous theoretical framework in a few places. The annihilation reactions (which play no role in the computability but in the complexity only) are implemented with a sufficiently large rate constant called *fast*, instead of with a large polynomial. The approximation error is not computed since we are not interested in the precision of the result and assume to know in advance some time horizon sufficient to get the results[6].

As a first example, let us consider the biochemical compilation of the oscillator defined by the cosine function $f = cos(t)$ as a function of time, itself defined by the PIVP $f'' = -f$ with $f(0) = 1$, i.e. $\{f' = z,\ z' = -f\}$ with $f(0) = 1$, $z(0) = 0$. This example compiles into the six elementary synthesis reactions below, where the first four reactions implement the PIVP, and the last two reactions the normalization reactions by mutual annihilation of the positive and negative variables.

```
biocham: compile_from_expression(cos, time, f).
  _ = [z2_p] => f_p.
  _ = [z2_m] => f_m.
  _ = [f_m] => z2_p.
  _ = [f_p] => z2_m.
  fast*z2_m*z2_p for z2_m+z2_p => _.
  fast*f_m*f_p for f_m+f_p => _.
  present (f_p, 1).
biocham: list_ode.
  d(f_p)/dt = z2_p-fast*f_m*f_p
  d(f_m)/dt = z2_m-fast*f_m*f_p
  d(z2_p)/dt = f_m-fast*z2_m*z2_p
  d(z2_m)/dt = f_p-fast*z2_m*z2_p
```

This reaction system, produced with initial concentration value $f_p = 1$ at time 0 (and 0 for all other variables), is designed for the differential semantics. Its robustness to extrinsic noise can be measured with respect to perturbations of the parameter values [42]. Such a reaction system can also be interpreted in the stochastic semantics [22], and simulated using Gillespie's SSA algorithm [24] to analyze its robustness to intrinsic noise. Figure 2 shows a differential simulation trace and one stochastic simulation trace.

4.2   Compilation of GPAC-Computable Functions

Let us first remark that a PIVP that *computes* the value of $y = f(x)$ at any point x can be derived from a PIVP that *generates* $f(t)$ as a function of time [39]. The idea is to replace the PIVP that generates $f(t)$ by a PIVP that generates $f(\gamma(t))$ where $\lim_{t\to\infty} \gamma(t) = x$, starting from a point x_0 such that $f(x)$ does not diverge along the trajectory $\gamma(t)$ [39]. Taking the trajectory $\gamma(t) = x + (x_0 - x)e^{-\lambda t}$ with $\lambda > 0$, we have $\gamma(t)' = -(x_0 - x)e^{-\lambda t} = x - \gamma(t)$.

Although not totally general since all GPAC-computable functions are not GPAC-generable, we limit ourselves to this method for compiling computable functions: with the following:

---

[6] Note also that the transformation to at most binary reactions is temporarily not included in our compiler.

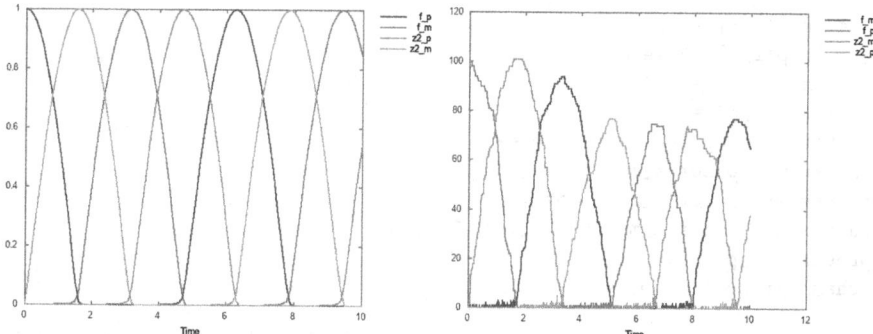

Fig. 2. Differential and stochastic simulation traces of the compiled reactions for generating the cosine function as a function of time.

Algorithm 1. Transformation of a PIVP that generates a function $f(t)$ in a PIVP that computes the function $f(x)$ for any x as $f(\gamma(t))$.

1. replace t by $\gamma(t)$ in the ODE that generates the function $f(t)$;
2. multiply all the terms of the ODE by $x - \gamma(t)$;
3. add the equation $\gamma' = x - \gamma$;
4. initialize γ to x_0 and the result variable to $f(x_0)$.

For instance, the compilation of the cosine function $cos(x)$ for any input concentration x generates the following elementary synthesis reaction system, where the first four reactions compute $\gamma(t)$ in g_p and g_m (with $\lambda = 1$), and the other reactions result from the multiplication by $x - \gamma$ of the ODE terms for $cos(t)$ which basically translates to the addition of catalysts x_p and g_m to the reactions for $cos(t)$:

```
biocham: compile_from_expression(cos, x, r).
  _ = [g_m]      => g_p.
  _ = [x_p]      => g_p.
  _ = [g_p]      => g_m.
  _ = [x_m]      => g_m.
  _ = [g_m+z4_p] => r_p.
  _ = [g_p+z4_m] => r_p.
  _ = [x_m+z4_m] => r_p.
  _ = [x_p+z4_p] => r_p.
  _ = [g_m+z4_m] => r_m.
  _ = [g_p+z4_p] => r_m.
  _ = [x_p+z4_m] => r_m.
  _ = [x_m+z4_p] => r_m.
  _ = [g_m+r_m]  => z4_p.
  _ = [g_p+r_p]  => z4_p.
  _ = [x_p+r_m]  => z4_p.
  _ = [x_m+r_p]  => z4_p.
```

```
_ = [g_m+r_p] => z4_m.
_ = [g_p+r_m] => z4_m.
_ = [x_m+r_m] => z4_m.
_ = [x_p+r_p] => z4_m.
fast*z4_m*z4_p for z4_m+z4_p => _.
fast*r_m*r_p for r_m+r_p => _.
fast*g_m*g_p for g_m+g_p => _.
fast*x_m*x_p for x_m+x_p => _.
present (r_p, 1).
biocham: present (x_p, 4).
```

This reaction system then computes $cos(x)$ by initializing the argument to the desired value, for instance $x_p = 4$ for which simulation traces are shown in Fig. 3.

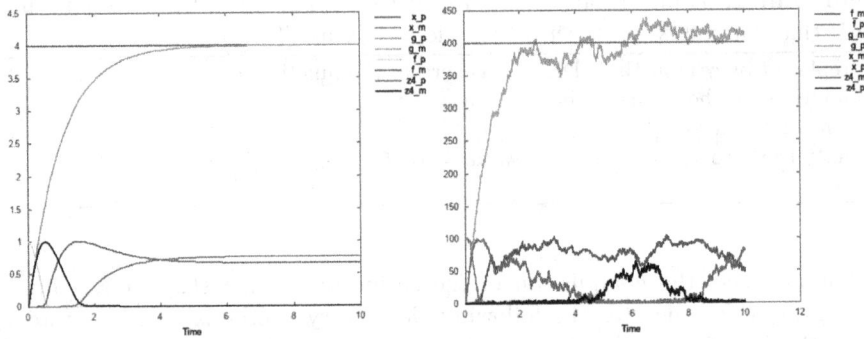

Fig. 3. Differential and stochastic simulation traces of the compiler-generated reactions for computing $cos(4)$.

5   Compilation of Sigmoid Functions

A sigmoid function is a bounded differentiable real function that is defined for all real input values and has a positive derivative at each point. Sigmoid functions have an "S" shape. They can be used to implement analog/digital converters which produce all-or-nothing outputs for a wide range of input levels. In biochemistry, Hill functions, of the form $x^n/(k + x^n)$, over $\mathbb{R}_+$ are examples of sigmoid functions that have been shown to approximate the input/output response of, first historically, cooperative allosteric enzymatic reactions [44], and more recently of the MAPK signaling network [28] for instance. In this section we study the biochemical compilation of various sigmoid functions which is key to the implementation of digital logic with molecular reactions [31,32].

5.1   Logistic, Hyperbolic Tangent, Arc Tangent and Hill Sigmoids

For the sake of simplicity, we restrict here to the generation of sigmoid functions as functions of time, with the idea of using Algorithm 1 for computing those functions as functions of some input variable. The logistic function $S(t) = 1/(1 + e^t)$ is a sigmoid function over $\mathbb{R}$ whose derivative can be written in terms of itself as $S'(t) = S(t) - S(t)^2$. It can be generated over $\mathbb{R}^+$ by two simple elementary reactions, one autocatalyzed synthesis and one autocatalyzed degradation:

```
S => 2*S.   S = [S] => _.   present(S,0.5).
```

The hyperbolic tangent $tanh(t)$ has also a simple derivative expression $tanh'(t) = 1 - tanh(t)^2$ which can be implemented with two elementary reactions:

```
_ => HT.   2*HT => HT.
```

The arc tangent $atan(t)$ has for derivative $atan'(t) = 1/(1 + t^2)$ which can be implemented by

```
_ => T.   1/(1+T^2) for /T => AT.
```

Note however that in this presentation, the second synthesis reaction uses T as reaction inhibitor, which is beyond the scope of this paper.

The Hill functions of degree d (resp. negative Hill functions) are defined by $H_d(t) = t^d/(k + t^d)$ (resp. $NH_d(t) = 1/(k + t^d)$) for some parameter $k \in \mathbb{R}$. One can easily check that they are solutions of the PIVP $H'_d = d * k * t^{d-1} * NH_d^2$, $NH'_d = -d * t^{d-1} * NH_d^2$ with $H_d(0) = 0$ and $NH_d(0) = 1/k$, which leads to the following (non elementary) reactions for their generation:

```
MA(d) for NHd = [(d-1)*T+NHd] => _. present(NHd,1/k).
MA(d*k) for _ = [(d-1)*T+2*NHd] => Hd.
```

5.2   Comparison to MAPK Signaling Circuits

MAPK (mitogen-activated protein kinases) signaling networks are very common biochemical reaction modules which are found in multiple copies in eukaryotic organisms. In these signaling cascades the proteins activated by phosphorylation are themselves kinases which catalyze in cascade other phosphorylations. Thus, the MAPK cascade has three stages of phosphorylation for a total of 30 elementary reactions: the entry E_1 of the cascade, directly linked to the membrane receptor, catalyses the phosphorylation of the kinase KKK of the first stage, which in turn phosphorylates the kinase KK of the second stage, which in this doubly phosphorylated form phosphorylates the protein K of the last stage of the cascade, which, when doubly phosphorylated in Kpp, is able to migrate into the nucleus and promote or inhibit gene transcription.

In [28] Huang and Ferrell have proposed an explanation for this structure by showing that the MAPK cascades exhibit a (stationary) response in the form of a Hill function which produces a nearly all-or-nothing response. That is, by denoting (u, y) the input-output relation of the system, they could approximate the dose-response diagram by an equation of the form $Y(u) \approx \lambda \frac{u^d}{c^d + u^d}$ with d in the order of 4.9 at the third level Kpp $d \sim 1.7$ to the second $KKpp$ and $d = 1$ at the first level $KKKp$.

The Hill function, as a function of an input, can be compiled in biochemical reactions by applying Algorithm 1 to the PIVP given in the previous section for the Hill function as a function of time. This leads to the following reaction system:

$$\left\{ \begin{array}{llll} \gamma & \rightarrow & - & y_2 + d + x + y_1 \rightarrow d + x + y_1 + 2y_2 \\ x & \rightarrow & x + \gamma & 2y_2 + d + x + y_1 \rightarrow d + x + y_1 + y_2 \\ 2y_1 + x & \rightarrow & y_1 + x & y_2 + d + \gamma + y_1 \rightarrow d + \gamma + y_1 \\ 2y_1 + \gamma & \rightarrow & 3y_1 + \gamma & 2y_2 + d + \gamma + y_1 \rightarrow d + \gamma + y_1 + 3y_2 \end{array} \right\}$$

with the initial conditions $(\gamma, y_1, y_2)_{t=0} = (1, 1, 1/2)$. This system satisfies $y_2 = \frac{x^d}{1+x^d}$ at steady state, and therefore constitutes a binary presence indicator: if $x \gg 1$, then $y_2 = 1$, and if $x \ll 1$, then $y_2 = 0$, the greater d, the greater the discrimination. Note that this value is given here by a fixed concentration of molecule but could be represented more simply by a kinetic constant. This converter, however, fails to create an intermediate value in $\frac{1}{\gamma}$ which gives an exponential amplitude for $x = 0$, and therefore an exponential computational complexity in the sense of the previous section. If we restrict ourselves to taking x in an interval of the form $[\varepsilon, +\infty[$, with $\varepsilon > 0$, then the complexity becomes polynomial. On the other hand, if we restrict to degree 2 and compile the expression $x^2/(1+x^2)$, the command `compile_from_expression(id*id/(1+id*id),x,y)` produces a system of 259 reactions over 23 species (70 reactions over 19 species for the function of time). However, the generated species for the possibly negative values, and their reactions, are useless in this example. Furthermore, our syntax-directed compilation strategy currently associates one variable per term occurrence, thus twice for the two occurrences of the expression x^2, and performs division in another variable. The computational complexity is polynomial, but with one component of amplitude x^2 which is computed in that strategy.

The natural MAPK circuit of 30 reactions [28] thus currently appears both more concise, and with a lesser computational complexity, than the system of reactions produced according to our first principles of compilation without any optimization.

6   Compilation of Sequentiality and Program Control Flows

The negative Hill sigmoid $\frac{c}{c+x^d}$ provides a binary absence indicator of higher quality than those proposed in [45] or even [29] for implementing sequentiality and program control flows, for which leakage phenomena may occur: even in the relative absence of the x species, the presence indicator remains at a sufficiently high concentration to catalyze certain reactions, or the opposite effect, the absence indicator may be too small. This is particularly visible in the sequentiality implementation: given the R_i reactions, if we want R_2 to be executed only once R_1 is completed, one can impose an indicator of the absence of one species consumed by R_1 as catalyst of R_2, ditto between R_2 and R_3, etc This leads however to the following phenomenon: the reactions are made all the more slowly as i is large, in other words, the reactions accumulate delay in their execution due to the retention of absence indicators.

With a sufficiently powerful absence indicator, it is possible to implement the sequentiality, the conditional instruction, and loop structures of algorithmic programming. It has been shown in [29] how to compile small imperative programs into a system of biochemical reactions wherein the molecular species are used as markers of the position of the program in a control flow graph. This was illustrated with the compilation of Euclidean division and greatest common divisor programs, and with strategies for species minimization in [30].

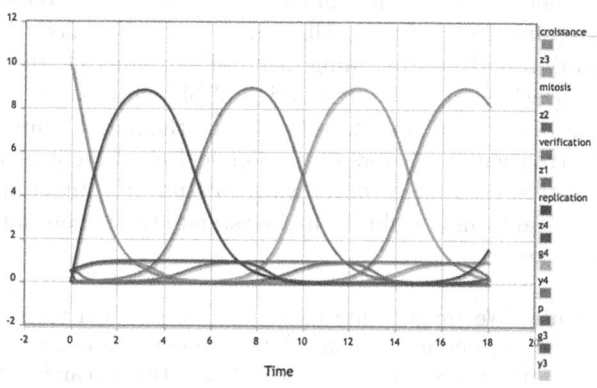

Fig. 4. ODE simulation trace of the generated reactions for the cell cycle loop.

Along the same lines, a minimalist specification of the cell division cycle can be specified by the program

```
while true do {Growth; Replication; Verification; Mitosis;}
```

The compilation of this program in elementary reactions implements the sequentiality of the four phases of the cycle by the degradation of the markers of each of the phases, depicted in Fig. 4. Interestingly, the resulting simulation curves are quite similar to the concentration curves obtained in cell cycle models [23] for the cyclin proteins, which appear here as necessary markers for implementing sequentiality with biochemical reactions.

7   Discussion and Perspectives

Though one lesson of Computer Science is that analog computation does not scale up, while digital computation does, the biological perspective provides a new impetus to the study of analog computation and mixed analog/digital parallel programs.

We have shown that recent results in computable analysis and theoretical complexity establish solid links between analog and digital computation, and can be used to compile analog specifications and mixed analog/digital programs into elementary biochemical reactions. This opens new research avenues to analyze natural protein interaction circuits not only from point of view of the size and the static complexity of the networks [1], but also from the computational complexity and robustness points of view [39], to revisit the important particular case of linear time invariant systems [14–16], to design reaction code optimizers, and compare natural circuits acquired by evolution to engineered and compiler-generated synthesized circuits.

The concept of biochemical computation and compilation can also be experimented *in vitro* and *in vivo*, either in Synthetic Biology, through the modification and reprogramming of living cells [18,35], or in Synthetic Biochemistry, through the creation and programming of non-living microfluidic vesicles [19], with various applications including the design of biomarkers [18].

Furthermore, the formal specification by mathematical functions of the input/output or transient behaviors of biochemical reaction systems under the differential semantics, establishes novel ways to study the functions of natural circuits mathematically, and on this route investigate their evolution history and evolution capabilities [47].

Acknowledgements. We are grateful to especially one reviewer for his expert proof-reading which helped us to improve the presentation of our results, and to the editors for providing us with the necessary extra space. Part of this research is funded by the ANR-MOST Biopsy project. The first author acknowledges fruitful discussions with Jie-Hong Jiang (NTU, Taiwan) on the compilation of program control flows with reactions, and motivating discussions with Frank Molina (CNRS, Sys2Diag, Montpellier) on the biochemical implementation by enzymatic reactions in microfluidic vesicles.

References

1. Barabási, A.L.: Network Science. Cambridge University Press, Cambridge (2016)
2. Berry, G., Boudol, G.: The chemical abstract machine. Theor. Comput. Sci. **96**, 217–248 (1992)
3. Bournez, O., Graça, D.S., Pouly, A.: Polynomial time corresponds to solutions of polynomial ordinary differential equations of polynomial length. J. ACM (2017, accepted)
4. Bournez, O., Campagnolo, M.L., Graça, D.S., Hainry, E.: Polynomial differential equations compute all real computable functions on computable compact intervals. J. Complex. **23**(3), 317–335 (2007). https://hal-polytechnique.archives-ouvertes.fr/inria-00102947
5. Bournez, O., Campagnolo, M.L., Graça, D.S., Hainry, E.: The general purpose analog computer and computable analysis are two equivalent paradigms of analog computation. In: Cai, J.-Y., Cooper, S.B., Li, A. (eds.) TAMC 2006. LNCS, vol. 3959, pp. 631–643. Springer, Heidelberg (2006). doi:10.1007/11750321_60

6. Bournez, O., Graça, D.S., Pouly, A.: Polynomial time corresponds to solutions of polynomial ordinary differential equations of polynomial length. The general purpose analog computer and computable analysis are two efficiently equivalent models of computations. In: 43rd International Colloquium on Automata, Languages, and Programming, ICALP 2016, Rome, Italy. LIPIcs, vol. 55, pp. 109:1–109:15. Schloss Dagstuhl - Leibniz-Zentrum fuer Informatik, 11–15 July 2016. http://drops.dagstuhl.de/opus/frontdoor.php?source_opus=6244

7. Bournez, O., Graça, D.S., Pouly, A.: On the functions generated by the general purpose analog computer. Inf. Comput. (2017, accepted under minor revision)

8. Buisman, H.J., ten Eikelder, H.M.M., Hilbers, P.A.J., Liekens, A.M.L.: Computing algebraic functions with biochemical reaction networks. Artif. Life 15(1), 5–19 (2009)

9. Busi, N., Gorrieri, R.: On the computational power of brane calculi. In: Priami, C., Plotkin, G. (eds.) Transactions on Computational Systems Biology VI. LNCS, vol. 4220, pp. 16–43. Springer, Heidelberg (2006). doi:10.1007/11880646_2

10. Cardelli, L., Zavattaro, L.: Turing universality of the biochemical ground form. Math. Struct. Comput. Sci. 20(1), 45–73 (2010)

11. Carothers, D.C., Parker, G.E., Sochacki, J.S., Warne, P.G.: Some properties of solutions to polynomial systems of differential equations. Electron. J. Differ. Eq. 40 (2005)

12. Chen, H.L., Doty, D., Soloveichik, D.: Rate-independent computation in continuous chemical reaction networks. In: Proceedings of the 5th Conference on Innovations in Theoretical Computer Science, ITCS 2014, pp. 313–326. ACM, New York (2014)

13. Chen, Y., Dalchau, N., Srinivas, N., Phillips, A., Cardelli, L., Soloveichik, D., Seelig, G.: Programmable chemical controllers made from DNA. Nat. Nanotechnol. 8, 755–762 (2013)

14. Chiang, H.J., Jiang, J.H., Fages, F.: Reconfigurable neuromorphic computation in biochemical systems. In: Proceedings of the 37th Annual International Conference of the IEEE Engineering in Medicine and Biology Society, EMBC (2015). http://lifeware.inria.fr/~fages/Papers/CJF15ieee.pdf

15. Chiang, K., Jiang, J.H., Fages, F.: Building reconfigurable circuitry in a biochemical world. In: BioCAS 2014: IEEE Biomedical Circuits and Systems Conference. IEEE, Lausanne, October 2014. http://lifeware.inria.fr/~fages/Papers/CJF14biocas.pdf

16. Chiu, T.Y., Chiang, H.J.K., Huang, R.Y., Jiang, J.H.R., Fages, F.: Synthesizing configurable biochemical implementation of linear systems from their transfer function specifications. PLoS ONE 10(9) (2015)

17. Cook, M., Soloveichik, D., Winfree, E., Bruck, J.: Programmability of chemical reaction networks. In: Condon, A., Harel, D., Kok, J.N., Salomaa, A., Winfree, E. (eds.) Algorithmic Bioprocesses, pp. 543–584. Springer, Heidelberg (2009). doi:10. 1007/978-3-540-88869-7_27

18. Courbet, A., Endy, D., Renard, E., Molina, F., Bonnet, J.: Detection of pathological biomarkers in human clinical samples via amplifying genetic switches and logic gates. Sci. Transl. Med. (2015)

19. Courbet, A., Amar, P., Fages, F., Renard, E., Molina, F.: Computer-aided biochemical programming of synthetic microreactors operating as logic-gated and multiplexed diagnostic devices (submitted)

20. Daniel, R., Rubens, J.R., Sarpeshkar, R., Lu, T.K.: Synthetic analog computation in living cells. Nature 497(7451), 619–623 (2013)

21. Fages, F., Gay, S., Soliman, S.: Inferring reaction systems from ordinary differential equations. Theor. Comput. Sci. **599**, 64–78 (2015). http://lifeware.inria.fr/~fages/Papers/FGS14tcs.pdf
22. Fages, F., Soliman, S.: Abstract interpretation and types for systems biology. Theor. Comput. Sci. **403**(1), 52–70 (2008). http://lifeware.inria.fr/~fages/Papers/FS07tcs.pdf
23. Gérard, C., Goldbeter, A.: Temporal self-organization of the cyclin/Cdk network driving the mammalian cell cycle. Proc. Natl. Acad. Sci. **106**(51), 21643–21648 (2009)
24. Gillespie, D.T.: General method for numerically simulating stochastic time evolution of coupled chemical-reactions. J. Comput. Phys. **22**, 403–434 (1976)
25. Gillespie, D.T.: Exact stochastic simulation of coupled chemical reactions. J. Phys. Chem. **81**(25), 2340–2361 (1977)
26. Graça, D., Costa, J.: Analog computers and recursive functions over the reals. J. Complex. **19**(5), 644–664 (2003)
27. Helmfelt, A., Weinberger, E.D., Ross, J.: Chemical implementation of neural networks and turing machines. PNAS **88**, 10983–10987 (1991)
28. Huang, C.Y., Ferrell, J.E.: Ultrasensitivity in the mitogen-activated protein kinase cascade. PNAS **93**(19), 10078–10083 (1996)
29. Huang, D.A., Jiang, J.H., Huang, R.Y., Cheng, C.Y.: Compiling program control flows into biochemical reactions. In: ICCAD 2012: IEEE/ACM International Conference on Computer-Aided Design, pp. 361–368. ACM, San Jose, November 2012. http://lifeware.inria.fr/~fages/Papers/iccad12.pdf
30. Huang, R.Y., Huang, D.A., Chiang, H.J.K., Jiang, J.H., Fages, F.: Species minimization in computation with biochemical reactions. In: IWBDA 2013: Proceedings of the Fifth International Workshop on Bio-Design Automation. Imperial College, London, July 2013. http://lifeware.inria.fr/~fages/Papers/HHCJF13iwbda.pdf
31. Jiang, H., Riedel, M., Parhi, K.K.: Digital signal processing with molecular reactions. IEEE Des. Test Comput. **29**(3), 21–31 (2012)
32. Jiang, H., Riedel, M., Parhi, K.K.: Digital logic with molecular reactions. In: ICCAD 2013: IEEE/ACM International Conference on Computer-Aided Design, pp. 721–727. ACM, November 2013
33. Lakin, M.R., Parker, D., Cardelli, L., Kwiatkowska, M., Phillips, A.: Design and analysis of DNA strand displacement devices using probabilistic model checking. J. Roy. Soc. Interface **9**(72), 1470–1485 (2012)
34. Magnasco, M.O.: Chemical kinetics is turing universal. Phys. Rev. Lett. **78**(6), 1190–1193 (1997)
35. Nielsen, A.A.K., Der, B.S., Shin, J., Vaidyanathan, P., Paralanov, V., Strychalski, E.A., Ross, D., Densmore, D., Voigt, C.A.: Genetic circuit design automation. Science **352**(6281) (2016)
36. Oishi, K., Klavins, E.: Biomolecular implementation of linear I/O systems. IET Syst. Biol. **5**(4), 252–260 (2011)
37. Arkin, P., Ross, J.: Computational functions in biochemical reaction networks. Biophys. J. **67**, 560–578 (1994)
38. Paun, G., Rozenberg, G.: A guide to membrane computing. Theor. Comput. Sci. **287**(1), 73–100 (2002)
39. Pouly, A.: Continuous models of computation: from computability to complexity. Ph.D. thesis, Ecole Polytechnique, July 2015
40. Qian, L., Soloveichik, D., Winfree, E.: Efficient turing-universal computation with DNA polymers. In: Sakakibara, Y., Mi, Y. (eds.) DNA 2010. LNCS, vol. 6518, pp. 123–140. Springer, Heidelberg (2011). doi:10.1007/978-3-642-18305-8_12

41. Rizik, L., Ram, Y., Danial, R.: Noise tolerance analysis for reliable analog and digital computation in living cells. J. Bioeng. Biomed. Sci. 6(186) (2016)
42. Rizk, A., Batt, G., Fages, F., Soliman, S.: Continuous valuations of temporal logic specifications with applications to parameter optimization and robustness measures. Theor. Comput. Sci. 412(26), 2827–2839 (2011). http://lifeware.inria.fr/ soliman/publi/RBFS11tcs.pdf
43. Sauro, H.M., Kim, K.: Synthetic biology: it's an analog world. Nature 497(7451), 572–573 (2013)
44. Segel, L.A.: Modeling Dynamic Phenomena in Molecular and Cellular Biology. Cambridge University Press, Cambridge (1984)
45. Senum, P., Riedel, M.: Rate-independent constructs for chemical computation. PLOS One 6(6) (2011)
46. Shannon, C.: Mathematical theory of the differential analyser. J. Math. Phys. 20, 337–354 (1941)
47. Valiant, L.: Probably Approximately Correct. Basic Books, New York (2013)
48. Weihrauch, K.: Computable Analysis: An Introduction. Springer, Heidelberg (2000). doi:10.1007/978-3-642-56999-9

A Scheme for Adaptive Selection of Population Sizes in Approximate Bayesian Computation - Sequential Monte Carlo

Emmanuel Klinger[1,2,3] and Jan Hasenauer[2,3(✉)]

[1] Department of Connectomics, Max Planck Institute for Brain Research,
60438 Frankfurt, Germany
emmanuel.klinger@brain.mpg.de
[2] Helmholtz Zentrum München - German Research Center for Environmental Health,
Institute of Computational Biology, 85764 Neuherberg, Germany
jan.hasenauer@helmholtz-muenchen.de, jan.hasenauer@tum.de
[3] Center for Mathematics, Chair of Mathematical Modeling of Biological Systems,
Technische Universität München, 85748 Garching, Germany

Abstract. Parameter inference and model selection in systems biology often requires likelihood-free methods, such as Approximate Bayesian Computation (ABC). In recent years, this approach has frequently been combined with a Sequential Monte Carlo (ABC-SMC) scheme. In this scheme, the approximation of the posterior distribution through a population of particles is iteratively improved by a sequential sampling strategy. However, it has been difficult to give general guidelines on how to choose the size of these populations. In this manuscript, we propose a method to adaptively and automatically select these population sizes. The method exploits the cross-validated approximation error of a kernel density estimate of the particles in the current population to select the number of particles for the subsequent population.

We found the proposed method to be robust to the initially chosen population size and to the number of posterior modes. We demonstrated that the method is applicable to parameter inference as well as to model selection. The study of a computationally demanding multiscale model confirmed the method's scalability. In conclusion, the proposed method is applicable to a wide range of parameter and model selection tasks. The method makes the influence of the population size on the approximation error explicit simplifying the application of ABC-SMC schemes.

Keywords: Parameter estimation · Likelihood-free inference · Approximate Bayesian Computation · Model selection · Sequential Monte Carlo · Population size

1 Introduction

Computer simulations have become an indispensable tool for scientific research. They facilitate to investigate regimes which are not analytically tractable anymore. It is often easy to simulate an experimental outcome for a model with given

© Springer International Publishing AG 2017
J. Feret and H. Koeppl (Eds.): CMSB 2017, LNBI 10545, pp. 128–144, 2017.
DOI: 10.1007/978-3-319-67471-1_8

parameters. However, it is often difficult to select the model and its parameters which are likely to explain a given experimental finding [1].

The Bayesian paradigm provides a natural framework to treat parameter estimation and model selection. Unfortunately, for stochastic models it is often impossible to calculate the likelihood efficiently, prohibiting a range of methods, such as e.g. variational inference [7], to approximate the Bayesian posterior. This has lead to the development of Approximate Bayesian Computation (ABC) schemes [2,14]. Amongst the different ABC schemes (see e.g. [1,24] and references therein) one particularly popular scheme uses a Sequential Monte Carlo (SMC) technique and is therefore called the Approximate Bayesian Computation - Sequential Monte Carlo (ABC-SMC) scheme [22,25,26]. In ABC-SMC, the posterior distribution of a parameter is approximated through a particle population. This population is sequentially refined from generation to generation improving the approximation. However, "it appears unfortunately difficult to give useful general guidelines how to select the population size as it is highly case depending" [16]. This is problematic as too small population sizes yield large approximation errors and might even hamper convergence, while too large population sizes result in an unnecessary computational burden.

We therefore investigated a method to select population sizes for ABC-SMC adaptively and automatically and describe it in this paper. The method is applied to examples with multimodal posteriors, model selection for Markov jump process models and multiscale, agent-based models. An implementation of the proposed method is provided as part of the pyABC framework (http:// pyabc.readthedocs.io/en/latest).

2   Methods

In the following, we introduce the ABC-SMC method and provide the corresponding algorithms. Transition kernels are discussed and related to kernel density estimation. Based on this relation, we suggest a scheme for the adaptive selection of population sizes.

2.1   ABC-SMC Algorithm

In ABC-SMC, populations of weighted parameter samples are sequentially constructed to approximate the posterior distribution of the parameter of interest. The ABC-SMC scheme considered in this study, and provided in Algorithm 1, is similar to the one from [25]. By $P = \{(w_i, \theta_i)\}_{i=1}^n$ we denote a population[1] of size n of weighted parameter samples with weights $w_i > 0$, $\sum_i w_i = 1$ and parameters $\theta_i \in \mathbb{R}^{d_{\mathrm{par}}}$ of dimension d_{par}. We denote the sequence of corresponding weighted distances $\delta_i \in \mathbb{R}^+$ by $D = \{(w_i, \delta_i)\}_{i=1}^n$. The distance δ_i is determined by evaluating the distance function $d = \mathrm{ComputeDistance} : \mathcal{S} \times \mathcal{S} \to \mathbb{R}^+$ for a

---

[1] A population is a sequence of pairs. Pairs (w, θ) can occur multiple times in a population. The curly braces {} denote a finite sequence (not a set).

pair of simulated data $s \in \mathcal{S}$, obtained by a stochastic simulation of the model for parameter θ_i via the function Simulate $: \mathbb{R}^{d_{\mathrm{par}}} \to \mathcal{S}$, and the observed data $s_{\mathrm{data}} \in \mathcal{S}$. In [25], a transition kernel was used to perturb samples of the previous population to generate proposals for the subsequent population. We reformulated the generation of parameter proposals using a generic (non-degenerate) density function K. This density is obtained from a kernel density estimator KDE $: P \mapsto K$ mapping a population P to a density function $K : \mathbb{R}^{d_{\mathrm{par}}} \to \mathbb{R}^+$, $\int_{\mathbb{R}^{d_{\mathrm{par}}}} K = 1$. The density K estimated on the current population serves as proposal distribution for the subsequent population. For the first population, the prior p_0 serves as proposal distribution. After each generation, the acceptance threshold ϵ is adapted via the function AdaptThreshold $: D \mapsto \epsilon$ and the population size n is adapted via the function AdaptPopulationSize $: (P, \mathrm{KDE}) \mapsto n$, which is described in Sect. 2.3 (Algorithm 3). Throughout this paper the acceptance threshold is adapted by setting the threshold for the subsequent population to the median of the weighted distances D of the previous population. The population size is initialized with $n_0 \in \mathbb{N}$. Sampling is stopped when either the maximum number of allowed generations t_{max} or the final acceptance threshold $\epsilon_{\mathrm{min}} > 0$ is reached.

Algorithm 1. ABC-SMC

Input: t_{max}, ϵ_{min}, n_0, s_{data}, p_0, KDE, Simulate, ComputeDistance
Output: P

$t \leftarrow 0$
$K \leftarrow p_0$
$\epsilon \leftarrow \infty$
$n \leftarrow n_0$
while $t < t_{\mathrm{max}}$ *and* $\epsilon > \epsilon_{\mathrm{min}}$ **do**
$\quad (P, D) \leftarrow$ SamplePopulation$(K, p_0, \epsilon, n, s_{\mathrm{data}},$ Simulate,
$\quad$ ComputeDistance$)$
$\quad K \leftarrow$ KDE(P)
$\quad n \leftarrow$ AdaptPopulationSize$(P,$ KDE$)$
$\quad \epsilon \leftarrow$ AdaptThreshold(D)
$\quad t \leftarrow t + 1$
end

The function SamplePopulation is described in Algorithm 2. There, a single candidate parameter θ is stochastically drawn from the density K by the function SampleSingleParameter $: K \mapsto \theta$. The model is then stochastically evaluated by the function Simulate $: \theta \mapsto s$ at this parameter θ, yielding the simulated data s. The parameter θ is added to the next population P only if the distance δ of the simulated data s to the observed data s_{data} is smaller than the current acceptance threshold ϵ. The two empty braces $\{\}$ denote the empty sequence with which the population P and the corresponding distances D are initialized. The + operator applied to sequences denotes the concatenation of these sequences.

For model selection, the parameter θ, the prior p_0, as well as the proposal density K can be decomposed into a component over the models and one over the model specific parameters. Assuming M models, the parameter θ is a sequence $\theta = (\theta_m)_{m=1}^M$ with $\theta_m \in \mathbb{R}^{d_{par,m}}$ and $d_{par,m}$ the dimension of the parameter space of model m. Moreover, the prior factorizes as $p_0(\theta) = p_0(\theta_m|m)p_0(m)$ and similarly the proposal density as $K(\theta) = K_{parameter}(\theta_m|m)K_{model}(m)$.

Algorithm 2. SamplePopulation

Input: K, p_0, ϵ, n, s_{data}, Simulate, ComputeDistance
Output: (P, D)

$P \leftarrow \{\}, D \leftarrow \{\}, Z \leftarrow 0$

while $|P| < n$ **do**
  repeat
    repeat
      | $\theta \leftarrow$ SampleSingleParameter(K)
    until $p_0(\theta) > 0$;
    $s \leftarrow$ Simulate(θ)
    $\delta \leftarrow$ ComputeDistance(s, s_{data})
  until $\delta < \epsilon$;
  $w \leftarrow p_0(\theta)/K(\theta)$
  $Z \leftarrow Z + w$
  $P \leftarrow P + \{(w,\theta)\}, D \leftarrow D + \{(w,\delta)\}$
end
$P \leftarrow \{(w/Z, \theta)|(w,\theta) \in P\}, D \leftarrow \{(w/Z, \delta)|(w,\delta) \in D\}$

The result of the ABC-SMC Algorithm 1 detailed above is an approximation of the posterior density, represented by a particle population (or by a KDE of it). The algorithm is implemented as part of the pyABC framework (http://pyabc.readthedocs.io/en/latest).

2.2 Kernel Density Estimation

In ABC-SMC the populations of weighted parameters are sequentially refined by decreasing the acceptance threshold from generation to generation. Of crucial importance in this process is the sampling of parameter proposals based on parameters accepted in the previous generation. This is commonly achieved by selecting an accepted parameter and perturbing it to generate a proposal [25]. This method is equivalent to sampling proposals from a non-parametric distribution approximation K, i.e. a kernel density estimate. In this study, the proposal densities K in Algorithms 1 and 2 were determined by kernel density estimators sharing the same general form.

General form: A density estimate K is expressed as sum of normally distributed kernels

$$K(\theta') = \sum_{i=1}^{n} w_i \, \mathcal{N}(\theta'|\theta_i, \Sigma(P, \theta_i)),$$

in which $P = \{(w_i, \theta_i)\}_{i=1}^{n}$ is a population of weighted parameters with weights $w_i \in \mathbb{R}^{+}$, $\sum_i w_i = 1$ and parameters $\theta_i \in \mathbb{R}^{d_{\text{par}}}$. The kernel $\mathcal{N}(\theta'|\theta, \Sigma)$ is a normal density with mean $\theta \in \mathbb{R}^{d_{\text{par}}}$ and covariance matrix $\Sigma \in \mathbb{R}^{d_{\text{par}} \times d_{\text{par}}}$, in the following referred to as bandwidth, evaluated at the parameter $\theta' \in \mathbb{R}^{d_{\text{par}}}$. Three different strategies were used to determine the bandwidth Σ.

Global bandwidth: The global bandwidth is a scaled covariance of the complete population. The population covariance matrix $\text{Cov}(P)$ is calculated from the population $P = \{(w_i, \theta_i)\}_{i=1}^{n}$, taking into account the sample weights:

$$\text{Cov}(P) = \sum_{i=1}^{n} w_i \, (\theta_i - \mu)(\theta_i - \mu)^t, \quad \mu = \sum_{i=1}^{n} w_i \, \theta_i.$$

The scaling factor b_{Silv} is estimated with Silverman's rule of thumb [21],

$$b_{\text{Silv}} = \left(\frac{4}{n_{\text{eff}}(d_{\text{par}} + 2)} \right)^{1/(d_{\text{par}}+4)}, \quad n_{\text{eff}} = \frac{1}{\sum_i w_i^2},$$

in which d_{par} denotes the parameter dimension, $\{w_i\}_{i=1}^{n}$ the sample weights and n_{eff} the effective population size. The kernel bandwidth is then $\Sigma = b_{\text{Silv}}^2 \text{Cov}(P)$. The bandwidth does thus not depend on the sample location θ and is therefore called "global".

Local bandwidth: The global bandwidth can be ill-suited for an accurate local approximation [19,21]. We therefore considered local bandwidths as well. The local bandwidths $\Sigma_{k,\text{nn}}(P, \theta_i)$ are constructed for each sample θ_i individually as twice the covariance matrix of the k nearest neighbors (in Euclidean distance) of sample θ_i. The overall density K is then given by

$$K(\theta') = \sum_{i=1}^{n} w_i \, \mathcal{N}(\theta'|\theta_i, \Sigma_{k,\text{nn}}(P, \theta_i)).$$

Similar bandwidths were examined before and were shown to yield good acceptance rates [6].

Cross-validated bandwidth: Since the scaling factor b_{Silv} is known to be too large for multimodal distributions [21], we also used cross-validated selection of the scaling factor for the population covariance matrix $\text{Cov}(P)$, according to the following scheme: the largest probed scaling factor is the Silverman scaling factor b_{Silv}. Five-fold cross-validation is used to determine the best of the downscaled factors $b_c = 2^{-e(c)} b_{\text{Silv}}$ with $e(c) = c/(2C)$, $c \in \{0, 1, \ldots, C\}$, $C = 4$.

The score function S of a scaling factor b_c with density K_c on a sub-population $\widetilde{P} = \{(w_i, \theta_i)\}_i$ is

$$S(K_c, \widetilde{P}) = \sum_i w_i \log K_c(\theta_i).$$

The score function is evaluated on the sub-population $\widetilde{P}$ of P which is not used for density estimation. The scaling factor b_c, whose corresponding density K_c yields the highest cumulative score summed over the five folds, is subsequently selected for the final density estimation on the complete population P.

2.3   Population Size Adaptation

The quality of an ABC-SMC scheme is determined by the accuracy and efficiency with which the posterior distribution is approximated. Besides the bandwidth selection strategy, a key parameter is the population size, i.e. the number of parameter samples of which a population consists. Not only the size of the last population but also the sizes of the intermediate populations have substantial influence. If intermediate populations are chosen too large, unnecessary computation is performed, rendering ABC-SMC inefficient. If intermediate populations are chosen too small, information about the posterior might get lost which cannot be efficiently recovered in the last population, rendering ABC-SMC inaccurate. For example, if the true posterior is multimodal, a small intermediate population might lack samples representing one of the posterior modes. This mode is unlikely to be recovered in the last population, unless the population size is chosen extraordinarily large. However, this would render ABC-SMC again inefficient and essentially equivalent to rejection sampling. Similarly, in model selection, one model with a small posterior probability might get completely extinct in an intermediate population. Hence, a consistent approximation quality across all intermediate populations is important for an accurate and efficient ABC-SMC scheme.

We thus developed an ABC-SMC scheme in which the population sizes are adaptively selected trying to match a specified target accuracy. We propose to express this accuracy in terms of the variation associated with kernel density estimates on a population (smaller variation corresponding to larger accuracy). To select the necessary population size to achieve the target variation, the effect of increasing or decreasing the population size (Fig. 1a) on the variation of the density estimate for the current population is determined with bootstrapped populations of varying sizes. By a parametric approximation to this population size dependent variation, a population size for the next generation is selected by interpolating to smaller population sizes if the current variation is too small and by extrapolating to larger population sizes if the current variation is too large.

Denoting by E_{CV} the desired target density variation, we propose the following scheme for adaptive population size selection: The number of particles in the initial population $t = 0$ is set to n_0. Given population $P_t = \{(w_i, \theta_i)\}_{i=1}^{n_t}$, $t \geq 0$, of size n_t, tentative population sizes $n_{t,q}^*$, evenly spaced between $\lfloor n_t/3 \rfloor$ and $2n_t$ with step size $\lfloor n_t/10 \rfloor$, are considered:

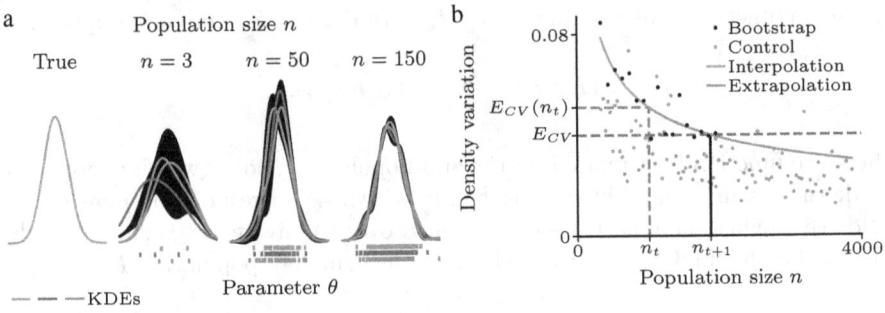

Fig. 1. Adaptive population size selection. (a) True density and kernel density estimates (KDEs) on populations of size n, sampled from the true density. The weighted differential density variation is computed by calculating the pointwise variation (at each parameter θ) and weighting it by the true density at this point. The density variation is the integrated weighted differential density variation and is higher for smaller population sizes n. **(b)** Fit of the density variation E_{CV} parametrized as $E_{CV}(n; \alpha, \beta) = \alpha n^{-\beta}$ on bootstrapped populations (black points) of tentative sizes $n_{t,q}^*$ (see (1)) and corresponding estimated variations $E_{CV}(n_{t,q}^*)$ (see (2)). Bootstrapped populations are drawn from the KDE of population t. The population size n_{t+1} of the subsequent population $t+1$ is selected to match the target variation E_{CV}, correcting the current variation $E_{CV}(n_t)$. Control estimates were directly obtained from populations of varying sizes n of the underlying true, unimodal normal distribution. The bandwidth was selected according to the Silverman rule (Sect. 2.2, global bandwidth).

$$n_{t,q}^* = \lfloor n_t/3 \rfloor + (q-1) \lfloor n_t/10 \rfloor, \ q \in \{1, \ldots, Q\}, \ Q = \max\{q | n_{t,q}^* \leq 2n_t\}. \quad (1)$$

To estimate the variation for each tentative population size $n_{t,q}^*$, a bootstrapped population $P_{t,q,b}$ of size $n_{t,q}^*$ is drawn from the density $K = \text{KDE}(P_t)$ for each bootstrap repetition $b \in \{1, \ldots, B\}$ (usually $B \approx 10$). Next, a density estimate, $K_{t,q,b} = \text{KDE}(P_{t,q,b})$ is calculated on each of the bootstrapped populations $P_{t,q,b}$. The density variation E_{CV} is then defined for each tentative population size $n_{t,q}^*$ according to

$$E_{CV}(n_{t,q}^*) = \sum_{i=1}^{n_t} w_i \ \text{CV}\left(\{K_{t,q,b}(\theta_i)\}_{b=1}^B\right), \quad (2)$$

with the coefficient of variation CV given by

$$\text{CV}(\{x_b\}_{b=1}^B) = \frac{\text{Std}(\{x_b\}_{b=1}^B)}{\text{Mean}(\{(x_b\}_{b=1}^B)},$$

computed from mean Mean and (biased) standard deviation Std:

$$\text{Mean}(\{x_b\}_{b=1}^B) = \frac{1}{B}\sum_{b=1}^B x_b, \ \text{Std}(\{x_b\}_{b=1}^B) = \sqrt{\frac{1}{B}\sum_{b=1}^B \left(x_b - \text{Mean}\left(\{x_b\}_{b=1}^B\right)\right)^2}.$$

For the inter- and extrapolation of E_{CV} a functional approximation f is used. The functional form of f is motivated by the scaling of the KDE mean squared error as function of the population size n [3]. Silverman [21] showed that the mean squared error of KDEs decreases with $\alpha\, n^{-\beta}$, depending on the properties of the distribution and the choice of the selected KDE. In this study, this functional form is employed on E_{CV}. The parameters α and β of the function $f(n; \alpha, \beta) = \alpha\, n^{-\beta}$ are fitted to the points $\{(n_{t,q}^*, E_{CV}(n_{t,q}^*))\}_{q=1}^{Q}$ (Fig. 1b, black points) with non-linear least squares and the Levenberg-Marquardt algorithm [11,15], yielding the optimized parameters α_t and β_t as well as the corresponding curve (Fig. 1b, interpolation and extrapolation). Finally, the size n_{t+1} of the subsequent population is selected such that the target variation is expected to be achieved: $n_{t+1} = \mathrm{round}(f^{-1}(E_{CV}; \alpha_t, \beta_t))$. In the case of multiple models, this scheme is performed on the joint parameter space. The pseudo-code is provided in Algorithms 3 and 4. There, Fit denotes fitting the function f, Round rounding to the nearest integer, Sample(K, n^*) drawing n^* samples from K and n_q^* corresponds to $n_{t,q}^*$ of (1).

Algorithm 3. AdaptPopulationSize

Input: P, KDE
Output: n
$N^* \leftarrow [n_1^*, \ldots, n_Q^*]$
$C^* \leftarrow [\mathrm{EstimateCV}(n^*, \mathrm{KDE}, P) \textbf{ for } n^* \textbf{ in } N^*]$
$f \leftarrow \mathrm{Fit}(N^*,\, C^*)$
$n \leftarrow \mathrm{Round}(f^{-1}(E_{CV}))$

Algorithm 4. EstimateCV

Input: n^*, KDE, P
Output: cv
$K \leftarrow \mathrm{KDE}(P)$
$K^* \leftarrow [\mathrm{KDE}(\mathrm{Sample}(K, n^*)) \textbf{ for } b \textbf{ in } \{1, ..., B\}]$
$\mathrm{cv} \leftarrow \sum_{(w,\theta)\in P} w\, \mathrm{CV}([K'(\theta) \textbf{ for } K' \textbf{ in } K^*])$

The population size is selected before the sampling of a population starts, instead of being continuously re-evaluated during sampling (after acceptance of each particle), to avoid potential bias towards distributions yielding lower variation for the same population size.

3   Results

To assess the proposed adaptation scheme, we applied it to problems with known true parameters and to problems of high practical relevance. We first assessed the appropriateness of the functional approximation, then examined the example of an analytical model with a multimodal posterior. We next applied the scheme to model selection of Markov Jump Process models and finally investigated a multiscale tumor growth model.

3.1 Appropriateness of the Functional Approximation of E_{CV} for Normal Distributions

We first asked if the chosen functional form $E_{CV}(n; \alpha, \beta) = \alpha n^{-\beta}$ did indeed capture the relation between population size and density variation reasonably well. A perfect approximation was not necessary, only an approximation which was good enough to ensure that the population size evolved towards the desired E_{CV} was required. For a first assessment we considered the case of a unimodal normal distribution. Indeed, the chosen functional form matched the relation between E_{CV} and n on the bootstrapped populations (Fig. 1b). Control samples from the true density revealed that in the extrapolated regime, the curve seemed to slightly overestimate E_{CV} but still captured the scaling behavior (Fig. 1b). We therefore continued with the first example.

3.2 Stability of the Population Size Adaptation for an Analytical Model

We considered a model, with a multimodal posterior (similar to [10]) to investigate the stability of the population sizes over the course of the generations, as well as a possible dependency on the number of posterior modes or the employed KDE. In this model, the simulated data $s \in \mathbb{R}^2$ were obtained by sampling from $s \sim \mathcal{N}(\mathrm{sq}(\theta, n_{\mathrm{modes}}), \sigma^2 I)$, in which I denotes the identity matrix in $\mathbb{R}^2$, $\sigma^2 > 0$, $n_{\mathrm{modes}} \in \{1, 2, 4\}$ denotes the number of posterior modes, and sq a squaring-like function squaring $\theta = (\theta_1, \theta_2)$ elementwise according to

$$\mathrm{sq}(\theta, n_{\mathrm{modes}}) = \begin{cases} (\theta_1, \theta_2) & \text{if } n_{\mathrm{modes}} = 1, \\ (\theta_1^2, \theta_2) & \text{if } n_{\mathrm{modes}} = 2, \\ (\theta_1^2, \theta_2^2) & \text{if } n_{\mathrm{modes}} = 4. \end{cases}$$

The form of the squaring function sq ensured that the number of posterior modes equaled n_{modes}. The parameter $\theta \in [-10, 10]^2$ was subject to posterior inference, with uniform prior $\theta \sim \mathcal{U}([-10, 10]^2)$ over the square $[-10, 10]^2$. The distance function d was $d(s, s_{\mathrm{data}}) = |s_1 - s_{\mathrm{data},1}| + |s_2 - s_{\mathrm{data},2}|$. For this model, $B = 10$ bootstrapped populations were used to estimate the density variation.

We performed ABC-SMC runs for $n_{\mathrm{modes}} = 1, 2, 4$ modes, with observed data $s_{\mathrm{data}} = (1, 1)$, $\sigma^2 = 0.5$ and $E_{CV} = 0.1$. The modes were correctly captured after a few generations for all scenarios (see Fig. 2a for $n_{\mathrm{modes}} = 4$). We then investigated how the population size evolved. To our surprise, even though the acceptance threshold decreased substantially (Fig. 2b$_1$), the population size and effective population size decayed only slightly (Fig. 2b$_2$). Runs with initial population sizes $n_0 = 10^1, 10^2, 10^3, 10^4$ converged within 3 generations to approximately the same population size (Fig. 2c). There was no further systematic dependency of the population sizes of later generations on the initial population size n_0 (Fig. 2c).

Furthermore, we found that the actually achieved variation matched the target E_{CV} well on average (Fig. 2d), confirming the adaptive population size

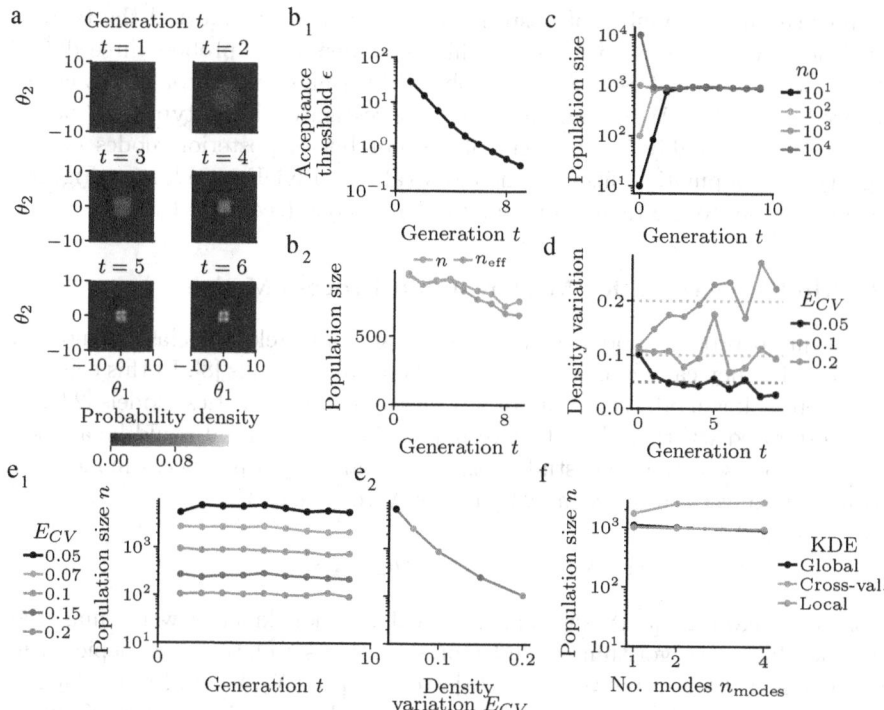

Fig. 2. Adaptive population size selection for multimodal posteriors. (a) Probability density of the first six generations of an ABC-SMC run with variation $E_{CV} = 0.1$, initial population size $n_0 = 500$, observed data $s_{\text{data}} = (1, 1)$, model parameters $\sigma^2 = 0.5$ and $n_{\text{modes}} = 4$. The bandwidth was selected according to the Silverman rule (global bandwidth, Sect. 2.2). **(b)** (b_1) Acceptance threshold ϵ; (b_2) population size n and effective population size n_{eff} for (a). At generation t, ϵ is set to the median of the observation–particle distances of generation $t - 1$. **(c)** Mean population size for initial population sizes $n_0 = 10^1, 10^2, 10^3, 10^4$, $E_{CV} = 0.1$, $n_{\text{modes}} = 4$, $\sigma^2 = 2$ and $s_{\text{data}} = (1, 1)$ averaged over 10 ABC-SMC runs. **(d)** Target E_{CV} (dashed) and actual density variation (solid) for $E_{CV} = 0.05, 0.1, 0.2$, $\sigma^2 = 0.5$, $s_{\text{data}} = (1, 1)$ and $n_{\text{modes}} = 4$. **(e)** Variation E_{CV} and population size n for $n_{\text{modes}} = 4$, $\sigma^2 = 0.5$ and $s_{\text{data}} = (1, 1)$. (e_1) Population size n over generation t for different variations E_{CV}. (e_2) Median population size n as function of E_{CV}. **(f)** Population size n as function of the number of posterior modes n_{modes} for global (Silverman), cross-validated and local bandwidth selection with $E_{CV} = 0.1$, $\sigma^2 = 2$ and $s_{\text{data}} = (1, 1)$.

selection. We then examined the dependency of the population size n on the variation E_{CV}. First, the population sizes remained approximately constant over the generations for each fixed E_{CV} (Fig. 2e$_1$). Second, the median population sizes increased with decreasing E_{CV} (Fig. 2e$_2$), as expected.

As quantitative changes of the distribution related to changes of the acceptance threshold did not influence the selected population size, we assessed its dependence on the number of posterior modes. Since we expected the type of kernel density estimator to influence this dependency, we probed three different estimators: a local estimator, a global (Silverman) estimator and a cross-validated estimator. We found that for all investigated KDE types the population size was roughly independent of the number of posterior modes n_{modes} (Fig. 2f). The population sizes for the cross-validated KDE were larger (Fig. 2f), as expected due to the (generally) smaller bandwidth (see Sect. 4).

3.3   Model Selection for Markov Jump Process Models

Markov jump process models constitute a practically relevant class of models. For example, they can be used to describe chemical reactions [8]. In this context, a common task is model selection for stochastic reaction kinetics models [23,27]. We investigated whether the adaptive population size method could be applied to such model selection and studied the two Markov jump process models m_1 and m_2 for conversion of (chemical) species X to species Y,

$$m_1 : X + Y \xrightarrow{k_1} 2Y, \quad m_2 : X \xrightarrow{k_2} Y,$$

considered before in [5,18,25]. The chemical reaction kinetics were simulated with the Gillespie algorithm [8] and representative simulations are depicted in Fig. 3a. The distance d between two trajectories $s_1 = (X_1, Y_1)$ and $s_2 = (X_2, Y_2)$ was defined as the absolute sum of concentration differences of species X evaluated at $N = 20$ time points:

$$d(s_1, s_2) = \sum_{n=1}^{N} |X_1(t_n) - X_2(t_n)|, \quad t_n = \frac{n}{N}T, \quad N = 20.$$

The simulation was run from $t_0 = 0$ until $T = 0.1$. The initial molecule numbers were $X(t_0) = 40$ and $Y(t_0) = 3$ for every simulation. The reference reaction rate, used for generation of the observed data s_{data}, was $k_1 = 2.1$ ($\log_{10} k_1 = 0.32$) [25]. The priors over the log-rates $\log_{10} k_1$ and $\log_{10} k_2$ were uniform priors $\log_{10} k_i \sim \mathcal{U}([-2, 2])$ for both rates $i = 1, 2$. The prior over the models was also uniform $p_0(m_1) = p_0(m_2) = 1/2$. The proposal density $K_{\text{model},t}(m)$ for model m at generation t with model probabilities p_t was given by $K_{\text{model},t}(m) = p_{\text{stay}}p_t(m) + (1 - p_{\text{stay}})p_t(m')$ if $p_t(m)p_t(m') > 0$ and $K_{\text{model},t}(m) = p_t(m)$ if $p_t(m)p_t(m') = 0$, in which m' denotes the other model $m' \neq m$ and $p_{\text{stay}} = 0.7$.

To assess the performance and reliability of the proposed scheme for adaptive selection of population sizes in model selection, we performed ABC-SMC inference for the generated artificial data s_{data} with $E_{CV} = 0.05$. While the posterior probability of m_1 was comparable to the one of m_2 for the first generations, we observed that as the generations progressed and the acceptance threshold decreased, the probability of m_1 increased as well (Fig. 3b). In the last generations, the posterior probability of m_1 was close to one, resulting in the selection of the true model. Accordingly, the parameter posterior distribution $p(\log_{10} k_1 | m = m_1)$ contained the parameter used to generate the artificial data

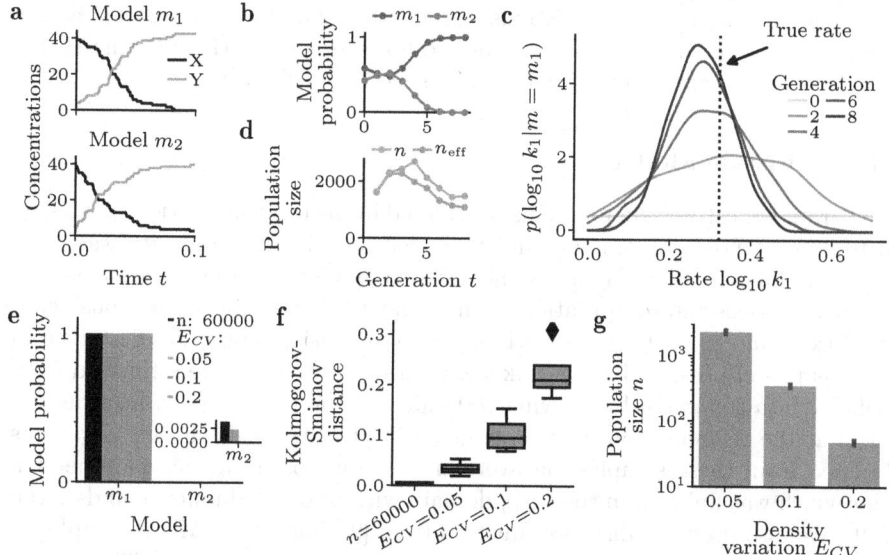

Fig. 3. Model comparison of two chemical reaction kinetics models. (a) Species concentrations X and Y over time t for a single realization of each of the two models m_1 and m_2 with rates $k_1 = 2.1$ and $k_2 = 30$. **(b–d)** Results of an ABC-SMC run for data generated from model m_1, $k_1 = 2.1$ ($\log_{10} k_1 = 0.32$) for $E_{CV} = 0.05$: (b) Model posterior distribution; (c) Parameter posterior distribution and true parameter (dashed line) used to generate the data; and (d) Population size n and effective population size n_{eff} for different generations t. **(e)** Model posterior distribution for adaptively selected population sizes and large, constant population size. In all cases sampling of new populations was stopped as soon as the acceptance threshold reached 1.5. Inset shows model m_2 only. **(f)** Kolmogorov-Smirnov distances of adaptive population size posteriors and large, constant population size posteriors for $\log_{10} k_1$, relative to the reference posterior. For each scenario, four independent runs were performed. **(g)** Mean population size n over all generations for different values of E_{CV}.

(Fig. 3c). The adaptation for E_{CV} yielded population sizes between 1458 and 2699 and effective population sizes between 1101 and 2284 particles (Fig. 3d). Unexpectedly, the population sizes decayed in the last generations (Fig. 3d).

We then examined the quality of the posterior approximation as a function of E_{CV}. As the posterior was not analytically accessible we used as reference an average ABC-SMC estimate obtained from four repetitions with the large and constant populations size of 60000 particles. We found that non-zero mass was attributed to both models by the reference posterior although the mass at model m_2 was small (Fig. 3e). For $E_{CV} = 0.2, 0.1$ model m_2 was completely extinct (Fig. 3e), only for smaller $E_{CV} = 0.05$ we obtained $p(m_2) > 0$. We quantified the mismatch between posterior approximation and reference posterior for the log-rate $\log_{10} k_1$ in terms of the Kolmogorov-Smirnov (KS) distance (Fig. 3f) between the posteriors obtained using adaptation of the population sizes and the mean reference posterior. This KS distance increased as E_{CV} increased (Fig. 3f). The number of required particles decreased however substantially for increasing E_{CV} (Fig. 3g), resulting in a decrease of the computation time.

3.4   Multiscale Models

The parameter estimation problems considered in the previous sections possessed up to two unknown parameters and their computational complexity was comparatively low. To assess if the method is also applicable to higher-dimensional parameter spaces and computationally more demanding problems, we considered a multiscale model for tumor growth on a two-dimensional plane, as described in [9]. This model possessed seven unknown parameters, which we estimated from artificial data generated by drawing 100 independent samples (Fig. 4a) from the model at the reference parameters as given in [9]. The artificial data s_{data} was obtained from these samples via averaging. We imposed on each parameter a prior which was uniform in the $\log_{10}$ domain with upper and lower bounds given in [9]. We also used the distance function from [9]. For ABC-SMC we employed a KDE with local bandwidth considering for each particle only the 20% nearest neighbors, measured in Euclidean distance (see Sect. 2.2).

The populations slowly contracted around the true (reference) parameters and clustered already for generation $t = 13$ around them (Fig. 4b). The last generation $t = 40$ showed that the posterior converged to the true parameters (Fig. 4b). To evaluate the quality of the posterior approximation, we drew 100 samples from the maximum a posteriori (MAP) parameters. We found that the distances (to the observed data) of samples from the MAP parameters were comparable to the distances of samples from the true parameters (Fig. 4c). This indicated that the population size adaptation method, paired with a local proposal distribution, worked successfully for the investigated multiscale model. We then examined the adaptation of the population sizes (Fig. 4d). The acceptance threshold (Fig. 4d$_1$) and the effective population size (Fig. 4d$_2$) decreased over the generations. Surprisingly, the population size increased instead (Fig. 4d$_2$). The estimation of E_{CV} took in this example a fraction of about $2.08 \cdot 10^{-5}$ of the total computation time and was therefore negligible.

4   Discussion

In many systems biology applications, model developers face the parameterization of complex computational models. While ABC-SMC algorithms are well suited for this task, the need of manually tuning population sizes – a task which requires substantial experience – limits their applicability. In this manuscript, we proposed a method to adaptively select population sizes based on the uncertainty of kernel density estimates. Our method complements existing methods for the adaptive choice of perturbation kernels [6], acceptance thresholds [20], and summary statistics [13,17]. We illustrated the method's applicability to parameter

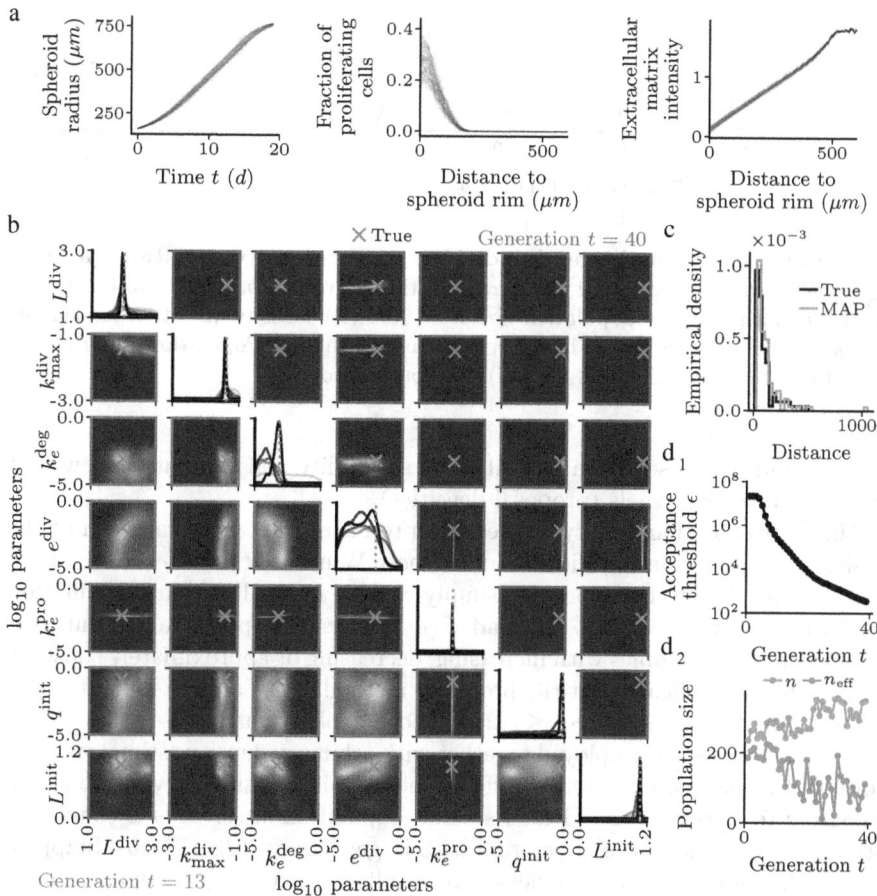

Fig. 4. Multiscale tumor growth model. (a) Tumor growth data. 100 samples from the reference parameters listed in [9]. The data were: (1) the spheroid radius over time, (2) the fraction of proliferating cells over distance to the spheroid rim, and (3) the extracellular matrix intensity over distance to the spheroid rim. **(b)** Posterior distributions obtained from an adaptive population size ABC-SMC run with $E_{CV} = 0.25$. Axis limits correspond to the support of the uniform prior. The purple crosses and dashed lines "True" indicate the reference parameters. Lower triangle: distribution of generation $t = 13$. Upper triangle: distribution of generation $t = 40$. Diagonal: distributions across the generations: light to dark: earlier to later generations. **(c)** Distribution of distances to the observed data s_{data} which was obtained as average of 100 samples from the reference parameter values [9]. True (black): Distances of the same 100 samples which were used to generate the observed data s_{data}. MAP (orange): Distances of 100 samples from the maximum a posteriori parameters. **(d)** (d_1) Acceptance threshold ϵ; (d_2) Population size n and effective population size n_{eff} for **(b)**. At generation t, ϵ is set to the median of the observation–particle distances of generation $t - 1$. (Color figure online)

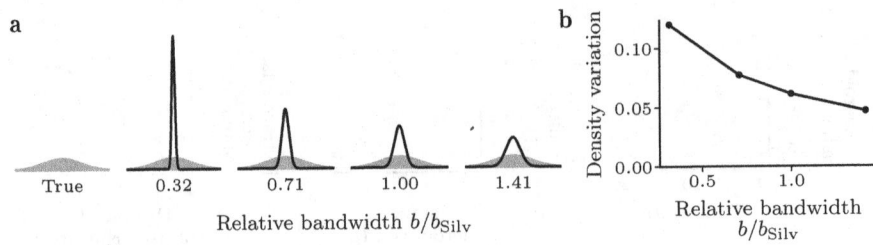

Fig. 5. Influence of the kernel bandwidth on the density variation. (a) True (unimodal normal) distribution and kernels with bandwidths b, relative to the Silverman bandwidth b_{Silv} for population size $n = 10^3$. **(b)** Density variation for kernels with bandwidths b, relative to the Silverman bandwidth b_{Silv}. Population size $n = 10^3$, drawn from the true (unimodal normal) distribution from (a).

inference and model selection as well as its scalability and compatibility with a range of transition kernels (proposal densities).

The approximation quality, expressed in terms of the target variation of the density E_{CV}, has to be specified in our method. While selecting E_{CV} adequately is important, the method does not simply replace manually tuning population sizes by manually tuning E_{CV}. Instead, E_{CV} is easier interpretable and thus easier to select. The examples with increasing, decreasing or approximately constant population sizes indicate that the proposed method is not a mere reparameterization. Empirically, $0.1 \leq E_{CV} \leq 0.2$ worked reliably in many cases.

Our method can be employed together with arbitrary density estimators. The choice of the estimator, however, affects the population sizes. Over-smoothing estimators (e.g. Silverman) yielded smaller population sizes. This is consistent with the lower variation of estimators with larger bandwidths (Fig. 5). Inappropriately chosen estimators can yield poor results with respect to approximation quality and computation time. For instance, cross-validated bandwidth selection can generate a notable computational overhead if the model simulation is fast.

We found no obvious difference between bootstrapping from the density or directly from the particle population; we therefore decided to bootstrap from the density. This avoids drawing the same particles of a population repeatedly as it would likely occur in bootstrapping from the population (with replacement). The assumed functional relation between the density variation and the population size was motivated by the Silverman rule [21] but might be further improved. For the considered applications, however, the approximation was sufficient and extrapolation of larger population sizes was facilitated.

The difficulty of choosing population sizes has been discussed in the literature before [16]. Our results suggest that probing population sizes over an order of magnitude, as done in some studies (e.g. [9]), can be avoided. To the best of our knowledge, this is the first attempt to adaptively and automatically select population sizes for ABC-SMC.

In the future, the interplay of density estimators and population sizes could be further explored. While the effect of density estimators on acceptance rates has been already investigated [6,10], it has not been related to population sizes yet. Alternative approaches to population size adaptation, for instance aiming for a constant effective population size, could be considered. Comparisons to methods requiring more specific problem structures than ABC-SMC, such as accelerated maximum likelihood [4] or the generalized method of moments [12], could be conducted where applicable. The adaptation scheme proposed here is compatible with virtually any ABC-SMC scheme. We expect our method to be applied to a wide range of model selection and parameter estimation tasks.

References

1. Beaumont, M.A.: Approximate Bayesian Computation in evolution and ecology. Annu. Rev. Ecol. Evol. Syst. **41**(1), 379–406 (2010)
2. Beaumont, M.A., Zhang, W., Balding, D.J.: Approximate Bayesian Computation in population genetics. Genetics **162**(4), 2025–2035 (2002)
3. Bowman, A.W., Azzalini, A.: Applied Smoothing Techniques for Data Analysis: The Kernel Approach with S-Plus Illustrations. Oxford University Press, Oxford (1997)
4. Daigle, B.J., Roh, M.K., Petzold, L.R., Niemi, J.: Accelerated maximum likelihood parameter estimation for stochastic biochemical systems. BMC Bioinform. **13**, 68 (2012)
5. Eigen, M.: Prionics or the kinetic basis of prion diseases. Biophys. Chem. **63**(1), A1–A18 (1996)
6. Filippi, S., Barnes, C.P., Cornebise, J., Stumpf, M.P.H.: On optimality of kernels for Approximate Bayesian Computation using Sequential Monte Carlo. Stat. Appl. Genet. Mol. Biol. **12**(1), 87–107 (2013)
7. Fox, C.W., Roberts, S.J.: A tutorial on variational Bayesian inference. Artifi. Intell. Rev. **38**(2), 85–95 (2012)
8. Gillespie, D.T.: Exact stochastic simulation of coupled chemical reactions. J. Phys. Chem. **81**(25), 2340–2361 (1977)
9. Jagiella, N., Rickert, D., Theis, F.J., Hasenauer, J.: Parallelization and high-performance computing enables automated statistical inference of multi-scale models. Cell Syst. **4**(2), 194–206 (2017)
10. Koutroumpas, K., Ballarini, P., Votsi, I., Cournède, P.H.: Bayesian parameter estimation for the WNT pathway: an infinite mixture models approach. Bioinformatics **32**(17), i781–i789 (2016)
11. Levenberg, K.: A method for the solution of certain non-linear problems in least squares. Q. Appl. Math. **2**(2), 164–168 (1944)
12. Lück, A., Wolf, V.: Generalized method of moments for estimating parameters of stochastic reaction networks. BMC Syst. Biol. **10**, 98 (2016)
13. Marin, J.M., Pillai, N.S., Robert, C.P., Rousseau, J.: Relevant statistics for Bayesian model choice. J. Roy. Stat. Soc. Ser. B (Stat. Methodol.) **76**(5), 833–859 (2014)
14. Marjoram, P., Molitor, J., Plagnol, V., Tavaré, S.: Markov Chain Monte Carlo without likelihoods. Proce. Natl. Acad. Sci. **100**(26), 15324–15328 (2003)
15. Marquardt, D.W.: An algorithm for least-squares estimation of nonlinear parameters. J. Soc. Ind. Appl. Math. **11**(2), 431–441 (1963)

16. Moral, P.D., Doucet, A., Jasra, A.: An adaptive Sequential Monte Carlo method for Approximate Bayesian Computation. Stat. Comput. **22**(5), 1009–1020 (2012)

17. Nunes, M.A., Balding, D.J.: On optimal selection of summary statistics for approximate bayesian computation. Stat. Appl. Genet. Mol. Biol. **9**(1), 34 (2010). doi:10.2202/1544-6115.1576

18. Prusiner, S.B.: Novel proteinaceous infectious particles cause Scrapie. Science **216**(4542), 136–144 (1982)

19. Salgado-Ugarte, I.H., Perez-Hernandez, M.A.: Exploring the use of variable bandwidth kernel density estimators. Stata J. **3**(2), 133–147 (2003)

20. Silk, D., Filippi, S., Stumpf, M.P.H.: Optimizing threshold-schedules for sequential Approximate Bayesian Computation: applications to molecular systems. Stat. Appl. Genet. Mol. Biol. **12**(5), 603–618 (2013)

21. Silverman, B.W.: Density Estimation for Statistics and Data Analysis, vol. 26. CRC Press, Boca Raton (1986)

22. Sisson, S.A., Fan, Y., Tanaka, M.M.: Sequential Monte Carlo without likelihoods. Proc. Natl. Acad. Sci. **104**(6), 1760–1765 (2007)

23. de Souza, L.G.M., Haida, H., Thévenin, D., Seidel-Morgenstern, A., Janiga, G.: Model selection and parameter estimation for chemical reactions using global model structure. Comput. Chem. Eng. **58**, 269–277 (2013)

24. Sunnåker, M., Busetto, A.G., Numminen, E., Corander, J., Foll, M., Dessimoz, C.: Approximate Bayesian Computation. PLOS Comput. Biol. **9**(1), e1002803 (2013)

25. Toni, T., Stumpf, M.P.H.: Simulation-based model selection for dynamical systems in systems and population biology. Bioinformatics **26**(1), 104–110 (2010)

26. Toni, T., Welch, D., Strelkowa, N., Ipsen, A., Stumpf, M.P.H.: Approximate Bayesian Computation scheme for parameter inference and model selection in dynamical systems. J. Roy. Soc. Interface **6**(31), 187–202 (2009)

27. Westerhuis, J.A., Boelens, H.F.M., Iron, D., Rothenberg, G.: Model selection and optimal sampling in high-throughput experimentation. Anal. Chem. **76**(11), 3171–3178 (2004)

Methods to Expand Cell Signaling Models Using Automated Reading and Model Checking

Kai-Wen Liang[1], Qinsi Wang[1], Cheryl Telmer[1], Divyaa Ravichandran[1],
Peter Spirtes[1], and Natasa Miskov-Zivanov[2(✉)]

[1] Carnegie Mellon University, Pittsburgh, PA 15213, USA
[2] University of Pittsburgh, Pittsburgh, PA 15213, USA
nmzivanov@pitt.edu

Abstract. Biomedical research results are being published at a high rate, and with existing search engines, the vast amount of published work is usually easily accessible. However, reproducing published results, either experimental data or observations is often not viable. In this work, we propose a framework to overcome some of the issues of reproducing previous research, and to ensure re-usability of published information. We present here a framework that utilizes the results from state-of-the-art biomedical literature mining, biological system modeling and analysis techniques, and provides means to scientists to assemble and reason about information from voluminous, fragmented and sometimes inconsistent literature. The overall process of automated reading, assembly and reasoning can speed up discoveries from the order of decades to the order of hours or days. Our framework described here allows for rapidly conducting thousands of *in silico* experiments that are designed as part of this process.

Keywords: Literature mining · Modeling Automation · Cancer

1 Introduction

Modeling, among many other advantages, facilitates explaining systems that we are studying, guides our data collection, illuminates core dynamics of systems, discovers new questions, or challenges existing theories [2]. However, the creation of models most often relies on intense human effort: model developers have to read hundreds of published papers and conduct numerous discussions with experts to understand the behavior of the system and to construct the model. This laborious process results in slow development of models, let alone validating the model and extending it with thousands of other possible component interactions that already exist in published literature. At the same time, research results are published at a high rate, and the published literature is voluminous, but often fragmented, and sometimes even inconsistent. There is a pressing need for automation of information extraction from literature, smart assembly into models, and model analysis, to enable researchers to re-use and reason about previously published work, in a comprehensive and timely manner.

© Springer International Publishing AG 2017
J. Feret and H. Koeppl (Eds.): CMSB 2017, LNBI 10545, pp. 145–159, 2017.
DOI: 10.1007/978-3-319-67471-1_9

In recent years, there has been an increasing effort to automate the process of explaining biological observations and answering biological questions. The goal of these efforts is to allow for rapid and accurate understanding of biological systems, treatment and prevention of diseases. To this end, several automated reading engines have been developed to extract interactions between biological entities from literature. These automated readers are capable of finding hundreds of thousands of such interactions from thousands of papers in a few hours [10]. However, in order to accurately and efficiently incorporate these pieces of knowledge into a model, we need a method to distinguish useful relationships from vast amounts of extracted information. The revised model often retains properties of the baseline model, but at the same time reflects new properties that the baseline model fails to satisfy, or suggests minimal interventions in the model that can lead to significant changes in outcomes.

To this end, the contributions of our work include: (i) Method to utilize previous research and published literature to validate existing knowledge about diseases, test hypotheses and raise new questions; (ii) Framework to rapidly conduct hundreds of *in silico* experiments via stochastic simulation and statistical model checking; (iii) Pancreatic cancer microenvironment case study that demonstrates the framework's effectiveness.

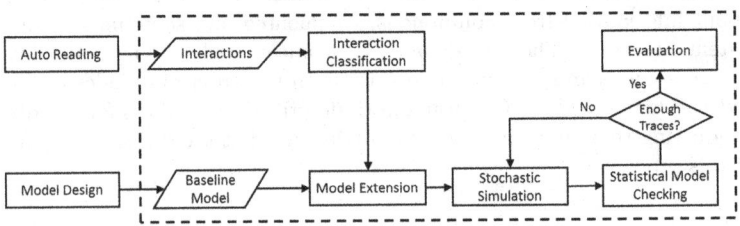

Fig. 1. Steps of our model extension approach.

Our framework is summarized in Fig. 1. The remainder of the paper is organized as follows. In Sect. 2 we provide details about the types of events extracted from literature. In Sect. 3, we outline methods to extend models. In Sect. 4, we describe model analysis methods. The results of applying our framework to pancreatic cancer microenvironment model are presented in Sect. 5. We discuss several important issues in Sect. 6 and conclude the paper with Sect. 7.

2   Events in Biomedical Literature

In this work, we focus on cellular pathways, that is, signal transduction, metabolic pathways and gene regulation. The literature that covers cellular pathways usually includes details such as molecular interactions, gene knock outs, inhibitors, stimulation with antigens. We conducted a brief exercise on a sample set of paragraphs from such published literature. The descriptions found

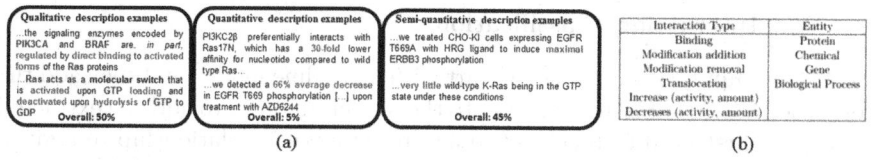

Fig. 2. Reading output: (a) Examples of the three types of interactions found in papers and the average number of occurrences of each type in a sample paragraph set; (a) Types of interactions and their arguments (entities).

in papers can be organized in three groups: qualitative, quantitative and semi-quantitative. Figure 2a shows examples of these three types of descriptions and the average number of occurrences for each type of interaction in the sample paragraph set. Automated reading engines [10] can extract events in the form of frames that contain an interaction with two entities (arguments). We list in Fig. 2b the interaction and entity types that are recognized by reading engines and that we use in this work. Here, we represent each interaction as a pair (u, v), where u is the regulator and v is the regulated element. For the first example sentence in Qualitative description in Fig. 2b, we can obtain two interaction pairs, $(Ras, PIK3CA)$, and $(Ras, BRAF)$.

2.1 Baseline Model Type

The interaction map of a model can be expressed as a *directed* graph $\mathcal{G} = (V, E)$. The set of vertices, V, represents model elements, $v_i \in V$, $i = 1..N$, where N is the number of elements in the model. The set of edges, E, $(v_j, v_i) \in E$, represents causal interactions between elements, that is, relationships of type affects/is-affected-by. The polarity of interactions (positive or negative) is also included in the interaction map.

In order to capture the type of information that most often occurs in published texts, as outlined in Fig. 2(a), we are using logical modeling approach. In logical models of cellular signaling, each element from the interaction map G has a corresponding Boolean variable $x_i \in \{0, 1\}$. The update rule for a variable x_i is a logic function of variables x_j's, where each x_j has a corresponding vertex $v_j \in V$, such that $(v_j, v_i) \in E$. That is, $f_i : \{0, 1\}^{k_i} \rightarrow \{0, 1\}$, where $k_i = |\{v_j : (v_j, v_i) \in E\}|$ is the in-degree of vertex v_i. For a logical model with n elements, there are 2^n possible configurations of variable values, and each configuration is called a state. The logical modeling approach works well with information extracted from text data, since the logical rules can be used to express the qualitative descriptions easily. For example, from the second example sentence in Qualitative description in Fig. 2a, we could extract two interactions (GTP, Ras) and $(!GDP, Ras)$, where '!' indicates negative regulation. We can implement all three elements, GTP, GDP, and Ras as Boolean variables, and write a logical rule for updating value of variable Ras as, for example, $Ras = GDP \ and \ not \ GTP$.

2.2 New Interaction Classification

Often, the computational modelers start with a baseline model, and they add the information extracted from literature to the model. In order to add the extracted events, they first need to be classified according to their relationship to a given model. The output from reading engines can be related to the model in several ways:

(i) *Corroborations*: The interaction from reading output matches an interaction already in the model. An example of corroboration is shown with green arrow in Fig. 3a.

(ii) *Extensions*: The interaction from reading output is not found in the baseline model. An example of extension is shown with blue arrow in Fig. 3a.

(iii) *Contradictions*: The interaction from reading output suggests a different mechanism from the model (for example, activation vs. inhibition). An example is shown with red arrow in Fig. 3a. In this work, we study *extensions only*, that is, new interactions that can be added to the model. Handling contradictions is part of our future work.

3 Model Extension

In Fig. 3b we show a toy example of model interaction map (solid arrows) and several extensions extracted by automated reading (dashed arrows). There are three kinds of model extensions (illustrated in Fig. 3b):

1. Interactions where both elements are already in the model (edges (E, D) and (F, D) in Fig. 3b). This kind of extension usually has a direct influence on the behavior of the model: when adding a new interaction between elements in the model, we are creating a new pathway or generating feed-forward or feedback loops. These structural changes may lead to a significant difference in the regulatory behavior.

2. Interactions where only one element is in the baseline model (for edge (H, A) in Fig. 3b the regulated element is in the baseline model, while the regulator is not; for edge (G, I) the regulator is in the baseline model while the regulated element is not). In cases where the regulated element is not in the baseline model, the regulated element will just hang from a pathway without having direct influence on the model. On the other hand, in extensions where the regulator is outside the baseline model, the regulator can act as a new model input, allowing for additional network control.

3. Interactions consist of elements outside the baseline model (edges (M, K), (K, J)). Such interactions alone do not affect the behavior of the model. However, when we are considering multiple extensions simultaneously, additional regulatory pathways may be constructed that will have effect on model behavior. The path $M \rightarrow K \rightarrow J \rightarrow H \rightarrow A$ in Fig. 3b is an example of newly formed pathway.

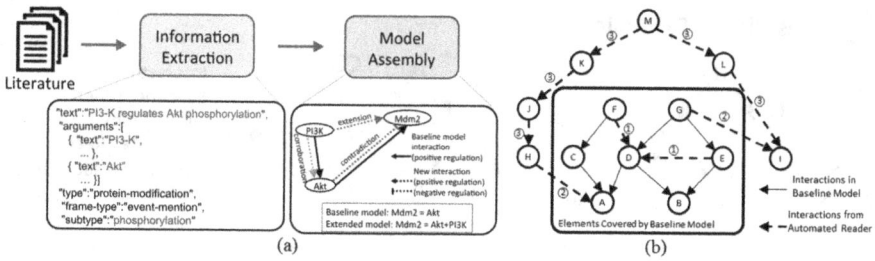

Fig. 3. Relationship between reading output and model. (a) The literature reading-assembly flow with example. (b) Example baseline model (solid arrows) and new interactions extracted by automated reading (dashed arrows). The circled numbers represent classification described in Sect. 2.2 (Color figure online)

3.1 Interaction Map Extension

Each interaction of extension type can be regarded as a candidate new edge in the model's interaction map. Let E^{ext} be the set of interactions provided by reading. Suppose the baseline model is $\mathcal{G} = (V, E)$. Each new candidate model can be obtained by adding a group of selected edges $E^{\text{new}} \in E^{\text{ext}}$ and its corresponding elements, that is, $\mathcal{G}' = (V', E')$, where $E' = E \cup E^{\text{new}}$. However, it is impossible to enumerate all configurations of whether or not to add a new edge, as the number of candidate models will become extremely large. For example, if there are 100 new interactions extracted by the automated reader, there are 2^{100} possible extensions of the model. This number is impossible to handle, therefore, we need heuristic methods to search for suitable configurations of model extensions.

A possible way to tackle the issue of large number of model extension configurations, is to list the elements of interest in the baseline model, and include, as an extension, interactions that are related to those elements. The set of model elements of interest can be defined by user, depending on the questions asked or hypotheses tested. Still, the extension configurations need to be constructed in a systematic manner. Here we introduce the concept of 'layer', where layer S_0 is the set of elements of interest. The next layer, S_1, is the set of direct parents of elements in S_0, and in general, S_i is the set of direct parents of elements in S_{i-1}. Elements in S_1 are direct regulators of S_0, and thus, the extensions including elements in S_1 may influence the model. Using this concept, we propose four different methods to create extension configurations.

Cumulative parent set with direct extensions (CD): In this method, we define the number of layer, n, that we want to consider, and include all new interactions that affect any element from layer 0 up to layer n. In other words, we add an extension $e = (u, v)$ to the model when at least one of the nodes u and v is mentioned in layers S_0 to S_n. Figure 4b demonstrates the result of this method where n equals 1. Starting from $S_0 = \{A, B, E\}$, we find its direct parents $S_1 = \{C, D\}$. The edges in the figure represent the union of layers S_0 and S_1.

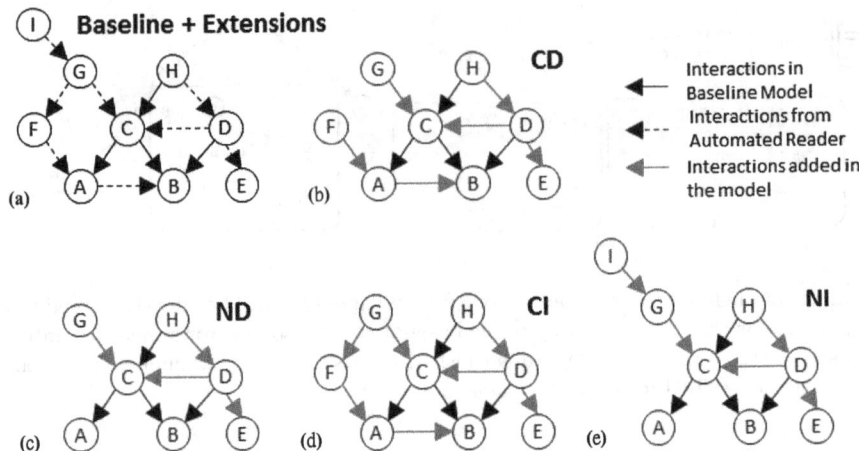

Fig. 4. Results from different extension methods: (a) The baseline model and the extensions from automated reading; (b) The result from method CD with $n = 1$; (c) The result from ND with $n = 1$; (d) The result from CI with $n = 1$; (e) The result from NI with $n = 1$ and $m = 1$.

The advantage for this method is that it includes as many relevant extensions as possible within a certain distance from the elements of interest. However, due to the large number of elements added, the behavior of the model may become intractable within a few layers, that is, if the behavior deviates from what we expect, it is hard to pin-point the source of the change.

Non-cumulative parent-set with direct extensions (ND): This method can be used when we want to know the influence only from the nth layer. When creating each layer, we exclude the elements that are already mentioned in the previous layer, and repeat this process for n times. As a result, from all elements in the ND set, layer 0 can be reached within n steps. We add extension $e = (u, v)$ to the model if and only if u or v is in the ND set. Figure 4c is an example with $n = 1$. After acquiring the layer S_1, we exclude the elements in the previous layer S_0, so we only include edges containing elements in S_1. Compared to the results of the CD method, ND method helps identify individual extension layers that may cause significant changes to the performance in different properties.

Cumulative parent-set with indirect extensions (CI): In the previous two methods, we find each layer only by looking for direct parents of previous layer, that is, the regulators that are already in the baseline model. In this method, we also look for indirect parents. In the example shown in Fig. 4d, we start from $S_0 = \{A, B, E\}$, which has as direct parents nodes C and D, and as an indirect parent node F. Therefore, if we consider indirect parents, S_1 includes $\{C, D, F\}$. The result in Fig. 4d is obtained by adding edges including elements in S_0 or S_1 into the model. This method incorporates more elements into the model, allowing us to examine the behavior of the model including all edges within certain layers. It also includes pathways outside the baseline model more often then the other methods. However, just like the first method, the behavior here may become intractable when n is large, especially when the network outside the baseline model is complicated.

Non-cumulative parent-set with indirect extensions (NI): This method is the combination of the two previous methods. The goal of this method is to provide information about influence on property values of m layers containing indirect edges, starting from the nth layer. In other words, we first look at the nth layer using the ND method, and perform the operation of CI for m times to find all the layers we are interested in. From Fig. 4e, we can see that using one ND step, we get the layer $S_1 = \{C, D\}$. Using CI for another time, we have the set $S_{1,1} = \{G, H\}$. Adding elements mentioned in S_1 and $S_{1,1}$ results in the structure in Fig. 4e. This method can be more comprehensive than ND, giving us a more thorough understanding of the extensions. However, it could also suffer from the issue of being intractable if m is large.

3.2   Executable Rule Updating

After choosing extension classification method and proper parameters for layer numbers, we create model extension sets. These sets extend the static interaction map of the model. Logical rules, on the other hand, allow for dynamic analysis of the model, as variable states change in time according to their update functions. Therefore, the set of logic update rules represents executable model. Incorporating new components into executable model rules can be done in several different ways. For example, if the original rule is $A = BorC$, and the extension interaction states that D positively regulates A, then the new update rule for A can be either $A = (BorC)andD$, or $A = BorCorD$. Other logic functions could be derived as well, but this largely depends on the information available in reading output about these interactions. Given that individual reading outputs only provide information of type 'participant a regulates participant b' (in our example, D positively regulates A), and no additional information about interactions with other regulators is given (in our example, that would be combined regulations of A by B, C and D), we use two naive approaches, which is to add new elements to update rules using either OR or AND operation.

4   Property Testing

After obtaining different extended models using the methods in Sect. 3, we evaluate the performance of each model by checking whether each extended model satisfies a set of biologically relevant properties. While simulations of logical models are known to be able to recapitulate certain experimental observations [5], verifying the results of the simulation against the properties manually is tedious and error-prone, especially when the number of models or properties becomes large. A feasible way to tackle this problem is to use formal methods.

We use statistical model checking that combines simulation and property checking on simulation traces to compute the probability for satisfying each property. We elaborate each framework component in the following subsections.

4.1 Stochastic Simulation

The original logical model and all the extended model versions are simulated using stochastic method. We identify initial states for all model elements, $\mathbf{x} = (x_1, ..., x_N)$, by assigning initial values to their corresponding Boolean variables $\mathbf{x}_0 = \{0, 1\}^n$. Next, we use update rules to compute new variable values, that is, new states of all model elements. The simulator we use is publicly available [6,8]. In the simulator, several different simulation schemes were designed to reflect different timing and element update approaches occurring in biological systems. The simulation scheme we use for this work is called *Uniform Step-Based Random Sequential (USB-RandSeq)*. In each simulation *step*, one model element is *randomly* chosen, its update function is evaluated, and the value of its corresponding variable is updated. At the beginning of simulation the number of these *sequential* steps is defined. In the case of *uniform* update approach, all variables have the same probability of being chosen. The variable values in each step, starting from the initial state, $\mathbf{x}_0$, are recorded in a trace $\sigma = (\mathbf{x}_0, \mathbf{x}_1, \ldots, \mathbf{x}_n)$. With the trace file at hand, we can use model checker to automatically verify whether or not the model meets several properties. Since the order of updating elements is random, when we run simulator to obtain multiple traces, the traces of variable values across different runs can vary.

4.2 Statistical Model Checking

The simulation of logical models is similar to discrete-time Markov chain, which means the verification problem is equivalent to computing the probability of whether a given temporal logic formula is satisfied by the system. One approach is to use numerical methods to compute the exact probability; however, this naive implementation suffers from the state explosion problem, and does not scale well to large-scale systems [14]. Statistical model checking provides an excellent solution to this problem, by estimating the probability using simulation and thus, avoiding a full state space search. To verify a model via statistical model checking against interesting properties, we first need to encode each property into temporal logic formulae. Here we use Bounded Linear Temporal Logic (BLTL) [3]. BLTL is a variant of Linear Temporal Logic [7], where the future condition of certain logic expressions is encoded as a formula with a time bound (see the supplementary material (http://ppt.cc/XlWF7) for BLTL's formal syntax and semantics). To verify whether a model satisfies the properties, statistical model checking treats it as a statistical inference problem for the model executions generated using the randomized sampling. For a stochastic system, the probability p that the system satisfies a property ϕ is unknown. Statistical model checking can handle two kinds of questions: (i) for a fixed $0 < \theta < 1$, determine whether $p \leq \theta$, and (ii) estimate the value of p. The first problem is solved using hypothesis testing methods, while the second is solved via estimation techniques. Statistical model checking assumes that, given a BLTL property ϕ, the behavior of a system can be modeled as a Bernoulli random variable M with parameter p, where p is the probability of the system satisfying ϕ. Statistical model checking first generates independent and identically distributed samples of M. Each sample σ is then checked against the property ϕ, and the yes/no answer corresponds to a 1/0 sample of the random variable M. The sample size does not need to be fixed, as the checking procedure will stop when it achieves the desired accuracy. This reduces the number of samples needed. The statistical model checking ha been applied in the past to the type of stochastic simulation that we use here, [11].

5   Results

The system that we studied is pancreatic cancer microenvironment, including pancreatic cancer cells (PCCs) and pancreatic stellate cells (PSCs). We adopted here the model created by Wang et al. [13], which has three major parts: (1) intracellular signaling network of PCC; (2) intracellular signaling network of PSC; (3) network located in extracellular space of the microenvironment, which contains mainly ligands of the receptors. In this model, several cellular functions, such as autophagy, apoptosis, proliferation, migration, are also implemented as elements of the model, which enables modeling of the system's behavior that can result from turning various signaling components ON or OFF. In total, there are 30 variables encoding intracellular PCC elements and 3 variables encoding PCC cellular function. For PSC, there are 24 variables for intracellular elements and 4 variables for PSC cellular function. In extracellular microenvironment, there are 8 variables encoding extracellular signaling elements with 1 environment function variable. Accordingly, there are 70 variables in the model that have associated update functions used to compute next state of those model elements. The interaction rules of this model are summarized in Table 1 in the Supplementary material (http://ppt.cc/XlWF7).

The framework is implemented in Python. The simulator described in Sect. 4.1 is implemented in Java [8]. We use PRISM [4] as our statistical model checker, which is a C++ tool for formal modeling and analysis of stochastic systems. Evaluating a model against one property, including running the simulations, takes about 10 min on a regular laptop (1.3 GHz dual-core Intel Core i5, 8GM LPDDR3 memory). The other components in the framework take less than 1 min. We used the REACH automated reading engine [9] output produced from 13,000 papers in publicly available domain. This output consists of 500,000 event files, with 170,000 possible extensions of our model (other events are corroborations or contradictions).

To demonstrate how our framework works, we identified elements of interest in the model (which were suggested by cancer experts), and defined a set of relevant properties reflecting important biological truths that the PCC-PSC model should satisfy [12]. In Table 1, we list 20 properties that we tested using

Table 1. System properties for model testing.

#	Property	Description
	Increased secretions of important growth factors:	
1	F[1000] (G[10000] VEGF)	
2	F[1000] (G[10000] bFGF)	Within 1000 time units, the concentration of VEGF (bFGF / PDGFBB / TGFβ1)
3	F[1000] (G[10000] PDGFBB)	in the tumor microenvironment will eventually reach a high amount, and stay in this high level for at least another 10000 time units.
4	F[1000] (G[10000] TGFβ1)	
	Over-expression of oncoproteins in PCCs and PSCs:	
5	F[1400] (G[10000] PCCHER2)	
6	F[1400] (G[10000] PCCRas)	Within 1400 time units, the concentration of HER2 in PCCs (/ RAS in PCCs / VEGFR in PSCs / ERK in PCCs) will eventually reach a high amount, and stay in
7	F[1400] (G[10000] PSCVEGFR)	this high level for at least another 10000 time units.
8	F[1400] (G[10000] PCCERK)	
	Inhibition of tumor suppressors in PCCs:	
9	F[1000] (PCCP21 ∧ F[2000] (G[10000] (! PCCP21)))	The concentration of P21 (/ PTEN / RB1 / P53) in PCCs will reach a high level
10	F[1000] (PCCPTEN ∧ F[2000] (G[10000] (! PCCPTEN)))	and act as a tumor suppressor within the first 1000 time units. Then, after at most
11	F[1000] (PCCRB1 ∧ F[2000] (G[10000] (! PCCRB1)))	2000 time units, its concentration will eventually drop to a low level, and stay in
12	F[1000] (PCCP53 ∧ F[2000] (G[10000] (! PCCP53)))	this low level for at least another 10000 time units.
	Cell functions of PCCs:	
13	F[1000] ((! PCCAutophagy) ∧ F[2000](G[10000] PCCAutophagy))	In the development of pancreatic cancer, apoptosis firstly overwhelms autophagy,
14	F[1000] ((PCCApoptosis) ∧ F[2000](G[10000] !PCCApoptosis))	and then autophagy takes the leading place after a certain time point.
15	F[1000] (G[10000] PCCProliferation)	Within 1000 time units, PCCs' proliferation will eventually be activated, and becomes a steady state for at least another 10000 time units.
16	!(! PCCP53 U[12000] PCCApoptosis)	It is not the case that, within 12000 steps, P53 in PCCs has to have a low concentration level until PCCs' Apoptosis being triggered.
	Cell functions of PSCs:	
17	F[1000] (G[10000] PSCActivation)	Within 1000 time units, PSCs' activation (/ migration) will eventually be
18	F[1000] (G[10000] PSCMigration)	activated, and becomes a steady state for at least another 10000 time units.
19	F[1000] (PSCApoptosis ∧ F[1000] (G[10000] (! PSCApoptosis)))	Within 1000 time units, PSCs' apoptosis will be triggered. Then, after at most 1000 time units, the initially functional apoptosis in PSCs will be inhibited and stay in inactive status for at least 10000 time units
20	F[12000] PSCProliferation	Within 12000 steps, PSCs' proliferation will eventually be triggered.

'∧' represents logic operator AND; '!' represents logic operator NOT, 'F' is eventually, 'G' is always, 'U' is until

statistical model checking. There are five major functions or phenomena that we are interested in: (1) increased secretion of important growth factors; (2) over-expresion of oncoproteins in PCC and PSCs; (3) inhibition of tumor suppressors in PCCs; (4) cell functions of PCCs; (5) cell functions of PSCs.

5.1 Impact of Proposed Extension Approaches on Model

The baseline model [13] has 70 elements and 114 regulatory interactions. Although there are 170,000 model extensions produced by reading, many of them are repetitions, and some of the reading outputs were missing one of the interaction participants. Therefore, in this work we used overall 1232 different interactions from reading output, which could lead to 2^{1232} possible models. Studying all possible model versions is impractical, and therefore, we used the four extension methods described in Sect. 3, to generate 46 different models. Using the CD method, we generated 2 models by having 1 or 2 layers. For ND, the number of layers we considered varied between 1 and 10, which resulted in 10 models. With CI, we used either 0, 1, 2 or 3 layers, which led to 4 different models. Finally, for NI, we have n ranges from 1 layer to 10 layer, and m ranges from 1 to 3, resulting in 30 models. We also test the model with all extensions being added to the baseline model.

Figure 5 summarizes results of our extension methods on 1232 interactions with respect to new node connections to the model:

(i) number of new nodes regulating baseline model elements, not regulated by baseline model elements (dark blue);

(ii) number of new nodes regulating baseline model elements, not regulated by any element, baseline or new (red);

(iii) number of new nodes regulated by baseline model elements, not regulating any elements in the baseline model (yellow);

(iv) number of new nodes regulated by baseline model elements, not regulating any element, baseline or new (purple);

(v) number of new nodes inserted into existing pathway - new regulators of baseline model elements that are also regulated by baseline model elements (green);

(vi) number of new nodes as intermediate elements of new pathways when multiple extensions are connected (light blue);

(vii) total number of all elements used in the extension method (dark red).

In Fig. 6(a), four different sections can be observed, and each section corresponds to one of the extension methods. Each method has its unique feature. For example, the ND method only includes relationships relevant to one layer, and this makes the number of new elements added to the model significantly smaller than other methods. Also, the light blue nodes indicate the number of newly added elements that are in a newly formed pathway. Since CD and ND do not include indirect parent interactions, we can see that the number of elements in new pathway is 0. While in CI and NI, we can tell that indirect interactions are included. The numbers within one method show higher similarity, but we can still observe some patterns. For example, the cumulative parent-set methods, CD and CI show an increase in the number of new nodes when more layers are considered. Furthermore, since NI has cumulative parents when they finish the noncumulative part, they also experience an increase when the step of noncumulative part is fixed. The numbers saturate at around 600, which is due to the limited size of baseline model and extensions we have. This is also the reason we choose to perform the cumulative approach for at most 3 steps.

In general, choosing the method to extend the model depends on the scenario a user is interested in. For example, if the focus is on the regulation of a specific element, one can track down each layer of parents using ND, and see the change of the model after modifying that specific layer. On the other hand, if the goal is to include as many new stimuli as possible with a fewer number of layers, cumulative methods such as CI or CD will fit better. We selected 20 elements as part of the base layer, since these elements appear in properties that we are testing, leading to relatively large base layer given the size of the baseline model. Therefore, by incorporating elements related to more than one layer, we capture almost all extensions related to the baseline model. Thus, the 'All In' method, which adds all extension interactions to the baseline model at once, does not change the counts shown in Fig. 6(a), when compared to many cases of CD, CI and NI methods.

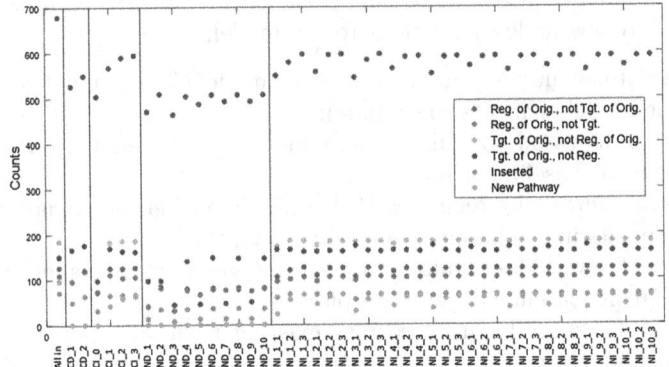

Fig. 5. Counts for newly added elements with certain structure (**Reg.** - regulator, **Tgt.** - regulated element, **Orig.** - baseline model elements, **New** - newly added element). All models studied are listed on x-axis, and y-axis is the count of new elements having certain structure. (Color figure online)

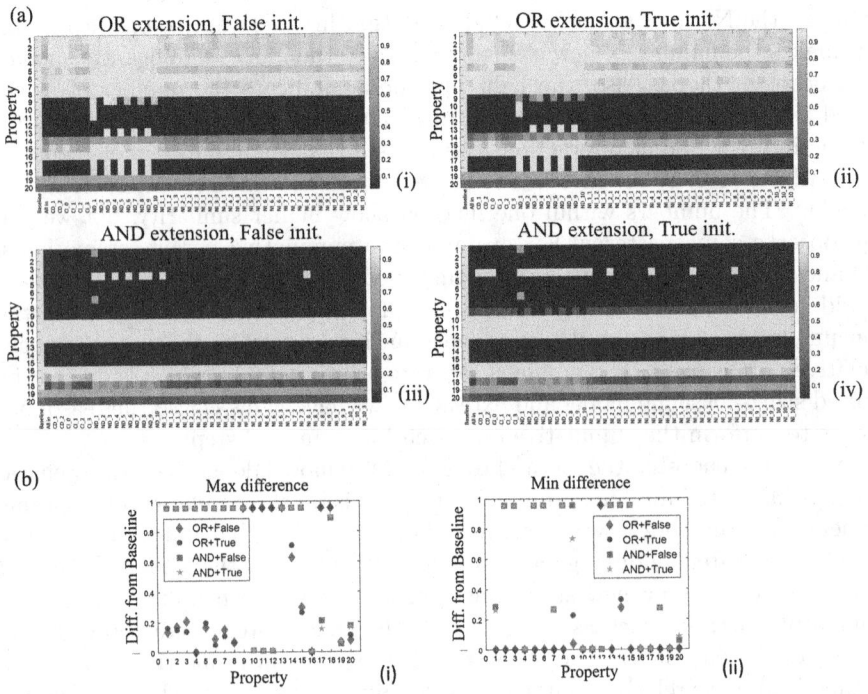

Fig. 6. (a) Results of statistical model checking of 20 properties in 68 different models. Each entity in x-axis is a model, and each row is the estimated probability for the corresponding property. (b) The Max and min difference from the baseline model of each property.

5.2 Impact of Model Extension on System Properties

Figure 6(a) shows the results of testing 48 models (baseline + All-In + 46 extended models) with different extension method (AND/OR) and different initialization of the newly added elements ($True/False$) against the 20 properties in Table 1. The values displayed are the estimated probabilities of each property. Just like the basic numbers of each model, different extension methods lead to different results of the properties. For example, we can see that the results from ND are different from other methods. The reason is that each ND method only deals with one layer at a time, and it will not insert new edges between elements mentioned in the properties. This leads to a more conservative extension. Also, for example, there are differences between OR-based ND models in properties 9 to 13 or property 4 in AND-based ND models, which are related to Inhibition of tumor suppressors and Autophagy in PCCs. By comparing the extension interactions added to those models, we found that the EGF (Epidermal Growth Factor) pathway plays the most important role. The p21 (regulator of cell cycle progression) pathway also influences the difference.

If we compare the models with different initialization of newly added nodes, we can see the results are actually quite similar. This means that the model is mostly influenced by the input elements in the baseline model, and to some degree, it emphasizes the robustness of the original model. On the other hand, if we compare extending the models with OR operations and those with AND operations, there is a huge difference. But the interesting part is that the behavior of models with the two types of extensions is opposite. They behave similarly only in properties 9, 13, 16, 19 and 20, while differently in all other 15 properties. This shows a drastic difference between AND-based and OR-based extension, and can be further designed according to the property we want to fit. Figure 6(b) shows the maximum/minimum difference compared to baseline that each model can achieve for each property. If a property probability is low in both max and min difference, it is relatively conservative to the extension interaction. An example is property 16, which depicts the relationship between p53 and Apoptosis. On the other hand, if a property probability is high in both max and min difference, it is a property susceptible to changes via extensions.

6 Discussion

The framework we describe here, although designed to extend an existing baseline model, can also be used to search for pathways or interactions that are vital to certain functions, and to suggest targets for drug development. For example, using the ND models and statistical model checker, we can study closely how each layer of elements influences the elements we are interested in. Then, we can pin-point the models that satisfy several properties that we desire, and we should be able to identify a few candidates that play important roles in the regulation. Or, by using NI method, we can further observe whether there is actually an upstream network that controls the behavior of the elements. This gives us a deeper understanding of the network and helps us in further model development.

One of our next steps is to improve the approach to incorporate new elements into logical rules. In this work we naively incorporate those rules using OR and AND operation. However, in reality the mutual relationships between the regulators are not necessary an AND or OR relationship. For example, a ligand and a receptor induce further response if they both exist, and there is another unrelated element activating the same target. This results in a format $A = B * C + D$. We are not able to capture this since the automated reader does not output this information, but from online databases such as UniProt [1], we are still able to gather pieces of knowledge about the true interaction between regulators. Also, the automated reader does not output the location of the interaction. For example, two types of cells, PCCs and PSCs, are in our baseline model, but we only extend the interactions to PCCs. More information of the location can also help us refine the extension method. As a future work, incorporating the on-line database should give us a more accurate extension of the model. But in the long run, if the automated reader can take into account these features, we should be able to construct a better model more easily. Finally, aside from extensions, the automated reader provides us with contradictions. In this work we ignore this kind of relationship and assume absolute correctness of interaction in the baseline model, but the contradictions serve as a great starting point to examine the validity of the baseline model, as well as to point to further improvements of reading engines.

7   Conclusion

We propose a framework that utilizes published work to collect extensions for existing models, and then analyzes these extensions using stochastic simulation and statistical model checking. With biological properties being formulated as temporal logic, model checker can use the trace generated by the simulator to estimate the probability that a certain property holds. This gives us an efficient approach (speed-up from decades to hours) to re-use previously published results and observations for the purpose of conducting hundreds of *in silico* experiments with different setups (models). Our methods and the framework that we have developed comprise a promising new approach to rapidly and comprehensively utilize published work for an increased understanding of biological systems, in order to identify new therapeutic targets for the design and improvement of disease treatments.

Acknowledgement. We would like to thank Mihai Surdeanu (REACH team) and Hans Chalupsky (RUBICON team) for providing output of their reading and assembly engines, and Michael Lotze for his guidance in studying cancer microenvironment. This work is supported by DARPA award W911NF-14-1-0422.

References

1. UniProt Consortium: UniProt: a hub for protein information. Nucl. Acids Res. **43**, D204–D212 (2014). doi:10.1093/nar/gku989
2. Epstein, J.M.: Why model? J. Artif. Soc. Soc. Simul. **11**(4), 12 (2008)
3. Jha, S.K., Clarke, E.M., Langmead, C.J., Legay, A., Platzer, A., Zuliani, P.: A bayesian approach to model checking biological systems. In: Degano, P., Gorrieri, R. (eds.) CMSB 2009. LNCS, vol. 5688, pp. 218–234. Springer, Heidelberg (2009). doi:10.1007/978-3-642-03845-7_15
4. Kwiatkowska, M., Norman, G., Parker, D.: PRISM 4.0: verification of probabilistic real-time systems. In: Gopalakrishnan, G., Qadeer, S. (eds.) CAV 2011. LNCS, vol. 6806, pp. 585–591. Springer, Heidelberg (2011). doi:10.1007/978-3-642-22110-1_47
5. Miskov-Zivanov, N., Turner, M.S., Kane, L.P., Morel, P.A., Faeder, J.R.: Duration of T cell stimulation as a critical determinant of cell fate and plasticity. Sci. Signaling **6**(300), ra97 (2013)
6. Miskov-Zivanov, N., Wei, P., Loh, C.S.C.: THiMED: time in hierarchical model extraction and design. In: Mendes, P., Dada, J.O., Smallbone, K. (eds.) CMSB 2014. LNCS, vol. 8859, pp. 260–263. Springer, Cham (2014). doi:10.1007/978-3-319-12982-2_22
7. Pnueli, A.: The temporal logic of programs. In: 18th Annual Symposium on Foundations of Computer Science, pp. 46–57. IEEE (1977)
8. Sayed, K., Kuo, Y.H., Kulkarni, A., Miskov-Zivanov, N.: Dish simulator: capturing dynamics of cellular signaling with heterogeneous knowledge. arXiv (2016). https://github.com/Yu-Hsin/simulator_java
9. Sprites & Palette. https://www.dropbox.com/s/mmks9xs0w4rjkcx/16K-fries_160331.tgz?dl=0
10. Valenzuela-Escárcega, M.A., Hahn-Powell, G., Hicks, T., Surdeanu, M.: A domain-independent rule-based framework for event extraction. In: Proceedings of the 53rd Annual Meeting of the Association for Computational Linguistics and the 7th International Joint Conference on Natural Language Processing of the Asian Federation of Natural Language Processing: Software Demonstrations (ACL-IJCNLP), pp. 127–132 (2015). http://www.aclweb.org/anthology/P/P15/P15-4022.pdf
11. Vardi, M.Y.: Automatic verification of probabilistic concurrent finite state programs. In: 26th Annual Symposium on Foundations of Computer Science, pp. 327–338. IEEE (1985)
12. Wang, Q.: Model checking for biological systems: languages, algorithms, and applications. Ph.D. thesis, Carnegie Mellon University (2016)
13. Wang, Q., Miskov-Zivanov, N., Liu, B., Faeder, J.R., Lotze, M., Clarke, E.M.: Formal modeling and analysis of pancreatic cancer microenvironment. In: Bartocci, E., Lio, P., Paoletti, N. (eds.) CMSB 2016. LNCS, vol. 9859, pp. 289–305. Springer, Cham (2016). doi:10.1007/978-3-319-45177-0_18
14. Younes, H.L.S., Simmons, R.G.: Probabilistic verification of discrete event systems using acceptance sampling. In: Brinksma, E., Larsen, K.G. (eds.) CAV 2002. LNCS, vol. 2404, pp. 223–235. Springer, Heidelberg (2002). doi:10.1007/3-540-45657-0_17

A Stochastic Model for the Formation of Spatial Methylation Patterns

Alexander Lück[1], Pascal Giehr[2], Jörn Walter[2], and Verena Wolf[1(✉)]

[1] Department of Computer Science, Saarland University, Saarbrücken, Germany
verena.wolf@uni-saarland.de
[2] Department of Biological Sciences, Saarland University, Saarbrücken, Germany

Abstract. DNA methylation is an epigenetic mechanism whose important role in development has been widely recognized. This epigenetic modification results in heritable changes in gene expression not encoded by the DNA sequence. The underlying mechanisms controlling DNA methylation are only partly understood and recently different mechanistic models of enzyme activities responsible for DNA methylation have been proposed. Here we extend existing Hidden Markov Models (HMMs) for DNA methylation by describing the occurrence of spatial methylation patterns over time and propose several models with different neighborhood dependencies. We perform numerical analysis of the HMMs applied to bisulfite sequencing measurements and accurately predict wild-type data. In addition, we find evidence that the enzymes' activities depend on the left 5' neighborhood but not on the right 3' neighborhood.

Keywords: DNA methylation · Hidden Markov model · Spatial stochastic model

1 Introduction

The DNA code of an organism determines its appearance and behavior by encoding protein sequences. In addition, there is a multitude of additional mechanisms to control and regulate the ways in which the DNA is packed and processed in the cell and thus determine the fate of a cell. One of these mechanisms in cells is DNA methylation, which is an epigenetic modification that occurs at the cytosine (C) bases of eukaryotic DNA. Cytosines are converted to 5-methylcytosine (5mC) by DNA methyltransferase (Dnmt) enzymes. The neighboring nucleotide of a methylated cytosine is usually guanine (G) and together with the GC-pair on the opposite strand, a common pattern is that two methylated cytosines are located diagonally to each other on opposing DNA strands. DNA methylation at CpG dinucleotides is known to control and mediate gene expression and is therefore essential for cell differentiation and embryonic development. In human somatic cells, approximately 70–80% of the cytosine nucleotides in CpG dyads are methylated on both strands and methylation near gene promoters varies considerably depending on the cell type. Methylation of promoters often correlates with low or no transcription [20] and can be used as a predictor of gene

© Springer International Publishing AG 2017
J. Feret and H. Koeppl (Eds.): CMSB 2017, LNBI 10545, pp. 160–178, 2017.
DOI: 10.1007/978-3-319-67471-1_10

expression [12]. Also significant differences in overall and specific methylation levels exist between different tissue types and between normal cells and cancer cells from the same tissue. However, the exact mechanism which leads to a methylation of a specific CpG and the formation of distinct methylation patterns at certain genomic regions is still not fully understood. Recently proposed measurement techniques based on hairpin bisulfite sequencing (BS-seq) allow to determine on both DNA strands the level of 5mC at individual CpGs dyads [15]. Based on a small hidden Markov model, the probabilities of the different states of a CpG can be accurately estimated (assuming that enough samples per CpG are provided) [1,13].

Mechanistic models for the activity of the different Dnmts usually distinguish de novo activities, i.e., adding methyl groups at cytosines independent of the methylation state of the opposite strand, and maintenance activities, which refers to the copying of methylation from an existing DNA strand to its newly synthesized partner (containing no methylation) after replication [10,17]. Hence, maintenance methylation is responsible for re-establishment of the same DNA methylation pattern before and after cell replication. A common hypothesis is that the copying of DNA methylation patterns after replication is performed by Dnmt1, an enzyme that shows a preference for hemimethylated CpG sites (only one strand is methylated) as they appear after DNA replication. Moreover, studies have shown that Dnmt1 is highly processive and able to methylate long sequences of hemimethylated CpGs without dissociation from the target DNA strand [10]. However, an exact transmission of the methylation information to the next cellular generation is not guaranteed. The enzymes Dnmt3a and Dnmt3b show equal activities on hemi- and unmethylated DNA and are mainly responsible for de novo methylation, i.e., methylation without any specific preference for the current state of the CpG (hemi- or unmethylated) [17]. However, by now evidence exists that the activity of the different enzymes is not that exclusive, i.e., Dnmt1 shows to a certain degree also de novo and Dnmt3a/b maintenance methylation activity [2]. The way how methyltransferases interact with the DNA and introduce CpG methylation was investigated in many *in vitro* studies. Basically, one can distinguish between two mechanisms. A distributive one, where the enzyme periodically binds and dissociates from the DNA, leaping more or less randomly from one CpG to another and a processive one in which the enzyme migrates along the DNA without detachment from the DNA [9,11,16], as illustrated in Fig. 1. Note that for Dnmt1, for instance, it is reasonable to assume that it is processive in 5' to 3' direction since it is linked to the DNA replication machinery. In particular for the Dnmt3's different hypotheses about the processivity and neighborhood dependence exist [3,5], but the detailed mechanisms remain elusive.

Several models that describe the dynamics of the formation of methylation patterns have been proposed. In the seminal paper of Otto and Walbot, a dynamical model was proposed that assumed independent methylation events for a single CpG. The main idea was to track the frequencies of fully, hemi- and unmethylated CpGs during several cell generations [18]. Later, refined models

allowed to distinguish between maintenance and de novo methylation on the
parent and daughter strands [7,19]. More sophisticated extensions of the original model of Otto and Walbot models have been successfully used to predict *in
vivo* data still assuming a neighbor-independent methylation process for a single
CpG site [2,8]. However, measurements indicate that methylation events at a
single CpG may depend on the methylation state of neighboring CpGs, which is
not captured by these models.

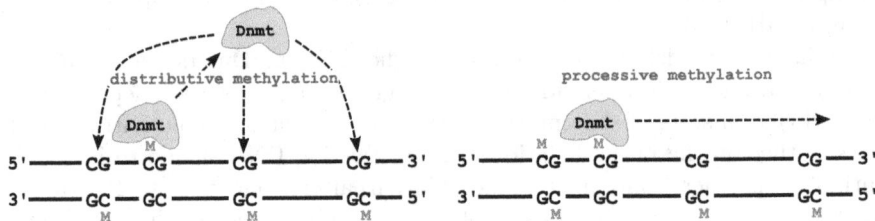

Fig. 1. Dnmts can methylate DNA in a distributive manner, "jumping" randomly from
one CpG to another or in a processive way where the enzyme starts at one CpG and
slides in 5' to 3' direction over the DNA.

Here, we follow the dynamical HMM approach proposed in [2] where knockout
data was used to train a model that accurately predicts wild-type methylation
levels for BS-seq data of repetitive elements from mouse embryonic stem cells. We
extend this model by describing the methylation state of several CpGs instead
of a single CpG and use similar dependency parameters as introduced in [4].
More specifically, we design different models by combining the activities of the
two types of Dnmts and test for both, maintenance and de novo methylation the
hypotheses illustrated in Fig. 1. The models vary according to the order in which
the enzymes act, whether they perform methylation in a processive manner or
not, and how much their action depends on the left/right CpG neighbor. We use
the same BS-seq data as in [2], i.e. data where Dnmt1 or Dnmt3a/b was knocked
out (KO) and learn the parameters of the different models. Then, similar as in
[2], we predict the behavior of the measured wild-type (WT), in which both
types of enzymes are active, by designing a combined model that describes the
activity of both enzymes and compare the results to the WT data.

We found that all proposed models show a similar behavior in terms of prediction quality such that no model can be declared as the best fit. However,
our results indicate that Dnmt1 works independently of the methylation state of
its neighborhood, which is in accordance to the current hypothesis that Dnmt1
is linked to the replication machinery and copies the methylation state on the
opposite strand. On the other hand, Dnmt3a/b shows a dependency to the left
but no dependency to the right, which supports hypotheses of processive or
cooperative behavior.

2   Preliminaries

Consider a sequence of L neighboring CpG dyads[1], which is represented as a lattice of length L and width two (for the two strands). Each cytosine in the lattice can either be methylated or not, leading to four possible states at each position l:

– *State 0*: Both sites are not methylated.
– *State 1*: The cytosine on the upper strand is methylated, the lower one not.
– *State 2*: The cytosine on the lower strand is methylated, the upper one not.
– *State 3*: Both cytosines are methylated.

A sequence of four CpGs, each of which is in one of the four possible states, is shown in Fig. 2.

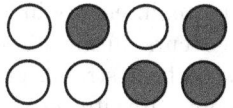

Fig. 2. A lattice of length $L = 4$ containing all possible states 0, 1, 2 and 3, forming the pattern 0123.

For a system of length L there are in total 4^L possibilities to combine the states of individual CpGs. These combinations are called *patterns* in the following. A pattern is denoted by a concatenation of states, e.g. 321, 0123 or 33221.

In order to represent the pattern distribution as a vector it is necessary to uniquely assign a reference number to each pattern. A pattern can be perceived as a number in the tetral system, such that converting to the decimal system leads to a unique reference number. After the conversion an additional 1 is added in order to start the referencing at 1 instead of 0.

Examples for $L = 3$:

$$
\begin{aligned}
000 &\longrightarrow & 1 \,(= 0 + 1) \\
123 &\longrightarrow & 28 \,(= 27 + 1) \\
333 &\longrightarrow & 64 \,(= 63 + 1)
\end{aligned}
$$

This reference number then corresponds to the position of the pattern in the respective distribution vector.

---

[1] The exact nucleotide distance between two neighboring dyads is not considered here, but we assume that this distance is small. For the BS-seq data that we consider, the average distance between two CpGs is 14 bp and the maximal distance is 46 bp.

3   Model

We describe the state of a sequence of L CpGs by a discrete-time Markov chain with pattern distribution $\pi(t)$, i.e., the probability of each of the 4^L patterns after t cell divisions. For the initial distribution $\pi(0)$, we use the distribution measured in the wild-type when the cells are in equilibrium. Note that other initial conditions gave very similar results, i.e., the choice of the initial distribution does not significantly affect the results. The reason is that also the KO data is measured after a relatively high number of cell divisions where the cells are almost in equilibrium. Transitions between patterns are triggered by different processes: First due to *cell division* the methylation on one strand is kept as it is (e.g. the upper strand), whereas the newly synthesized strand (the new lower strand) does not contain any methyl group. Afterwards, methylation is added due to different mechanisms. On the newly synthesized strand a site can be methylated if the cytosine at the opposite strand is already methylated (*maintenance*). It is widely accepted that maintenance in form of Dnmt1 is linked to the replication machinery and thus occurs during/directly after the synthesis of the new strand. Furthermore, CpGs on both strands can be methylated independent of the methylation state of the opposite site (*de novo*). The transition matrix P is defined by composition of matrices for cell division, maintenance and de novo methylation of each site.

3.1   Cell Division

Depending on which daughter cell is considered after cell replication, the upper $(s = 1)$ or lower $(s = 2)$ strand is the parental one after cell division. Then, the new pattern can be obtained by applying the following state replacements:

$$
s = 1: \begin{cases} 0 & \longrightarrow & 0 \\ 1 & \longrightarrow & 1 \\ 2 & \longrightarrow & 0 \\ 3 & \longrightarrow & 1 \end{cases} \qquad s = 2: \begin{cases} 0 & \longrightarrow & 0 \\ 1 & \longrightarrow & 0 \\ 2 & \longrightarrow & 2 \\ 3 & \longrightarrow & 2 \end{cases} \tag{1}
$$

Given some initial pattern with reference number i, applying the transformation (1) to each of the L positions leads to a new pattern with reference number j (notation: $i \overset{(1)}{\leadsto} j$). The corresponding transition matrix $D_s \in \{0,1\}^{4^L \times 4^L}$ has the form

$$
D_s(i,j) = \begin{cases} 1, & \text{if } i \overset{(1)}{\leadsto} j, \\ 0, & \text{else.} \end{cases} \tag{2}
$$

3.2   Maintenance and De Novo Methylation

For maintenance and de novo methylation, the single site transition matrices are built according to the following rules:

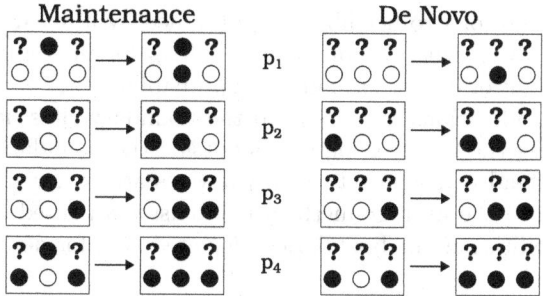

Fig. 3. Possible maintenance and de novo transitions depicted for the lower strand, where ○ denotes an unmethylated, ● a methylated site and ? a site where the methylation state does not matter. Note that the same transitions can occur on the upper strand.

Consider at first the (non-boundary) site $l = 2, \ldots, L - 1$ and its left and right neighbor $l - 1$ and $l + 1$ respectively. The remaining sites do not change and do not affect the transition. The probabilities of the different types of transitions in Fig. 3 have the form

$$p_1 = 0.5 \cdot (\psi_L + \psi_R)x, \tag{3}$$

$$p_2 = 0.5 \cdot (\psi_L + \psi_R)x + 0.5 \cdot (1 - \psi_L), \tag{4}$$

$$p_3 = 0.5 \cdot (\psi_L + \psi_R)x + 0.5 \cdot (1 - \psi_R), \tag{5}$$

$$p_4 = 1 - 0.5 \cdot (\psi_L + \psi_R)(1 - x), \tag{6}$$

where $x = \mu$ is the maintenance probability, $x = \tau$ is the de novo probability and $\psi_L, \psi_R \in [0, 1]$ the dependency parameters for the left and right neighbor. A dependency value of $\psi_i = 1$ corresponds to a total independence on the neighbor whereas $\psi_i = 0$ leads to a total dependence. Hence, μ and τ can be interpreted as the probability of maintenance and de novo methylation of a single cytosine between two cell divisions assuming independence from neighboring CpGs. Moreover, all CpGs that are part of the considered window of the DNA have the same value for the parameters μ, τ, ψ_L, and ψ_R, since in earlier experiments only very small differences have been found between the methylation efficiencies of nearby CpGs [2].

In order to understand the form of the transition probabilities consider at first a case with only one neighbor. The probabilities then have the form ψx if the neighbor is unmethylated and $1 - \psi(1 - x)$ if the neighbor is methylated. Note that both forms evaluate to x for $\psi = 1$, meaning that a site is methylated with probability x, independent of its neighbor. For $\psi = 0$ the probabilities become 0 and 1, meaning that if there is no methylated neighbor the site cannot be methylated or will be methylated for sure if there is a methylated neighbor respectively.

The probabilities for two neighbors are obtained by a linear combination of the one neighbor cases, with ψ_L for the left and ψ_R for the right neighbor, and an additional weight of 0.5 to normalize the probability.

The same considerations also apply to the boundary sites however there is no way of knowing the methylation states outside the boundaries (denoted by ?). Therefore instead of a concrete methylation state (o for unmethylated, • for methylated site) the average methylation density ρ is used to compute the transition probabilities at the boundaries (depicted here for de novo):

$$? \circ \circ \to ? \bullet \circ \qquad \tilde{p}_1 = (1 - \rho) \cdot p_1 + \rho \cdot p_2, \tag{7}$$

$$? \circ \bullet \to ? \bullet \bullet \qquad \tilde{p}_2 = (1 - \rho) \cdot p_3 + \rho \cdot p_4, \tag{8}$$

$$\circ \circ ? \to \circ \bullet ? \qquad \tilde{p}_3 = (1 - \rho) \cdot p_1 + \rho \cdot p_3, \tag{9}$$

$$\bullet \circ ? \to \bullet \bullet ? \qquad \tilde{p}_4 = (1 - \rho) \cdot p_2 + \rho \cdot p_4. \tag{10}$$

Note that the same considerations hold for maintenance at the boundaries if the opposite site of the boundary site is already methylated.

For each position l, there are four transition matrices: two for maintenance and two for de novo, namely one for the upper and one for the lower strand in each process. In order to construct these matrices consider the three positions $l-1$, l and $l+1$, where the transition happens at position l. Only the transitions depicted in Fig. 3 can occur. Furthermore the transitions are unique, i.e. for a given reference number i the new reference number j is uniquely determined. For patterns not depicted in Fig. 3 no transition can occur, i.e. the reference number does not change.

The matrix describing a maintenance event at position l and strand s has the form

$$M_s^{(l)}(i, j) = \begin{cases} 1, & \text{if } i = j \text{ and } \not\exists j' : i \leadsto j', \\ 1 - p, & \text{if } i = j \text{ and } \exists j' : i \leadsto j', \\ p, & \text{if } i \neq j \text{ and } i \leadsto j, \\ 0, & \text{else,} \end{cases} \tag{11}$$

where the probability p is given by one of the Eqs. (3)–(10) that describes the corresponding case and $x = \mu$. Note that $M_s^{(l)}$ depends on s and l since it describes a single transition from pattern i to pattern j, which occurs on a particular strand and at a particular location with probability p. We define matrices $T_s^{(l)}$ for de novo methylation according to the same rules except that $x = \tau$ and the possible transitions are as in Fig. 3, right.

The advantage of defining the matrices position- and process-wise is that different models can be realized by changing the order of multiplication of these matrices.

It is important to note that 5mC can be further modified by oxidation to 5-hydroxymethyl- (5hmC), 5-formyl- (5fC) and 5-carboxyl cytosine(5caC) by Tet enzymes. These modifications are involved in the removal of 5mC from the DNA and can potentially interfere with methylation events. However, our data does not capture these modifications and therefore we are not able to consider these modifications in our model.

3.3   Combination of Transition Matrices

For all subsequent models it is assumed that first of all cell division happens and maintenance methylation only occurs on the newly synthesized strand given by s, whereas de novo methylation happens on both strands. Given the mechanisms in Fig. 1, the two different kinds of methylation events, and the two types of enzymes, there are several possibilities to combine the transition matrices. We consider the following four models, which we found most reasonable based on the current state of research in DNA methylation:

1. first processive maintenance and then processive de novo methylation

$$P_s = \prod_{l_1=1}^{L} M_s^{(l_1)} \prod_{l_2=1}^{L} T_1^{(l_2)} \prod_{l_3=1}^{L} T_2^{(l_3)}, \tag{12}$$

2. first processive maintenance and then de novo in arbitrary order

$$P_s = \frac{1}{(L!)^2} \prod_{l_1=1}^{L} M_s^{(l_1)} \left(\sum_{\sigma_1 \in S_L} \prod_{l_2=1}^{L} T_1^{(\sigma_1(l_2))} \right) \left(\sum_{\sigma_2 \in S_L} \prod_{l_3=1}^{L} T_2^{(\sigma_2(l_3))} \right), \tag{13}$$

3. maintenance and de novo at one position, processive

$$P_s = \prod_{l=1}^{L} M_s^{(l)} T_1^{(l)} T_2^{(l)}, \tag{14}$$

4. maintenance and de novo at one position, arbitrary order

$$P_s = \frac{1}{L!} \sum_{\sigma \in S_L} \prod_{l=1}^{L} M_s^{(\sigma(l))} T_1^{(\sigma(l))} T_2^{(\sigma(l))}, \tag{15}$$

where S_L is the set of all possible permutations for the numbers $1, \ldots, L$.

Note that the de novo events on both strands are independent, i.e. the de novo events on the upper strand do not influence the de novo events on the lower strand and vice versa, such that $[T_1^{(l)}, T_2^{(l')}] = 0$ independent of ψ_i^2. Obviously it is important whether maintenance or de novo happens first, since the transition probabilities and the transitions themselves depend on the actual pattern. Furthermore in the case $\psi_i < 1$ (dependency on right and/or left neighbor) the order of the transitions on a strand matters, i.e. $[M_s^{(l)}, M_s^{(l')}] \neq 0$ and $[T_s^{(l)}, T_s^{(l')}] \neq 0$ for $l \neq l'$. The total transition matrix is then given by a combination of the cell division and maintenance/de novo matrices.

---

[2] $[A, B] = AB - BA$ is the commutator of the matrices A and B.

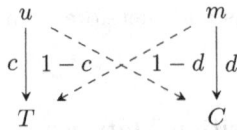

Fig. 4. Conversions of the unobservable states u, m to observable states T, C with respective rates.

Recall that we consider two different types of Dnmts, i.e., Dnmt1 and Dnmt3a/b. If only one type of Dnmt is active (KO data) the matrix has the form

$$P = 0.5 \cdot (D_1 \cdot P_1 + D_2 \cdot P_2) \qquad (16)$$

and if all Dnmts are active (WT data)

$$P = 0.5 \cdot (D_1 \cdot P_1 \cdot \tilde{P}_1 + D_2 \cdot P_2 \cdot \tilde{P}_2), \qquad (17)$$

where P_s and $\tilde{P}_s$ have one of the forms (12)–(15). This leads to four different models for one active enzyme or 16 models for all active enzymes respectively. In the second case P_s represents the transitions caused by Dnmt1 and $\tilde{P}_s$ the transitions caused by Dnmt3a/b. Note that if $\psi_L = \psi_R = 1$ all models are the same within each case.

3.4 Conversion Errors

The actual methylation state of a C cannot be directly observed. During BS-seq, with high probability every unmethylated C (denoted by u) is converted into Thymine (T) and every 5mC (denoted by m) into C. However, conversion errors may occur and we define their probability as $1 - c$ and $1 - d$, respectively, as shown by the dashed arrows in Fig. 4. It is reasonable that these conversion errors occur independently and with approximately identical probability at each site and thus the error matrix for a single CpG takes the form

$$\Delta_1 = \begin{pmatrix} c^2 & c(1-c) & c(1-c) & (1-c)^2 \\ c(1-d) & cd & (1-c)(1-d) & d(1-c) \\ c(1-d) & (1-c)(1-d) & cd & d(1-c) \\ (1-d)^2 & d(1-d) & d(1-d) & d^2 \end{pmatrix}. \qquad (18)$$

Due to the independency of the events this matrix can easily be generalized for systems with $L > 1$ by recursively using the Kronecker-product

$$\Delta_L = \Delta_1 \otimes \Delta_{L-1} \qquad \text{for } L \geq 2. \qquad (19)$$

Hence, Δ_L gives the probability of observing a certain sequence of C and T nucleotides for each given unobservable methylation pattern. In order to compute the likelihood $\hat{\pi}$ of the observed BS-seq data, we therefore first compute the transient distribution $\pi(t)$ of the underlying Markov chain at the corresponding time instant[3] t by solving

$$\pi(t) = \pi(0) \cdot P^t \tag{20}$$

and then multiply the distribution of the unobservable patterns with the error matrix.

$$\hat{\pi} = \pi(t) \cdot \Delta_L. \tag{21}$$

Note that this yields a hidden Markov model with emission probabilities Δ_L. In the following the values for c were chosen according to [2]. Since the value for d was not determined in [2], we measured the conversion rate $d = 0.94$ in an independent experiment under comparable conditions (data not shown).

3.5 Maximum Likelihood Estimator

In order to estimate the parameters $\theta = (\mu, \psi_L, \psi_R, \tau)$, we employ a Maximum (Log)Likelihood Estimator (MLE)

$$\hat{\theta} = \arg \max_{\theta} \ell(\theta), \quad \ell(\theta) = \sum_{j=1}^{4^L} \log(\hat{\pi}_j(\theta)) \cdot N_j, \tag{22}$$

where $\hat{\pi}$ is the pattern distribution obtained from the numerical solution of (20) and (21) for a given time t and N_j is the number of occurrences of pattern j in the measured data. The parameters $\theta = \hat{\theta}$ are chosen in such a way that ℓ is maximized. Visual inspection of all two dimensional cuts of the likelihood landscapes showed only a single local maximum.

We employ the MLE twice in order to estimate the parameter vector $\hat{\theta}_1$ for Dnmt1 from the 3a/b DKO (double knockout) data and the vector $\hat{\theta}_{3a/b}$ for Dnmt3a/b from the Dnmt1 KO data, where transition matrix (16) is used. The corresponding time instants are $t = 26$ for the 3a/b DKO data and $t = 41$ for the 1KO data.

We approximate the standard deviations of the estimated parameters $\hat{\theta}$ as follows: Let $\mathcal{I}(\hat{\theta}) = \mathbb{E}[-\mathcal{H}(\hat{\theta})]$ be the expected Fisher information, with the Hessian $\mathcal{H}(\hat{\theta}) = \nabla\nabla^{\mathsf{T}}\ell(\hat{\theta})$. The inverse of the expected Fisher information is a lower bound for the covariance matrix of the MLE such that we can use the approximation $\sigma(\hat{\theta}) \approx \sqrt{\mathrm{diag}(-\mathcal{H}(\hat{\theta}))}$.

A prediction for the wild-type can be computed by combining the estimated vectors such that in the model both types of enzymes are active. For this, we insert $\hat{\theta}_1$ in P_s and $\hat{\theta}_{3a/b}$ in $\tilde{P}_s$ in (17) to obtain the transition matrix for the wild-type.

---

[3] The number of cell divisions is estimated from the time of the measurement since these cells divide once every 24 hours.

4   Results

For our analysis we focused at the single copy genes Afp (5 CpGs) and Tex13 (10 CpGs) as well as the repetitive elements IAP (intracisternal A particle) (6 CpGs), L1 (Long interspersed nuclear elements) (7 CpGs) and mSat (major satellite) (3 CpGs). Repetitive elements occur in multiple copies and are dispersed over the entire genome. Therefore they allow capturing an averaged, more general behavior of methylation dynamics. If a locus contains more than three CpGs, the analysis is done for all sets of three adjacent sites independently, in order to keep computation times short and memory requirements low. In the sequel, we mainly focus on the estimated dependency parameters ψ_L and ψ_R and on the prediction quality of the different models.

The estimates for all the available KO data and all suggested models obtained using the transition matrix in Eq. (16) are summarized as histograms in Fig. 5. Because of the different possibilities to combine the four different models in Eqs. (12)–(15) and because of the different loci considered, in total there are 84 estimates for each KO data set. We plot the number of occurrences N of ψ_L (left) and ψ_R (right) in different ranges for both sorts of KO data (Dnmt1KO and Dnmt3a/b DKO).

The estimates of ψ_L spread over the whole interval $[0, 1]$ while in the case of ψ_R, nearly all estimates are larger than 0.99 and only in a few cases the dependency parameter is significantly smaller than 1. Hence, in most cases the methylation probabilities are independent of the right neighbor for both Dnmt1KO and Dnmt3a/b DKO. For ψ_L the dependency parameter in the Dnmt3a/b DKO case occurs more often close to 1, meaning that the transitions induced by Dnmt1 have little to no dependency on the left neighbor. On the other hand for Dnmt1KO the dependency parameter occurs more often at smaller values giving evidence that there is a dependency on the left neighbor for the activity of Dnmt3a/b. Note that all models show a similar behavior in terms of the dependency parameters for a given locus or position within a locus respectively, i.e. either $\psi_i \approx 1$ or $\psi_i < 1$ for all models. The difference between the behaviors at different loci and positions may be explained by explicitly including the distances between the CpGs and is planned as future work.

Since ψ_R is usually close to 1 a smaller model with only three parameters $\theta = (\mu, \psi, \tau)$ can be proposed, where ψ is a dependency parameter for the left neighbor. This model can either be obtained by fixing $\psi_R = 1$ in the original model and setting $\psi = \psi_L$ or by redefining the transition probabilities to ψx if the left neighbor is unmethylated and $1 - \psi(1 - x)$ if the left neighbor is methylated. In that case ψ and ψ_L are related via $\psi = 0.5(\psi_L + 1)$. Note that both versions yield the same results.

In order to check whether there is a significant difference in the original and the smaller model, we performed a Likelihood-ratio test with the null hypothesis that the smaller model is a special case of the original model. Since the original model with more parameters is always as least as good as the smaller model, our goal is to check in which cases the smaller model is sufficient. Indeed if ψ_R was estimated to be approximately 1 the Likelihood-ratio test indicates that the

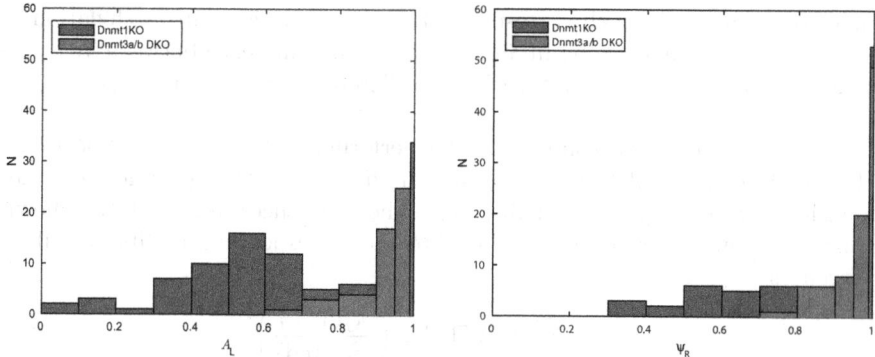

Fig. 5. Histograms for the estimated dependency parameters ψ_L and ψ_R for all sets of three adjacent CpGs in all loci and for all suggested models.

smaller model is sufficient (p-value ≈ 1). On the other hand, for the few cases where ψ_R differs significantly from 1 the original model has to be used (p-value < 0.01).

As a next step we used the estimated parameters from the KO data to predict the WT data. The models from Eqs. (12)–(15) are referred to as *Models 1–4*. For the prediction, the notation (x, y) is used to refer to Model x for the Dnmt3a/b DKO (only Dnmt1 active) and Model y for the Dnmt1KO case (only Dnmt3a/b active). One instance of the prediction, for which Model 1 was used for both Dnmt1KO and Dnmt3a/b DKO, i.e. $(1, 1)$, are shown in Fig. 6. Note that all wild-type predictions yielded a very similar accuracy. We list the corresponding estimations for the parameters for an example of a single copy gene (Afp) and a repetitive element (L1) in Table 1. While the standard deviation of the estimated parameters for μ is always of the order 10^{-2} and for τ of order 10^{-3}, it is usually of order 10^{-2} for ψ_i. Depending on the model, locus and position, standard deviations up to order 10^{-1} may occur for the dependency parameters in a few cases.

In Fig. 6 the predictions for the pattern distribution together with the WT pattern distribution and a prediction from the neighborhood independent model ($\psi_L = \psi_R = 1$) for all loci are shown in the main plot. As an inset the distributions are shown on a smaller scale to display small deviations. With the exception

Table 1. Estimated parameters for the KO data and model based on Eq. (12) for the loci Afp and L1 with sample size n.

KO	μ	ψ_L	ψ_R	τ	n	Locus
Dnmt1	0.452 ± 0.062	0.383 ± 0.076	1.000 ± 0.094	0.091 ± 0.016	134	Afp
Dnmt3a/b	0.990 ± 0.003	0.984 ± 0.011	1.000 ± 0.006	$10^{-10} \pm 0.011$	186	Afp
Dnmt1	0.334 ± 0.051	0.576 ± 0.067	1.000 ± 0.122	0.038 ± 0.004	1047	L1
Dnmt3a/b	0.789 ± 0.037	1.000 ± 0.038	0.984 ± 0.045	$10^{-10} \pm 0.002$	805	L1

of patterns 0 and 64 (which corresponds to no methylation/full methylation of all sites) in L1 and pattern 64 in all loci, where the difference between WT and the numerical solution is about 10%, the difference is always small ($< 5\%$) as seen in the insets.

In general all 16 models show a similar performance for all loci and positions in terms of accuracy of the prediction. On the large scale the differences are not visible and even for the smaller scale the differences are small, as shown for mSat in Fig. 7. This is in accordance to the corresponding Kullback-Leibler divergences

$$KL = \sum_{j=1}^{4^L} \pi_j(\text{WT}) \log \left(\frac{\pi_j(\text{WT})}{\pi_j(\text{pred})} \right) \tag{23}$$

that we list in Table 2. The difference in KL between the "best" and the "worst" case is about 0.01. The mean and standard deviation for KL was obtained via bootstrapping of the wild-type data (10.000 bootstrap samples for each model). Since no confidence intervals of the parameters are included, this standard deviation can be regarded as a lower bound. However, even with these lower bounds the intervals of KL overlap for all models, such that no model can be favorized.

Table 2. Kullback-Leibler divergence KL for the 16 models.

Model	$(1,1)$	$(1,2)$	$(1,3)$	$(1,4)$
KL	0.1398 ± 0.0134	0.1398 ± 0.0134	0.1398 ± 0.0134	0.1337 ± 0.0127
Model	$(2,1)$	$(2,2)$	$(2,3)$	$(2,4)$
KL	0.1438 ± 0.0137	0.1439 ± 0.0136	0.1439 ± 0.0137	0.1374 ± 0.0133
Model	$(3,1)$	$(3,2)$	$(3,3)$	$(3,4)$
KL	0.1399 ± 0.0134	0.1399 ± 0.0134	0.1398 ± 0.0133	0.1337 ± 0.0127
Model	$(4,1)$	$(4,2)$	$(4,3)$	$(4,4)$
KL	0.1410 ± 0.0137	0.1411 ± 0.0136	0.1409 ± 0.0135	0.1349 ± 0.0130

5   Related Work

In [4] location- and neighbor-dependent models are proposed for single-stranded DNA methylation data in blood and tumor cells. The (de-)methylation rates depend on the position of the CpG relative to the 3' or 5' end and/or on the methylation state of the left neighbor only. The dependency is realized by the introduction of an additional parameter. In our proposed models we use double-stranded DNA and can therefore include hemi-methylated sites and even distinguish on which strand the site is methylated. Furthermore we allow dependencies on both neighbors by introducing two different dependency parameters. In contrast [6] copes with the neighborhood dependency indirectly by allowing different parameter values for different sites. In order to reduce the dimensionality

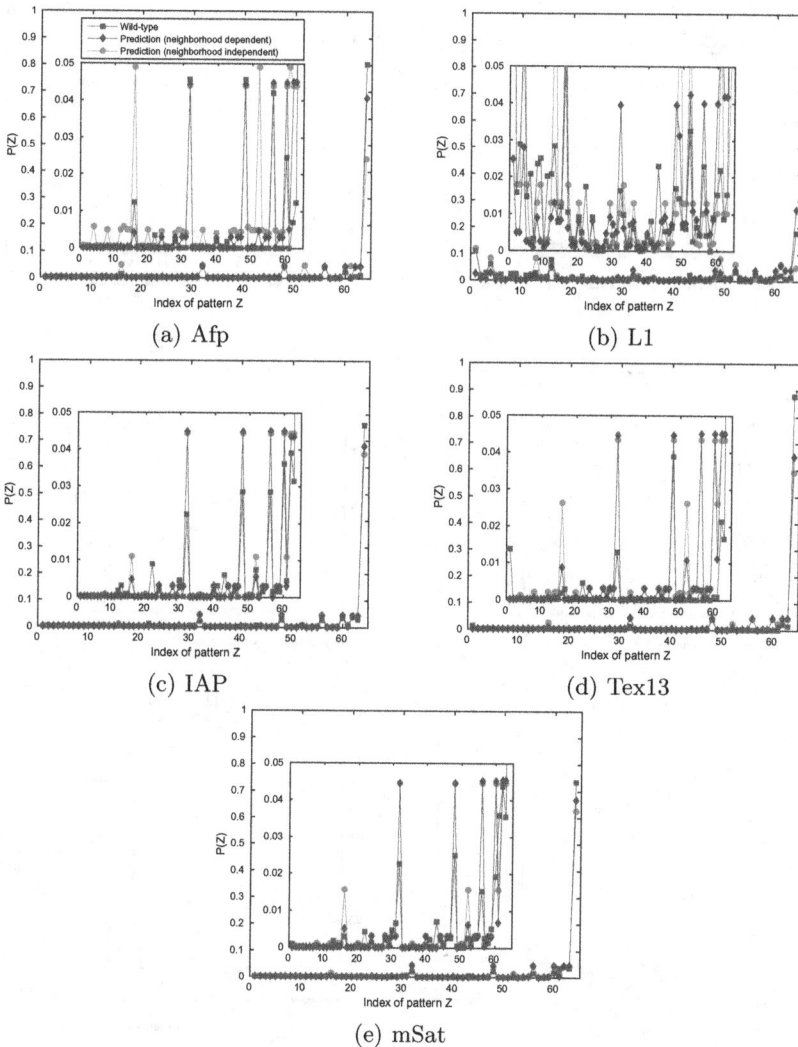

Fig. 6. The figures show an example for the predicted (neighborhood dependent and neighborhood independent) and the measured pattern distribution for each locus. The inset shows a zoomed in version of the distribution.

of the parameter vector, a hierarchical model based on beta distributions is proposed. Another difference to our model is the distinction between de novo rates for parent and daughter strand. However, this can easily be included in future work. A density-dependent Markov model was proposed [14]. In this model, the probabilities of (de-)methylation events may depend on the methylation density in the CpG neighborhood. In addition, a neighboring sites model has been developed, in which the probabilities for a given site are directly influenced by

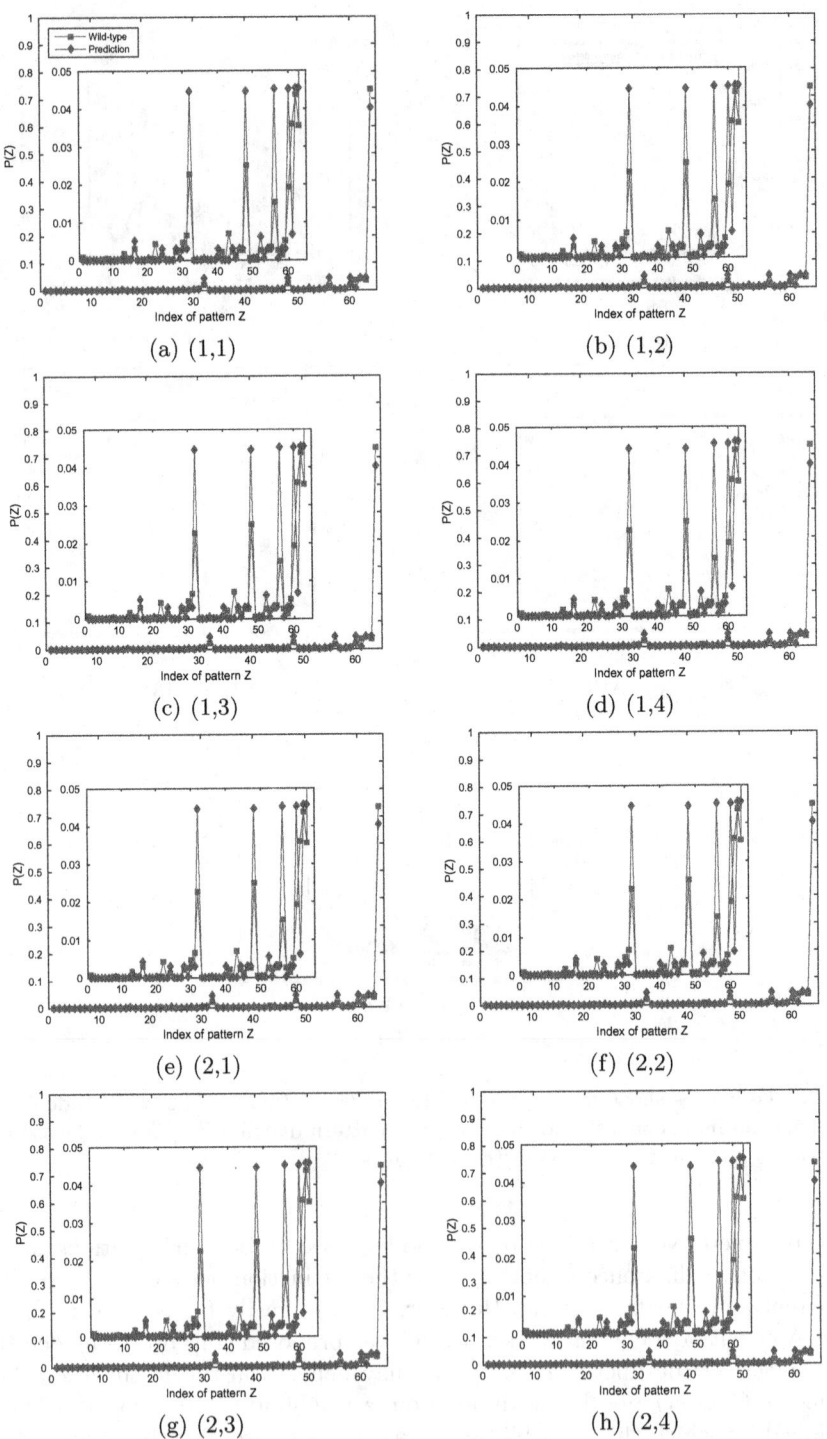

Fig. 7. The figures show the predicted and the measured pattern distribution for all 16 models for mSat. The inset shows a zoomed in version of the distribution.

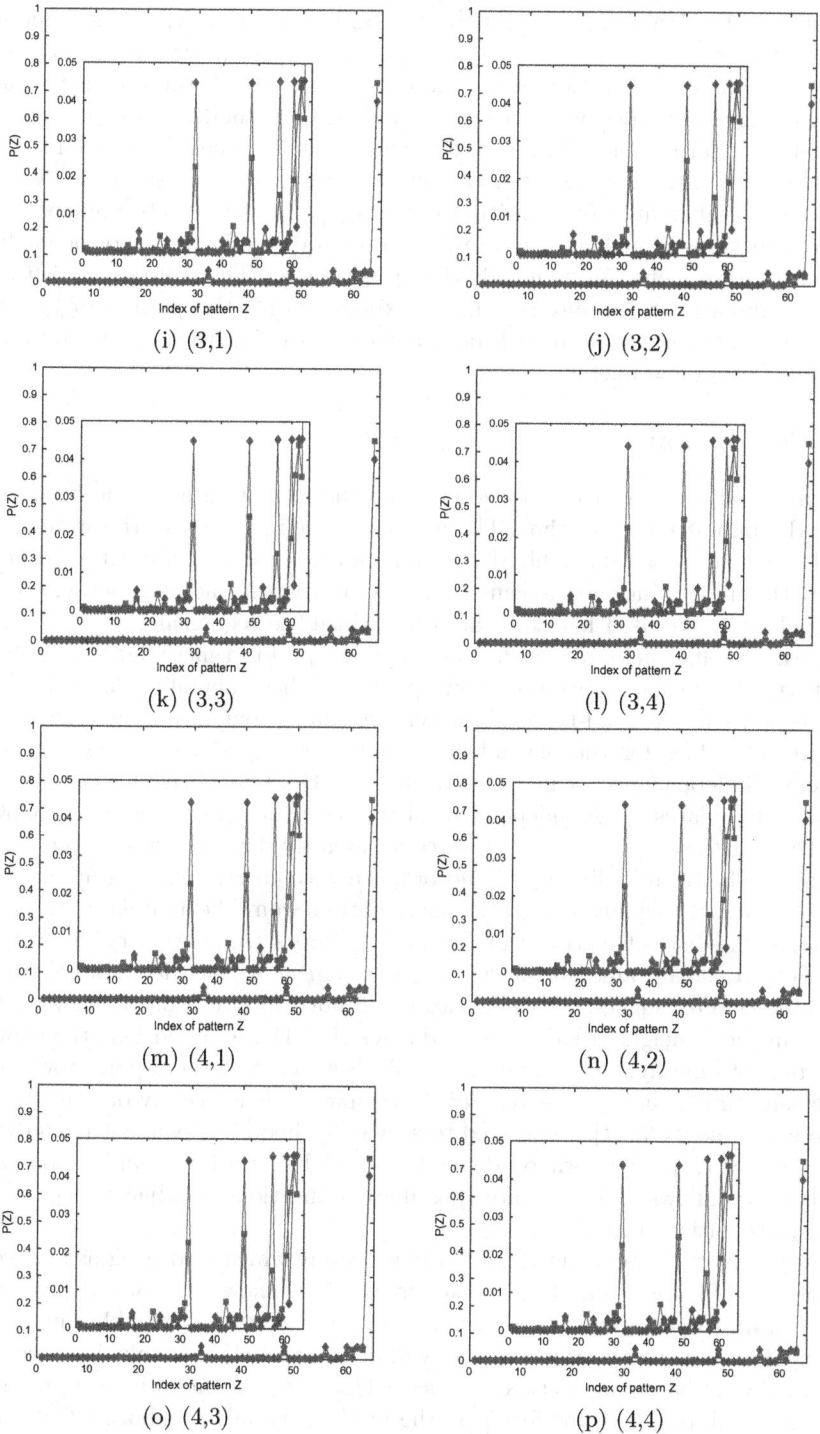

Fig. 7. (*continued*)

the states of neighboring sites to the left and right [14]. When these models were tested on double-stranded methylation patterns from two distinct tandem repeat regions in a collection of ovarian carcinomas, the density-dependent and neighboring sites models were superior to independent models in generating statistically similar samples. Although this model also includes the dependence on the methylation state on the left and right neighbor for double-stranded DNA the approach is different. The transition probabilities of the neighbor-independent model are transformed into a transition probability of a neighbor-dependent model by introducing only one additional parameter. The state of the left and right neighbor are taken into account by exponentiating this parameter by some norm. In addition, this approach does not allow the intuitive interpretation of the dependency parameter.

6   Conclusion

We proposed a set of stochastic models for the formation and modification of methylation patterns over time. These models take into account the state of the CpG sites in the spatial neighborhood and allow to describe different hypotheses about the underlying mechanisms of methyltransferases adding methyl groups at CpG sites. We used knockout data from bisulfite sequencing at several loci to learn the efficiencies at which these enzymes perform methylation. By combining these efficiencies, we accurately predicted the probability distribution of the patterns in the wild-type. Moreover, we found that in all cases the models predict values for the dependency parameters ψ_L and ψ_R close to 1 and therefore independence of methylation for the Dnmt3a/b DKO meaning that Dnmt1 methylates CpGs independent of the methylation of neighboring CpGs. For Dnmt3a/b on the other hand we could identify dependencies on the neighboring CpGs. Both findings are in accordance with current existing mechanistic models: Dnmt1 reliably copies the methylation from the template strand to maintain the distinct methylation patterns, whereas Dnmt3a/b try to establish and keep a certain amount of CpG methylation at a given loci. Interestingly, our models only suggest dependencies of de novo methylation activity on the CpGs in the 5' neighborhood. This indicates that Dnmt3a and Dnmt3b show a preference to methylate CpGs in a 5' to 3' direction and could point towards a processive or cooperative behavior of these enzymes like recently described in *in vitro* experiments [5,11]. Compared to a neighborhood independent model with $\psi_L = \psi_R = 1$, a neighborhood dependent model shows better predictions and furthermore allows to investigate (possible) connections of adjacent CpGs and their methylation states.

As future work, we plan to investigate models in which we distinguish between the actions of Dnmt3a and Dnmt3b and in which we allow a diagonal dependency for de novo methylation, i.e., a dependency on the state of neighboring CpGs on the opposite strand. Moreover, we will design models that take into account the number of base pairs between adjacent CpG sites. To investigate a potential impact of oxidized cytosine forms on the methylation at neighboring CpG sites we further plan to include the CpG states 5hmC, 5fC and 5caC in our model.

References

1. Äijö, T., Huang, Y., Mannerström, H., Chavez, L., Tsagaratou, A., Rao, A., Lähdesmäki, H.: A probabilistic generative model for quantification of DNA modifications enables analysis of demethylation pathways. Genome Biol. **17**(1), 49 (2016)
2. Arand, J., Spieler, D., Karius, T., Branco, M.R., Meilinger, D., Meissner, A., Jenuwein, T., Xu, G., Leonhardt, H., Wolf, V., et al.: In vivo control of CpG and non-CpG DNA methylation by DNA methyltransferases. PLoS Genet. **8**(6), e1002750 (2012)
3. Baubec, T., Colombo, D.F., Wirbelauer, C., Schmidt, J., Burger, L., Krebs, A.R., Akalin, A., Schübeler, D.: Genomic profiling of DNA methyltransferases reveals a role for DNMT3B in genic methylation. Nature **520**(7546), 243–247 (2015)
4. Bonello, N., Sampson, J., Burn, J., Wilson, I.J., McGrown, G., Margison, G.P., Thorncroft, M., Crossbie, P., Povey, A.C., Santibanez-Koref, M., et al.: Bayesian inference supports a location and neighbour-dependent model of DNA methylation propagation at the MGMT gene promoter in lung tumours. J. Theor. Biol. **336**, 87–95 (2013)
5. Emperle, M., Rajavelu, A., Reinhardt, R., Jurkowska, R.Z., Jeltsch, A.: Cooperative DNA binding and protein/DNA fiber formation increases the activity of the Dnmt3a DNA methyltransferase. J. Biol. Chem. **289**(43), 29602–29613 (2014)
6. Fu, A.Q., Genereux, D.P., Stöger, R., Laird, C.D., Stephens, M.: Statistical inference of transmission fidelity of DNA methylation patterns over somatic cell divisions in mammals. The Annals of Applied Statistics **4**(2), 871 (2010)
7. Genereux, D.P., Miner, B.E., Bergstrom, C.T., Laird, C.D.: A population-epigenetic model to infer site-specific methylation rates from double-stranded DNA methylation patterns. PNAS **102**(16), 5802–5807 (2005)
8. Giehr, P., Kyriakopoulos, C., Ficz, G., Wolf, V., Walter, J.: The influence of hydroxylation on maintaining CpG methylation patterns: a hidden Markov model approach. PLoS Comput. Biol. **12**(5), e1004905 (2016)
9. Gowher, H., Jeltsch, A.: Molecular enzymology of the catalytic domains of the Dnmt3a and Dnmt3b DNA methyltransferases. J. Biol. Chem. **277**(23), 20409–20414 (2002)
10. Hermann, A., Goyal, R., Jeltsch, A.: The Dnmt1 DNA-(cytosine-c5)-methyltransferase methylates DNA processively with high preference for hemimethylated target sites. J. Biol. Chem. **279**(46), 48350–48359 (2004)
11. Holz-Schietinger, C., Reich, N.O.: The inherent processivity of the human de novo methyltransferase 3A (DNMT3A) is enhanced by DNMT3L. J. Biol. Chem. **285**(38), 29091–29100 (2010)
12. Kapourani, C.A., Sanguinetti, G.: Higher order methylation features for clustering and prediction in epigenomic studies. Bioinformatics **32**(17), i405–i412 (2016)
13. Kyriakopoulos, C., Giehr, P., Wolf, V.: H(O)TA: estimation of DNA methylation and hydroxylation levels and efficiencies from time course data. Bioinformatics (2017, to appear)
14. Lacey, M.R., Ehrlich, M., et al.: Modeling dependence in methylation patterns with application to ovarian carcinomas. Stat Appl Genet Mol Biol **8**(1), 40 (2009)
15. Laird, C.D., Pleasant, N.D., Clark, A.D., Sneeden, J.L., Hassan, K.A., Manley, N.C., Vary, J.C., Morgan, T., Hansen, R.S., Stöger, R.: Hairpin-bisulfite PCR: assessing epigenetic methylation patterns on complementary strands of individual DNA molecules. PNAS **101**(1), 204–209 (2004)

16. Norvil, A.B., Petell, C.J., Alabdi, L., Wu, L., Rossie, S., Gowher, H.: Dnmt3b methylates DNA by a noncooperative mechanism, and its activity Is unaffected by manipulations at the predicted dimer interface. Biochemistry (2016). http://dx.doi.org/10.1021/acs.biochem.6b00964

17. Okano, M., Bell, D.W., Haber, D.A., Li, E.: DNA methyltransferases Dnmt3a and Dnmt3b are essential for de novo methylation and mammalian development. Cell **99**(3), 247–257 (1999)

18. Otto, S.P., Walbot, V.: DNA methylation in eukaryotes: kinetics of demethylation and de novo methylation during the life cycle. Genetics **124**(2), 429–437 (1990)

19. Sontag, L.B., Lorincz, M.C., Luebeck, E.G.: Dynamics, stability and inheritance of somatic DNA methylation imprints. J. Theor. Biol. **242**(4), 890–899 (2006)

20. Suzuki, M.M., Bird, A.: DNA methylation landscapes: provocative insights from epigenomics. Nat. Rev. Genet. **9**(6), 465–476 (2008)

Temporal Reprogramming of Boolean Networks

Hugues Mandon[1,2](✉), Stefan Haar[1], and Loïc Paulevé[2]

[1] LSV, ENS Cachan, INRIA, CNRS, Université Paris-Saclay, Cachan, France
hugues.mandon@inria.fr
[2] CNRS, LRI UMR 8623, Univ. Paris-Sud, Université Paris-Saclay, Orsay, France

Abstract. Cellular reprogramming, a technique that opens huge opportunities in modern and regenerative medicine, heavily relies on identifying key genes to perturb. Most of computational methods focus on finding mutations to apply to the initial state in order to control which attractor the cell will reach. However, it has been shown, and is proved in this article, that waiting between the perturbations and using the transient dynamics of the system allow new reprogramming strategies. To identify these *temporal* perturbations, we consider a qualitative model of regulatory networks, and rely on Petri nets to model their dynamics and the putative perturbations. Our method establishes a complete characterization of temporal perturbations, whether permanent (mutations) or only temporary, to achieve the existential or inevitable reachability of an arbitrary state of the system. We apply a prototype implementation on small models from the literature and show that we are able to derive temporal perturbations to achieve trans-differentiation.

1 Introduction

Regenerative medicine is gaining traction with the discovery of cell reprogramming, a way to change a cell phenotype to another, allowing tissue or neuron regeneration techniques. After proof that cell fate decisions could be reversed [17], scientists need efficient and trustworthy methods to achieve it. Instead of producing induced pluripotent stem cells and force the cell to follow a distinct differentiation path, new methods focus on *trans-differentiating* the cell, without necessarily going (back) through a multipotent state [8,9].

This paper addresses the formal prediction of perturbations for cell reprogramming from computational models of gene regulation. We consider qualitative models where the genes and/or the proteins, notably transcription factors, are nodes with an assigned value giving the level of activity, e.g., 0 for inactive and 1 for active, in a Boolean abstraction. The value of each node can then evolve in time, depending on the value of its regulators.

This research was supported by Labex DigiCosme (project ANR-11-LABEX-0045-DIGICOSME) operated by ANR as part of the program "Investissement d'Avenir" Idex Paris-Saclay (ANR-11-IDEX-0003-02); by ANR-FNR project "AlgoReCell" (ANR-16-CE12-0034); and by CNRS PEPS INS2I 2017 "FoRCe".

© Springer International Publishing AG 2017
J. Feret and H. Koeppl (Eds.): CMSB 2017, LNBI 10545, pp. 179–195, 2017.
DOI: 10.1007/978-3-319-67471-1_11

The attractors, or long term dynamics, of qualitative models typically correspond to differentiated and stable states of the cell [13,18]. In such a setting, cell reprogramming can be interpreted as triggering a change of attractor: starting within an initial attractor, perform perturbations which would de-stabilize the network and lead the cell to a different attractor.

Current experimental settings and computational models mainly consider cell reprogramming by applying the set of perturbations simultaneously in the initial state. However, as suggested in [14] and as we will demonstrate formally in this paper, considering *temporal* reprogramming, i.e., the application of perturbations in particular moments in time, and in a particular *ordering*, brings new reprogramming strategies, potentially requiring fewer interventions.

Contribution. This paper establishes the formal characterization of all possible reprogramming paths between two states of asynchronous Boolean networks by the means of a bounded number of either permanent (mutations) or temporary perturbations. Solutions account both for perturbations applied only in the initial state, and perturbations applied in a specific ordering and in specific states. Moreover, the solutions can guarantee that the target state *may* be reached, or will be reached inevitably.

Our method relies on a Petri net modelling jointly the asynchronous dynamics of the Boolean network and the candidate perturbations. The reprogramming solutions are identified from the state transition graph of the resulting model. We apply our approach on biological networks from the literature, and show that the *temporal* application of perturbations brings new reprogramming solutions.

Related work. The computational prediction for reprogramming of Boolean networks has been addressed mainly by considering mutations to be applied in the initial state only, letting then the system stabilize itself in the targeted attractor [1,6,7,15,16,19]. Our method includes temporal perturbations, which none of these methods do: perturbations which takes into account the latent dynamics of the system for the reprogramming, allowing more solutions to be found, and possibly some needing fewer nodes to be perturbed.

Other approaches consider stochastic frameworks for exploring by simulation potential reprogramming event in Boolean networks, such as [10] for stochastic transitions between cell cycles. Statistical methods are also used to extract combinations of transcription factors that are key for cellular differentiation from gene expression data [3,14]. In [14], starting from expression data, they derive a continuous dynamical model from which control strategies for reprogramming can be computed. They show that time-dependent perturbations can provide potential reprogramming strategies.

Most of mentioned methods provide incomplete or non-guaranteed results. Our aim is to provide a formal framework for the complete and exact characterisation of the initial state and temporal reprogramming of Boolean networks.

Outline. Section 2 details an example of Boolean network which motivates temporal reprogramming. Section 3 introduces our model of temporal reprogramming, Sect. 4 establishes the identification of temporal reprogramming strategies, and Sect. 5 applies it to biological networks from the literature. Section 6 concludes the paper.

Notations: The set $\{1, ..., n\}$ is noted $[n]$; Given $x \in \{0,1\}^n$ and $i \in [n]$, $\bar{x}^{\{i\}} \in \{0,1\}^n$ is such that for all $j \in [n]$, $\bar{x}_j^{\{i\}} \triangleq \neg x_j$ if $j = i$ and $\bar{x}_j^{\{i\}} \triangleq x_j$ if $j \neq i$.

2 Background and Motivating Example

This section illustrates the benefit of *temporal* reprogramming on a small Boolean network in order to trigger a change of attractor. We consider both perturbations to be applied solely in the initial state, and perturbations to be applied in a specific sequence in specific states. We show that, in the first setting, 3 perturbations are always required for the reprogramming, whereas the temporal approach necessitates only 2.

2.1 Boolean Networks

A Boolean Network is a tuple of Boolean functions giving the future value of each node with respect to the global state of the network.

Definition 1 (Boolean Network (BN)). *A Boolean Network of dimension n is a function f such that:*

$$f : \{0,1\}^n \to \{0,1\}^n$$
$$x = (x_1, ..., x_n) \mapsto f(x) = (f_1(x), ..., f_n(x))$$

The dynamics of a Boolean network f are modelled by *transitions* between its states $x \in \{0,1\}^n$. In the scope of this paper, we consider the *asynchronous semantics* of Boolean networks: a transition updates the value of only one node $i \in [n]$. Thus, from a state $x \in \{0,1\}^n$, there is one transitions for each vertex i such that $f_i(x) \neq x_i$. The *transition graph* (Definition 2) is a digraph where vertices are all the possible states $\{0,1\}^n$, and edges correspond to asynchronous transitions. The transition graph of a Boolean network f can be noted as STG(f).

Definition 2 (Transition Graph). *The transition graph (also known as state graph) is the graph having $\{0,1\}^n$ as vertex set and the edges set $\{x \to \bar{x}^{\{i\}} \mid x \in \{0,1\}^n, i \in [n], f_i(x) = \neg x_i\}$. A path from x to y is noted $x \to^* y$.*

The terminal strongly connected components of the transition graph can be seen as the long-term dynamics or "fates" of the system, that we refer to as *attractors*. An attractor may model sustained oscillations (*cyclic attractor*) or a unique state, referred to as a *fixpoint*, $f(x) = x$.

2.2 Cell Reprogramming: The Advantage of Temporal Perturbations

Let us consider the following Boolean Network:

$$f_1(x) = x_1 \qquad f_2(x) = x_2 \qquad f_3(x) = x_1 \wedge \neg x_2 \qquad f_4(x) = x_3 \vee x_4$$

Figure 1 gives the transition graph of this Boolean Network, and the different perturbation techniques. To understand the benefit of temporal perturbations, let us consider the perturbations to apply in the fixpoint 0000 in order to reach the fixpoint 1101.

Because 0000 is a fixpoint, there exists no sequence of transitions from 0000 to 1101. It can also be seen that if one or two vertices are perturbed at the same time, by affecting them new values, 1101 is not reachable, as shown in Fig. 1(top). However, if two vertices are perturbed, but the system is allowed to follow its own dynamics between the changes, 1101 can be reached, as shown in Fig. 1(bottom), by using the path $0000 \xrightarrow{x_1=1} 1000 \to 1010 \to 1011 \xrightarrow{x_2=1} 1111 \to 1101$, i.e. we first force the activation of the first node, then wait until the system reaches (by itself) the state 1011 before activating node 2. From the perturbed state, the system is guaranteed to end up in the wanted fixpoint, 1101.

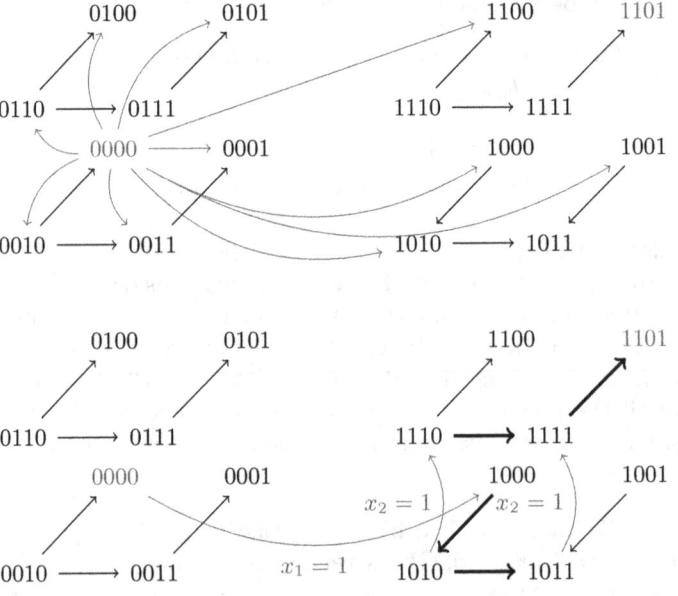

Fig. 1. Transition graph of f and candidate perturbations (magenta) for the reprogramming from 0000 to 1101: (top) none of candidate perturbations of one or two nodes in the initial state allow to reach 1101; (bottom) sequences of two temporal perturbations allow to reach 1101.

Inevitable and existential reprogramming. Thus, this example shows that some attractors may be reached by changing the values of vertices in a particuliar order and using the transient dynamics. We remark that there exists another reprogramming path, where node 2 is perturbed when the system reach 1010. Note that, in this case, after the second perturbation, the system can reach 1101, but it is not guaranteed. We say that, in the first reprogramming path, the reprogramming is *inevitable*, whereas it is only *existential* in the second case.

Permanent and temporary solutions. The previous example shows the difference between what we will call temporal and initial reprogramming. How perturbations are made has also to be considered. The model can either only be slightly perturbed, by changing the value of a vertex i for a time (setting i to 0 or 1), or the change can be permanent, by changing the function of the vertex (setting f_i to 0 or 1). On the example above, making permanent changes would not change the solutions found. However, if the initial state is 1011 and the target state is 1100, then it has different solutions (Fig. 2).

Indeed, if the objective is to go from 1011 to 1100 in the same transition graph using only permanent perturbations, then their order does not matter. Perturbating x_2 and x_4 from the initial state is enough to make 1100 the only reachable state. On the other hand, if the perturbations are temporary, x_2 has to be perturbed first, then when 1101 is reached, x_4 can be perturbed. If this order is not followed, 1101 is reachable as well as 1100.

In most case, the perturbations done in permanent reprogramming and the ones done in temporary reprogramming can be on different nodes.

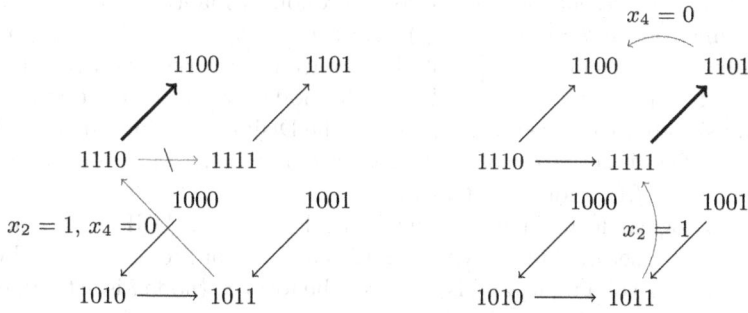

Fig. 2. Right part of the transition graph of f from initial state 1011 to 1100, with permanent perturbations (left) and temporary ones (right)

3   Modelling Temporal Reprogramming with Petri Nets

In this section, we introduce a new model for the temporal reprogramming of Boolean Networks (BNs) using Safe (1-bounded) Petri nets [2]. We take advantage of the transition-centred specification of Petri nets and their ability to define specific coupled transitions (the simultaneous change of value of several components) to model the candidate perturbations.

Definition 3 (Safe Petri Net). *A Petri net is a tuple* (P, T, A, M_0) *where P and T are sets of nodes, called* places *and* transitions *respectively, and* $A \subseteq (P \times T) \cup (T \times P)$ *is a* flow relation *whose elements are called* arcs. *A subset* $M \subseteq P$ *of the places is called a* marking, *and* M_0 *is a distinguished* initial *marking.*

For any node $u \in P \cup T$, *we call* pre-set *of u the set* $^{\bullet}u = \{v \in P \cup T \mid (v, u) \in A\}$ *and* post-set *of u the set* $u^{\bullet} = \{v \in P \cup T \mid (u, v) \in A\}$.

A transition $t \in T$ *is* enabled *at a marking M if and only if* $^{\bullet}t \subseteq M$. *The application of such a transition leads to the new marking* $M' = (M \setminus {}^{\bullet}t) \cup t^{\bullet}$, *and is denoted by* $M \xrightarrow{t} M'$. *A marking* M' *is* reachable *if there exists a sequence of transitions* $t_1, \ldots, t_k$ *such that* $M_0 \xrightarrow{t_1} \ldots \xrightarrow{t_k} M'$.

A Petri net is safe *if and only if any reachable marking M is such that for any* $t \in T$ *that can fire from M leading to* M', *the following property holds:* $\forall p \in M \cap M', p \in {}^{\bullet}t \cap t^{\bullet} \vee p \notin {}^{\bullet}t \cup t^{\bullet}$.

Less formally, a safe Petri Net is a Petri Net where in all reachable markings from the initial marking, all places have at most one token. A subset of places $\{p_1, \ldots, p_k\} \subseteq P$ is *mutually exclusive* if every reachable marking M contains at most one these place.

3.1   Encoding Asynchronous Boolean Networks

The equivalent representation of the asynchronous semantics of a Boolean network of dimension n $f = (f_1, \cdots, f_n)$ in Petri net has been addressed in [4,5]. Essentially, to each node $i \in [n]$ of the Boolean network f is associated two places, i_0 and i_1, acting respectively for the node i being inactive and active. Then, transitions are derived from clauses of the Disjunctive Normal Form (DNF; disjunction of conjunctive clauses) representation of $[\neg x_i \wedge f_i(x)]$ for i activation, and from $[x_i \wedge \neg f_i(x)]$ for i inactivation.

Given a logical formula $[e]$, we write DNF$[e]$ for its DNF representation. DNF$[e]$ is thus a set of clauses, where clauses are sets of literals. A literal correspond to the state of a node, and is either of the form x_i (node i is active), or $\neg x_i$ (node i is inactive). Given such a literal l, place(l) associates the corresponding Petri net place: place($[x_i]$) $\triangleq i_1$ and place($[\neg x_i]$) $\triangleq i_0$.

The safe Petri net encoding the asynchronous semantics of a Boolean network f is defined as follows.

Definition 4 (PN(f)). *Given a Boolean network f of dimension n and an initial state* $x \in \{0, 1\}^n$, $\mathrm{PN}(f) = (P_f, T_f, A_f, M_0)$ *is the corresponding Safe Petri Net such that:*

- $P_f = \bigcup_{i \in [n]} \{i_0, i_1\}$ *is the set of places,*
- T_f *and* A_f *are the smallest sets which satisfy, for each* $i \in [n]$,

- *for each clause $c \in \text{DNF}[\neg x_i \wedge f_i(x)]$, there is a transition $t_{i,c} \in T_f$ with A_f such that $^\bullet t_{i,c} = \{\text{place}(l) \mid l \in c\}$ and $t_{i,c}^\bullet = \{i_1\} \cup {}^\bullet t_{i,c} \setminus \{i_0\}$;*
- *for each clause $c \in \text{DNF}[x_i \wedge \neg f_i(x)]$, there is a transition $t_{\neg i,c} \in T_f$ with A_f such that $^\bullet t_{\neg i,c} = \{\text{place}(l) \mid l \in c\}$ and $t_{\neg i,c}^\bullet = \{i_0\} \cup {}^\bullet t_{\neg i,c} \setminus \{i_1\}$,*
- $M_0 = \{i_{x_i} \mid i \in [n]\}$ *is the initial marking.*

Note that [5] also extends the encoding to multi-valued networks into 1-bounded Petri nets (contrary to the encoding of multi-valued networks of [4] which does not result in a safe Petri net). For the sake of simplicity, we restrict the presentation to Boolean networks. However, our encoding of temporal perturbations can be easily extended to multi-valued networks.

Example 1. Figure 3 gives the resulting Petri net encoding of the Boolean function $f_3(x) = x_1 \wedge \neg x_2$. In this case, $\text{DNF}[\neg x_3 \wedge (x_1 \wedge \neg x_2)] = \{\{\neg x_3, x_1, \neg x_2\}\}$ and $\text{DNF}[x_3 \wedge (\neg x_1 \vee x_2)] = \{\{x_3, \neg x_1\}, \{x_3, x_2\}\}$.

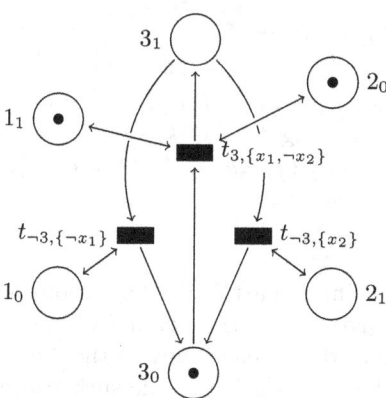

Fig. 3. Safe Petri net encoding of $f_3(x) = x_1 \wedge \neg x_2$. Places are drawn as circles and transitions as rectangles. Marked places have a dot.

3.2 Encoding Temporal Perturbations

Perturbations are modelled as additional transitions which modify the state of nodes of the BN f. These perturbations can be performed at any time during the transient dynamics, and independently of the current state of the network.

In the scope of this paper, we consider two kinds of perturbations: *temporary perturbations* induce a state change of nodes, but these nodes can later be updated according to their Boolean function. Such perturbations can model, for instance, the transient activation of transcription factor through a signalling pathway. *Permanent perturbations* induce a permanent state change of nodes. These perturbations model mutations (loss or gain of functions).

In both cases, we consider a limited amount of allowed perturbations: only up to k successive perturbations can be performed.

Temporary Perturbations. In addition to the places for the BN node values, we add k mutually exclusive places $c_1, \ldots, c_k$ and two mutually exclusive places p_0 and p_1. Essentially, c_j is marked if the next perturbation is the j-th; and p_0 is marked if the j-th perturbation is yet to be performed, and p_1 is marked if the j-th perturbation has been performed.

The transitions are the same as in $PN(f)$, with additional transitions $t_{i,0}$ and $t_{i,1}$ for each node $i \in [n]$ which set their value to 0 and 1 respectively. To be enabled, these transitions need p_0 to be marked, and after the transition, p_1 is marked. Finally, a transition t_{cj} re-enabling p_0 is defined for each c_j, $j \in [k-1]$, which moves the marking of c_j to c_{j+1}.

Definition 5. *Given a Boolean network f of dimension n, the Petri net (P, T, A, M_0) modelling its k temporary perturbations is given by*

- $P = P_f \cup \{p_0, p_1, c_1, \ldots, c_k\}$,
- T *and* A *are the smallest sets which satisfy*
  (a) **BN transitions** $T_f \subseteq T$, $A_f \subseteq A$;
  (b) **Perturbation transitions** *for* $i \in [n]$,
      $t_{i,0} \in T$ *with* $^\bullet t_{i,0} = \{i_1, p_0\}$ *and* $t_{i,0}^\bullet = \{i_0, p_1\}$
      $t_{i,1} \in T$ *with* $^\bullet t_{i,1} = \{i_0, p_0\}$ *and* $t_{i,1}^\bullet = \{i_1, p_1\}$;
  (c) **Perturbation enabling** *for* $j \in [k-1]$,
      $t_{cj} \in T$ *with* $^\bullet t_{cj} = \{p_1, c_j\}$ *and* $t_{cj}^\bullet = \{p_0, c_{j+1}\}$,
- $M_0 = \{i_{x_i} \mid i \in [n]\} \cup \{p_0, c_1\}$,

where $(P_f, T_f, A_f, M_0') = PN(f)$.

Example 2. Figure 4(top) shows part of the transitions added by the modelling of $k = 2$ temporary perturbations in the example of Fig. 3. In the given marking, the perturbation are enabled, therefore, any of the 3 shown perturbation transitions can be applied. The application of one such transition disable the other perturbation transitions (as p_0 is no longer marked). By applying the transition t_{c1}, the perturbations transitions are then re-enabled, allowing a second (and last) one to be applied.

Permanent Perturbations (mutations). Contrary to temporary perturbations, once a node has been (permanently) perturbed, its state should no longer change. This is modelled by *locks*: if the i-th lock is active the node i cannot perform any transition. In addition to the places introduced for temporary perturbations, our encoding add mutually exclusive places $lock_{i0}, lock_{i1}$ for each each node $i \in [n]$, $lock_{i0}$ being marked if the node i has not been perturbed, $lock_{i1}$ being marked otherwise.

The transitions of the BN are then modified so that a transition changing the state of node i requires the place $lock_{i0}$ to be marked. For each node i, 4 perturbations transitions are defined: two for the value changes (0 to 1 and 1 to 0) also inducing the marking of $lock_{i1}$; and two for the marking of $lock_{i1}$ without value change: indeed, a mutation does not necessarily have to change the current value of the node, but it prevents any further evolution of it.

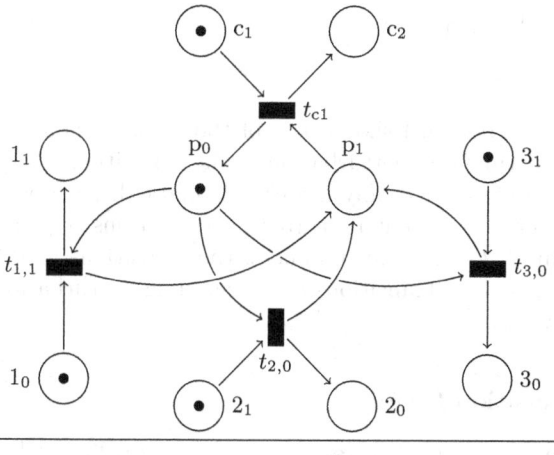

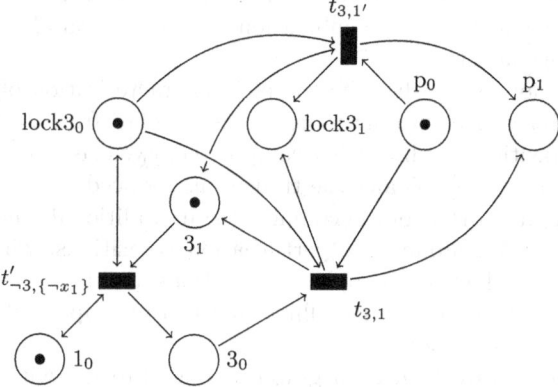

Fig. 4. (top) Excerpt of the encoding of temporary perturbations. (bottom) Excerpt of the encoding of permanent perturbations.

Definition 6. *Given a Boolean network* f *of dimension* n*, the Petri net* (P, T, A, M_0) *modelling its* k *permanent perturbations is given by*

- $P = P_f \cup \{\mathrm{p}_0, \mathrm{p}_1, \mathrm{c}_1, \ldots, \mathrm{c}_k\} \cup \bigcup_{i \in [n]} \{\mathrm{lock}i_0, \mathrm{lock}i_1\}$
- T *and* A *are the smallest sets which satisfy*

  BN transitions $\forall t_{l,c} \in T_f$, *with* $l = i$ *or* $l = \neg i$, $i \in [n]$,
  $\quad t'_{l,c} \in T$ *with* $^\bullet t'_{l,c} = {}^\bullet t_{l,c} \cup \{\mathrm{lock}i_0\}$ *and* $t'_{l,c}{}^\bullet = t_{l,c}{}^\bullet \cup \{\mathrm{lock}i_0\}$

  Perturbation transitions *for* $i \in [n]$,
  $\quad t_{i,0} \in T$ *with* $^\bullet t_{i,0} = \{i_1, \mathrm{p}_0, \mathrm{lock}i_0\}$ *and* $t_{i,0}{}^\bullet = \{i_0, \mathrm{p}_1, \mathrm{lock}i_1\}$
  $\quad t_{i,0'} \in T$ *with* $^\bullet t_{i,0'} = \{i_0, \mathrm{p}_0, \mathrm{lock}i_0\}$ *and* $t_{i,0'}{}^\bullet = \{i_0, \mathrm{p}_1, \mathrm{lock}i_1\}$
  $\quad t_{i,1} \in T$ *with* $^\bullet t_{i,1} = \{i_0, \mathrm{p}_0, \mathrm{lock}i_0\}$ *and* $t_{i,1}{}^\bullet = \{i_1, \mathrm{p}_1, \mathrm{lock}i_1\}$
  $\quad t_{i,1'} \in T$ *with* $^\bullet t_{i,1'} = \{i_1, \mathrm{p}_0, \mathrm{lock}i_0\}$ *and* $t_{i,1'}{}^\bullet = \{i_1, \mathrm{p}_1, \mathrm{lock}i_1\}$

  Perturbation enabling *for* $j \in [k-1]$,
  $\quad t_{cj} \in T$ *with* $^\bullet t_{cj} = \{\mathrm{p}_1, \mathrm{c}_j\}$ *and* $t_{cj}{}^\bullet = \{\mathrm{p}_0, \mathrm{c}_{j+1}\}$

- $M_0 = \{i_{x_i} \mid i \in [n]\} \cup \{p_0, c_1\}$

where $(P_f, T_f, A_f, M_0') = \mathrm{PN}(f)$.

Example 3. Figure 4(bottom) shows part of the transitions added by the modelling of $k = 2$ permanent perturbations. The transition $t_{3,\{\neg x_1\}}$ of Fig. 3 is modified so that it is enabled only if lock3$_0$ is marked, i.e., the node 3 has not been perturbed yet. Permanent perturbation transitions $t_{3,1}$ and $t_{3,1'}$ lock the node 3 to its value 1. Once applied, none of the transitions modifying the value of node 3 can be enabled. Transitions for re-enabling perturbations are identical to the temporary case.

3.3   State Transition Graph

Given a BN f and an initial state x, the above modelling allows to compute all the states reachable by any combination and succession of k perturbations, temporary or permanent.

The next section establishes the complete characterisation of perturbations for the existential and inevitable reprogramming of f from x. It relies on an explicit state transition graph which is composed of two classes of transitions: the transitions induced by BN f, and the transitions induced by the perturbation.

It can be remarked that our encoding uses an additional kind of transition: the transitions for re-enabling the perturbation transitions, when strictly less than k perturbations have been applied (transitions noted $t_{cj}, j \in [k-1]$). These transitions are artefacts of the modelling, and can be skipped during the state transition graph construction.

Let us define a state transition graph among states S with two classes of transitions $\mathcal{E}$ (induced by f) and $\mathcal{M}$ (induced by the perturbations), as the smallest digraph $(S, \mathcal{E}, \mathcal{M})$ such that $M_0 \in S$, and for each $M \in S$, for each $t \in T$ such that ${}^\bullet t \subseteq M$, let $M' = (M \setminus {}^\bullet t) \cup t^\bullet$,

- if $p_1 \in M$ and $c_k \notin M$, then $\exists j \in [k] : {}^\bullet t_{cj} \in M'$; let $M'' = (M' \setminus {}^\bullet t_{cj}) \cup t_{cj}{}^\bullet$, $M'' \in S$ and $(M, M'') \in E$,
- otherwise, $M' \in S$, and if $t = t_{l,c}$, then $(M, M') \in \mathcal{E}$, else $(M, M') \in \mathcal{M}$.

Given any marking $M \in S$ of the resulting state transition graph, the number of perturbations applied to reach M is given by $j + b$ where $c_j \in M$ and $p_b \in M$.

4   Complete Identification of Temporal Reprogramming Strategies

This section explains how, from the transition graph obtained by the model of Sect. 3, the complete set of reprogramming solutions leading to a set of final states $\mathcal{F} \subseteq S$ can be identified.

The *Perturbation Transition Graph* (Definition 7) gathers the transitions of the Boolean network from the state x, and the perturbation transitions with a label specifying the performed perturbation. Each node of the original transition graph have multiple copies, given how many perturbations are used to reach it: thus, a state of the perturbation transition graph is composed of the state of the Boolean network and a perturbation counter. A transition of the Boolean network is necessarily between two states with the same counter; a perturbation transition is necessarily between a state with counter i to a state with counter $i + 1$. This Perturbation Transition Graph can be directly computed from the Petri net of Sect. 3. Given a subset M of perturbation transitions, a *Perturbation Path* (Definition 8) is a sequence of Boolean networks transitions and transitions in M.

Definition 7 (Perturbed Transition Graph). *Given a Boolean network f of dimension n and a maximum number of allowed perturbations k, the Perturbed Transition Graph is a tuple $(\mathcal{S}_0, \mathcal{E}_0, \mathcal{M}_0)$ where*

- $\mathcal{S}_0 = \{0,1\}^n \times \{0, .., k\}$ *is the set of states;*
- $\mathcal{E}_0 \subseteq \{(s,i) \rightarrow (s',i) \mid i \in [0;k], (s \rightarrow s') \in \mathcal{E}_f\}$, *where* $\mathrm{STG}(f) = (\{0,1\}^n, \mathcal{E}_f)$, *is the set of* normal *transitions, which corresponds to a subset of the asynchronous transitions of the Boolean network f;*
- $\mathcal{M}_0 \subseteq \{(s,i) \rightarrow (s',i+1) \mid i \in [0;k-1], s, s' \in \{0,1\}^n\} \times L$ *is a set of perturbation transitions, where L is the set of labels describing the perturbation.*

Definition 8 (Perturbation path $(\rightarrow_M^*)$). *Given a Perturbation State Graph $(\mathcal{S}, \mathcal{E}, \mathcal{M})$ and a set of perturbation transitions $M \subseteq \mathcal{M}$, $\rightarrow_M^* \subseteq \mathcal{S} \times \mathcal{S}$ is a binary relation such that*

$$(s,i) \rightarrow_M^* (s',i') \overset{\Delta}{\Leftrightarrow} (s,i) = (s',i') \ or \ \exists (s'',i'') \in \mathcal{S} \ with$$
$$(s,i) \rightarrow (s'',i'') \in \mathcal{E} \cup M \ and \ (s'',i'') \rightarrow_M^* (s',i')$$

4.1   Complete Identification of Reprogramming Solutions

In the scope of this paper, we consider two classes of reprogramming solutions: the reprogramming solutions which build a path that reaches one of the final states, referred to as *existential reprogramming* (Definition 9); and the reprogramming solutions which ensure that a final state is always reached, referred to as *inevitable reprogramming* (Definition 10).

Definition 9 (Existential Reprogramming). *Given a Perturbation Transition Graph $(\mathcal{S}, \mathcal{E}, \mathcal{M})$ of a Boolean network f, a state $(s_0, i_0) \in \mathcal{S}$ has an existential reprogramming to a set of states $\mathcal{F} \subseteq \mathcal{S}$ if and only if there exists a set of perturbation transitions $M \subseteq \mathcal{M}$ such that there is a path from (s_0, i_0) to a state $w \in \mathcal{F}$ using only $\mathcal{E}$ and M transitions, i.e., $(s_0, i_0) \rightarrow_M^* w$.*

Definition 10 (Inevitable Reprogramming). *Given a Perturbation Transition Graph $(\mathcal{S}, \mathcal{E}, \mathcal{M})$ of a Boolean network f, a state $(s_0, i_0) \in \mathcal{S}$ has an inevitable reprogramming to a set of states $\mathcal{F} \subseteq \mathcal{S}$ if and only if there exists a set of perturbation transitions $M \subseteq \mathcal{M}$ such that from any state $(s,i) \in \mathcal{S}$*

reachable from (s_0, i_0) using $\mathcal{E}$ and M transitions, there exists a path from (s, i) to a state in $\mathcal{F}$ using $\mathcal{E}$ and M transitions: $\forall (s, i) \in \mathcal{S}$ with $(s_0, i_0) \rightarrow_M^* (s, i)$, $\exists w \in \mathcal{F}$ such that $(s, i) \rightarrow_M^* w$.

Given a reprogramming property, the set of nodes that verify it (called "valid nodes" below) can be computed iteratively, by browsing the transition graph in a reverse topological order of the strongly connected components. As a topological order is used, the complexity is linear in the number of states in the Perturbed Transition Graph.

It can be noted that all strongly connected components have the same value of perturbations counter, as there are no edges that decrease the counter. As a consequence, all edges between two strongly connected components are either only normal edges, or only perturbation edges.

In the following part, we only consider the condensed graph $G = (\mathcal{S}, \mathcal{E}, \mathcal{M})$ of the perturbed transition graph. For a graph $G_0 = (\mathcal{S}_0, \mathcal{E}_0, \mathcal{M}_0)$, the condensed transition graph G, is defined by:

- A set of strongly connected components of the states. $\forall u \in \mathcal{S}_0, \exists s \in \mathcal{S}, u \in s$. $\mathcal{S}$ is a partition of $\mathcal{S}_0$.
- A set of normal edges between the strongly connected components: $\mathcal{E} = \{((s, i) \rightarrow (s', i)) \mid s, s' \in \mathcal{S}, \exists s_0 \in s, s'_0 \in s' \text{ such that } ((s_0, i) \rightarrow (s'_0, i)) \in \mathcal{E}_0\}$
- A set of perturbation edges between the strongly connected components: $\mathcal{M} = \{((s, i) \xrightarrow{l} (s', i+1)) \mid s, s' \in \mathcal{S}, \exists s_0 \in s, s'_0 \in s' \text{ such that } ((s_0, i) \xrightarrow{l} (s'_0, i+1)) \in \mathcal{M}_0\}$

Given the construction of the graph, G is a Perturbed Transition Graph as well.

Existential Reprogramming: In the case of existential reprogramming, a node is valid if it is one of the final nodes or if it has an edge (a normal edge or a perturbation edge) that leads to a valid node.

Definition 11. *Given a Perturbation Transition Graph $(\mathcal{S}, \mathcal{E}, \mathcal{M})$, the set of valid nodes for existential reprogramming $V_E \subseteq \mathcal{S}$ is defined by:*
$V_E = \{(u, i) \in \mathcal{S} \mid \exists M \subseteq \mathcal{M}, \exists (v, j) \in \mathcal{F}, (u, i) \rightarrow_M^* (v, j)\}$

Inevitable Reprogramming: In the case of inevitable reprogramming, a valid node is either: (a) a final node, (b) a node from which all children through normal edges are valid nodes, or (c) a node that reaches a valid node through one perturbation edge.

Definition 12. *Given a Perturbation Transition Graph $(\mathcal{S}, \mathcal{E}, \mathcal{M})$, the set of valid nodes for inevitable reprogramming $V_I \subseteq \mathcal{S}$ is defined by:*
$V_I = \{(u, i) \in \mathcal{S} \mid \exists M \subseteq \mathcal{M}, \exists (v, j) \in \mathcal{F}, (u, i) \rightarrow_M^* (v, j) \text{ and } \forall (u', i') \in \mathcal{S} \text{ verifying } (u, i) \rightarrow_M^* (u', i'), \exists (v', j') \in \mathcal{F}, (u', i') \rightarrow_M^* (v', j')\}$

Validity of the Initial Node: If the initial node is not valid as defined above, then there is no reprogramming solution given the settings. Otherwise, there exist one or more paths that correspond to reprogramming solutions. This will be illustrated on examples from the literature in Sect. 5.

4.2 Example

Applied on the example of Sect. 2 for the inevitable reprogramming from 0000 to 1101 with $k = 2$, the algorithm returns the graph of Fig. 5, with nodes verifying the reprogramming property in black and the other ones in gray.

The temporal reprogramming path identified in Sect. 2 is the only strategy for inevitable reprogramming.

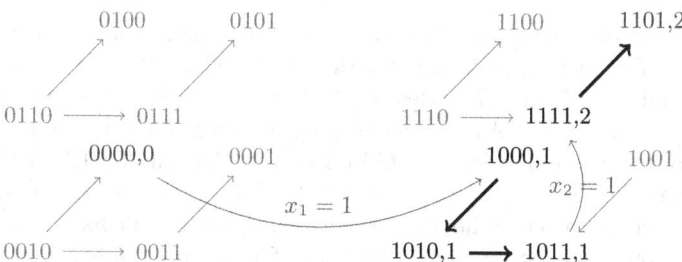

Fig. 5. The perturbation path returned by the algorithm on the example of Sect. 2

4.3 Initial Reprogramming Vs Temporal Reprogramming

In most other works, perturbations are performed only in the initial state. Our method allows finding temporal perturbations paths, which accounts for the transient dynamics of the system between the perturbations. We also capture perturbations of the sole initial state: they correspond to paths where all the first edges are perturbation edges, only followed by normal edges.

We consider that temporal reprogramming can return new reprogramming strategies when the perturbations act on different nodes than perturbations of the initial state only. Given the Perturbation Transition Graph, one can first compute the reprogramming solutions for the initial state, and then enumerate the perturbation paths that use different sets of perturbations.

5 Case Studies

5.1 Identifying Reprogramming Paths

The set of reprogramming paths can be summarized by the perturbations they involve and their ordering. These perturbations can be extracted from the valid node computation introduced in Sect. 4 as follows.

To each *valid* node $u \in V_E$ or V_I of the Peturbation Transition Graph, we associate a set S_u of sequences of perturbations, specified by the label of perturbation transitions. S_u gathers all possible perturbations to get from the node u to a final state in $\mathcal{F}$.

If $u \in \mathcal{F}$, $S_u = \{\emptyset\}$, i.e., no perturbation is necessary. Otherwise, S_u consists of the union of S_v for every children v where $(u \rightarrow v) \in \mathcal{E}$ and of the union of $\{l \oplus s \mid s \in S_v\}$ for every children v where $u \xrightarrow{l} v \in \mathcal{M}$, and $l \oplus s$ is the sequence starting with l and followed by s.

To get a minimal set of temporal perturbations, every perturbation sequence that is equal or a superset of initial perturbations are removed, and only the smallest sub-sequences (in terms of sequence inclusion) are kept.

5.2   T-Helper Cells

We applied a prototype implementation of our algorithm[1] on the model of the multi-valued T helper regulatory network introduced in [12].

The initial model has 17 nodes, with 2 or 3 possible values for each. We applied the identification of *inevitable* reprogramming of the initial state where all the nodes are inactive, except GATA3, IL4, IL4R and STAT6 that have an initial value of 1, to any attractor where Tbet is active, using at most 2 *permanent* perturbations. The Perturbed Transition Graph has 21,647 nodes, and 20,941 connected components. The set of temporal reprogramming paths uses the following perturbations:

– IFNg $= 2$, then, after several transitions, IFNgR $= 0$
– IFNg $= 2$, then, after several transitions, STAT1 $= 0$
– IFNgR $= 2$, then, after several transitions, STAT1 $= 0$

The graph in Fig. 6 gives an example of a possible perturbation path that uses INFg $= 2$ and STAT1 $= 0$:

From the initial state, a permanent perturbation (INFg $= 2$) is performed. The new perturbed state, 1, has several possible futures, one of which leads to the state 4 in the graph. From this state, the system can continue to follow its usual dynamics, or can be perturbed again with STAT1 $= 1$ to go to the state 5, that will always reach the final state. It can be seen that there are branching paths: our method guarantees that from each reachable node there is perturbation path leading to the final state, using one the three perturbation paths given above.

If one applies these perturbations (IFNg $= 2$ and STAT $= 1$) directly in the initial state, the attractor where Tbet is active is not reachable. Therefore, this perturbation path gives a new reprogramming strategy. Moreover, the temporal reprogramming solutions returned by our method are complete.

---

[1] Scripts and models available at http://www.lsv.fr/~mandon/CMSB2017.zip.

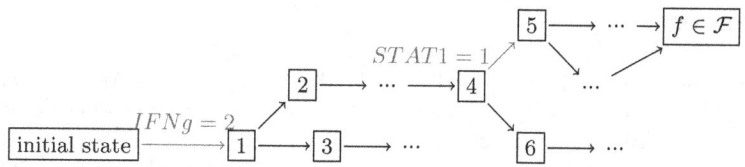

Fig. 6. Simplification of a perturbation path for T-helper cells

5.3 Cardiac Gene Regulatory Network

The same algorithm has been applied to the Boolean model of the cardiac gene regulatory network built in [11]. The Boolean network has 15 nodes. Its Perturbed Transition Graph with at most 3 permanent perturbations has around 60,000 reachable states.

In this example, we computed the fixpoints of the Boolean network and identified reprogramming solutions to change from one fixpoint to another.

For some cases, we observe that temporal reprogramming provides solutions requiring only two perturbations when at least three perturbations are required when applied only in the initial state.

For instance, let us consider the inevitable reprogramming from the fixpoint where all nodes are active except Bmp2, Fgf8, Tbx5, exogen_BMP2_I, and exogen_BMP2_II to the fixpoint where all nodes are inactive but Bmp2, exogen_BMP2_I, and exogen_BMP2_II. Our method identifies 1 set of 3 perturbations to apply in the initial state; and 14 sequences of temporal perturbations, one of which requires only 2 perturbations (the loss of function of exogen_CanWnt_I, followed later by the gain of function of exogen_BMP2_I).

6 Discussion

Temporal reprogramming consists in applying perturbations in a specific order and in specific states of the system to trigger and control an attractor change.

This paper establishes the complete characterization of temporal perturbations for Boolean networks reprogramming. Perturbations can be applied at the initial state, and during the transient dynamics of the system. This later feature allows to identify new strategies to reprogram regulatory networks, by providing solutions with different targets and possibly requiring less perturbations than when applied only in the initial state.

Our method relies on a Petri net modelling the combination of Boolean network asynchronous transitions with perturbation transitions. The identification of temporal reprogramming solutions then relies on a explicit exploration of the resulting state transition graph. Our framework can handle temporary (e.g., through signalling) and permanent (e.g., mutations) perturbations for the existential and inevitable reprogramming to the targeted state.

Future work will focus on increasing the scalability of temporal reprogramming predictions. Notably, we aim at using partial order exploration and unfolding of the Petri net model in order to exploit the concurrency of transitions.

References

1. Abou-Jaoudé, W., Monteiro, P.T., Naldi, A., Grandclaudon, M., Soumelis, V., Chaouiya, C., Thieffry, D.: Model checking to assess t-helper cell plasticity. Frontiers Bioeng. Biotechnol. **2** (2015)
2. Bernardinello, L., De Cindio, F.: A survey of basic net models and modular net classes. In: Rozenberg, G. (ed.) Advances in Petri Nets 1992. LNCS, vol. 609, pp. 304–351. Springer, Heidelberg (1992). doi:10.1007/3-540-55610-9_177
3. Chang, R., Shoemaker, R., Wang, W.: Systematic search for recipes to generate induced pluripotent stem cells. PLoS Comput. Biol. **7**(12), e1002300 (2011)
4. Chaouiya, C., Naldi, A., Remy, E., Thieffry, D.: Petri net representation of multi-valued logical regulatory graphs. Nat. Comput. **10**(2), 727–750 (2011)
5. Chatain, T., Haar, S., Jezequel, L., Paulevé, L., Schwoon, S.: Characterization of reachable attractors using petri net unfoldings. In: Mendes, P., Dada, J.O., Smallbone, K. (eds.) CMSB 2014. LNCS, vol. 8859, pp. 129–142. Springer, Cham (2014). doi:10.1007/978-3-319-12982-2_10
6. Cohen, D.P.A., Martignetti, L., Robine, S., Barillot, E., Zinovyev, A., Calzone, L.: Mathematical modelling of molecular pathways enabling tumour cell invasion and migration. PLOS Comput. Biol. **11**(11), 1–29 (2015)
7. Crespo, I., Perumal, T.M., Jurkowski, W., del Sol, A.: Detecting cellular reprogramming determinants by differential stability analysis of gene regulatory networks. BMC Syst. Biol. **7**(1), 140 (2013)
8. del Sol, A., Buckley, N.J.: Concise review: a population shift view of cellular reprogramming. Stem Cells **32**(6), 1367–1372 (2014)
9. Graf, T., Enver, T.: Forcing cells to change lineages. Nature **462**(7273), 587–594 (2009)
10. Hannam, R., Annibale, A., Kuehn, R.: Cell reprogramming modelled as transitions in a hierarchy of cell cycles. ArXiv e-prints, December 2016
11. Herrmann, F., Groß, A., Zhou, D., Kestler, H.A., Kühl, M.: A boolean model of the cardiac gene regulatory network determining first, second heart field identity. PLOS ONE **7**(10), 1–10 (2012)
12. Mendoza, L.: A network model for the control of the differentiation process in th cells. Biosystems **84**(2), 101–114 (2006). Dynamical Modeling of Biological Regulatory Networks
13. Morris, M.K., Saez-Rodriguez, J., Sorger, P.K., Lauffenburger, D.A.: Logic-based models for the analysis of cell signaling networks. Biochemistry **49**(15), 3216–3224 (2010). PMID: 20225868
14. Ronquist, S., Patterson, G., Brown, M., Chen, H., Bloch, A., Muir, L., Brockett, R., Rajapakse, I.: An Algorithm for Cellular Reprogramming. ArXiv e-prints, March 2017
15. Sahin, O., Frohlich, H., Lobke, C., Korf, U., Burmester, S., Majety, M., Mattern, J., Schupp, I., Chaouiya, C., Thieffry, D., Poustka, A., Wiemann, S., Beissbarth, T., Arlt, D.: Modeling ERBB receptor-regulated G1/S transition to find novel targets for de novo trastuzumab resistance. BMC Syst. Biol. **3**(1), 1–20 (2009)
16. Samaga, R., Von Kamp, A., Klamt, S.: Computing combinatorial intervention strategies and failure modes in signaling networks. J. Comput. Biol. **17**(1), 39–53 (2010)
17. Takahashi, K., Yamanaka, S.: A decade of transcription factor-mediated reprogramming to pluripotency. Nat. Rev. Mol. Cell Biol. **17**(3), 183–193 (2016)

18. Wang, R.-S., Saadatpour, A., Albert, R.: Boolean modeling in systems biology: an overview of methodology and applications. Phys. Biol. **9**(5), 055001 (2012)
19. Zañudo, J.G.T., Albert, R.: Cell fate reprogramming by control of intracellular network dynamics. PLOS Comput. Biol. **11**(4), 1–24 (2015)

Detecting Toxicity Pathways with a Formal Framework Based on Equilibrium Changes

Benjamin Miraglio[1]([envelope]), Gilles Bernot[1], Jean-Paul Comet[1],
and Christine Risso-de Faverney[2]

[1] Université Côte d'Azur, CNRS, I3S, Sophia Antipolis, France
{miraglio,bernot,comet}@unice.fr
[2] Université Côte d'Azur, CNRS, ECOMERS, Nice, France

Abstract. Toxicology aims at studying the adverse effects of exogenous chemicals on organisms. As these effects mainly concern metabolic pathways, reasoning about toxicity would involve metabolism modeling approaches. Usually, metabolic network models approaches are rule-based and describe chemical reactions, indirectly depicting equilibria as results of competing rule kinetics. By altering these kinetics, an exogenous compound can shift the system equilibria and induce toxicity. As equilibria are kept implicit, the identification of possible toxicity pathways is hindered as they require a fine understanding of chemical reactions dynamics to infer possible equilibria disruptions. Paradoxically, the toxicity pathways are based on a succession of very abstract (coarse grained) events. To reduce this mismatch, we propose a more abstract framework making equilibria first-class citizens. Our rules describe qualitative equilibrium changes and the chaining of rules is controlled by constraints expressed in extended temporal logic. This higher abstraction level fosters the detection of toxicity pathways, as we will show through an example of endocrine disruption of the thyroid hormone system.

Keywords: Discrete dynamic systems · Rule-based modeling · Temporal logic · Computational toxicology

1 Introduction

The purpose of toxicology is to study the adverse effects caused by chemical substances on living organisms. In this perspective, the central paradigm of the discipline assumes that the more an organism is exposed to a compound, the greater the effects of this compound will be.

This dose-response relationship underpins toxicity studies, where toxicologists aim at determining the threshold of toxicity of a compound (*i.e.* the lowest exposure from which an induced toxicity is observable). These studies also aim at identifying how a chemical disrupts physiological equilibria, and how these disruptions propagate in an organism, linking the exposure to a chemical to its observable toxicity. This causal chain of equilibrium changes, also known as

© Springer International Publishing AG 2017
J. Feret and H. Koeppl (Eds.): CMSB 2017, LNBI 10545, pp. 196–213, 2017.
DOI: 10.1007/978-3-319-67471-1_12

pathway of toxicity, is widely used by regulating authorities to assess the toxicity of a compound.

Indeed, as our exposure to chemical products is becoming an area of great concern for society, authorities are implementing increasingly strict regulations. As a consequence, chemical manufacturers must now conduct extensive toxicity studies to demonstrate the innocuousness of their products, skyrocketing the development cost of such products.

Related works. This context provides ground for modeling toxicity, and so far, most of these modeling approaches are quantitative [9]. They aim at either inferring the toxic threshold of a chemical substance or confirming its specific pathway of toxicity. These objectives require a lot of biological data, which can be restrictive given the current acquisition cost of such data. An alternative approach consists in shifting the focus from toxic thresholds to toxicity pathways. Indeed, describing these pathways in a *qualitative* manner would allow to focus only on equilibrium changes and would therefore require comparatively less biological data. Moreover, such an approach would allow to use automated reasoning tools.

Several generic formalisms have already been developed to qualitatively model biological processes [3,5,15,16,19]. These formalisms use formal methods to reason about these standard processes. However, expressing toxicology problems in terms manageable for the formalism is frequently troublesome. Several specificities of toxicology make these environments not optimal. As an example, Biocham [7] is based on rules able to qualitatively model many biochemical processes thanks to either Boolean or discrete semantics. The transformation of A into B thanks to the catalyst C can for instance be written A =[C]=> B. However, this formalism does not allow to express intuitively the possibility for this process to be further enhanced by an entity E, or conversely, to be stopped by the presence of an inhibitor I, two very common situations in toxicology.

In addition, these formalisms describe chemical reactions, only depicting equilibria as indirect results of competing rule kinetics. Yet, toxicity pathways are sequences of equilibrium changes. As such, keeping equilibria implicit while building a toxicological model can thus prove confusing for toxicologists, hindering the identification of possible toxicity pathways. In this mind, several aspects of automata networks and René Thomas' theory, especially its asynchronicity or the continuity of its variables, fit nicely with toxicology. However, it is common in toxicology to see cases where two entities A and B affect the level of a third one. This influence is classically linked to the concentration of both A and B, with the less concentrated entity *limiting the influence of both entities*. This concept is actually poorly handled by René Thomas' formalism, such cases leading to an explosion in the number of model parameters.

A two layers formalism. To solve these limitations, we present in this article a domain-oriented formalism directly describing qualitative equilibrium changes thanks to two layers. First, a rule-based language allows to express the different equilibrium changes present in a biological system. Then, the chaining of rules can be corseted thanks to constraints expressed in an extended temporal logic. These constraints are usually based on toxicological observations regarding specific conditions of the system. Finally, automated reasoning tools can be used on the resulting dynamics to detect possible toxicity pathways, providing useful insights to improve the experimental strategies of toxicity studies.

As our formalism is presented alongside examples inspired from the thyroid hormone system, the next section sketches an overview of this system. In Sect. 3, we explain how to use the new formalism to describe the equilibrium changes of a system. In Sect. 4, we show how to integrate toxicological knowledge in the system using an extended temporal logic. Finally, this formalism is applied to a model of the thyroid hormone system in Sect. 5.

2   The Thyroid Hormone System in a Nutshell

The thyroid hormone system plays a crucial role in the organism homeostasis. For example, alteration of thyroid hormones (TH) levels leads to troubles in the energy metabolism and in the adaptive thermogenesis in adults. This crucial role is even further highlighted during the organism development, where a slight disruption of the thyroid hormone homeostasis can lead to severe adverse effects such as neuronal defects, deafness or impaired bone and muscle formation [22,23].

Consequently, as most endocrine systems, the thyroid hormone homeostasis is maintained by a complex regulation network involving a central control carried out by cerebral regions. However, this regulation is unusually strengthened peripherally by dedicated enzymes, the *deiodinases*. Indeed, contrarily to most endocrine systems, the blood circulating form of TH, tetraiodothyronine (T_4) is inactive and must be 5'-deiodinated into triiodothyronine (T_3) to act on its target receptors.

Another metabolite of T_4, reverse triiodothyronine (rT_3), can be obtained through 5-deiodation. Similarly to T_4, rT_3 is not able to activate thyroid hormone receptors and is thus considered to be inactive. It should be noted that recent experiments suggest that both T_4 and rT_3 have other biological activities [14]. However, these actions need further investigations and will not be developed through this article.

TH Synthesis. As deiodinases are also present in the thyroid gland, the gland produces both thyroid hormone forms (T_3 and T_4). However, T_4 still accounts for roughly 90% of the gland production [11]. The synthesis process in itself starts with thyroid follicular cells extracting large quantities of iodide from the blood. This import is carried out thanks to dedicated iodide transporters. Thyroid iodide is then used by an enzyme, thyroid peroxidase (TPO), to assemble TH in the follicle [10]. Finally, TH are released in the blood, where they are associated with transporter proteins, as neither T_3 nor T_4 are soluble in water.

Central regulation. The thyroid hormone synthesis, from iodide uptake to TPO activity, can be stimulated by thyroid-stimulating hormone (TSH) [12]. TSH synthesis is performed in the pituitary gland when triggered by thyrotropin-releasing hormone (TRH), itself produced in the hypothalamus [23]. Both TSH and TRH synthesis are down-regulated by high concentrations of T_3, creating a negative feedback loop described as the thyroid hormone *central regulation*, or hypothalamo-pituitary-thyroid axis (HPT axis, see Fig. 1).

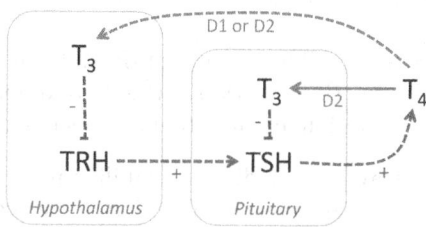

Fig. 1. Representation of the HPT axis integrating deiodinases action. Plain arrows show the deiodination of T_4 in T_3 by D_1 or D_2. Dashed arrows represent positive or negative regulations.

Peripheral regulation. Only a minor part of circulating T_3 is synthesized by the thyroid gland, the remaining part is produced by deiodinases directly in tissues sensitive to thyroid hormone [11,22]. The activation *in situ* of T_4 places deiodinases as key actors in thyroid hormone level regulation. This regulation is even more fine tuned thanks to three types of deiodinases, performing either 5'- or 5-deiodations, respectively activating or inactivating TH.

Type 3 deiodinase (D_3) is the main TH inactivator [6]. By catalyzing 5-deiodations, D_3 converts T_4 in rT_3 and T_3 in 3,3'-diiodothyronine, both inactive compounds. D_3 physiological role is to protect tissues from a local hyperthyroidism. As such, high concentrations of T_3 increase D_3 activity and conversely, the activity of the enzyme is reduced in hypothyroidism conditions.

Diametrically opposed to D_3, type 2 deiodinase (D_2) is the main TH activator [22]. D_2 catalyzes the 5'-deiodination of T_4 into T_3 and is down-regulated by T_3. As such, hyperthyroidism inhibits D_2 while low levels of T_3 increases D_2 activity. Interestingly enough, D_2 also plays a crucial role in the HPT axis (see Fig. 1). Indeed, D_2 is required to transform T_4 into T_3 in the pituitary gland [22], making D_2 necessary to complete the negative feedback of T_3 on TSH production.

Finally, type 1 deiodinase (D_1) has several roles. This enzyme is able to catalyze both 5- and 5'-deiodations but is extremely inefficient when compared to D_2 or D_3 [22]. Despite this inefficiency, D_1 is able to mitigate the effects of the absence of D_2 by converting enough T_4 into T_3, preventing any global hypothyroidism. On top of this, D_1 primary role actually concerns iodine recycling [11]. Indeed, D_1 is highly affine with sulfated TH (*i.e.* hormones about to be eliminated, see next paragraph). As the thyroid hormone system is extremely dependent on iodine intake, D_1 role is then to recycle as much iodine as possible from sulfated TH before their excretion.

TH Metabolism. TH metabolism is mainly carried out by hepatic enzymes. These enzymes are referred to as detoxifying enzymes since they are apt to inactivate a vast range of compounds (either exogenous or endogenous). This inactivation involves the conjugation of the compound with a specific residue, marking the compound for excretion. For instance, the action of the hepatic enzyme sulfotransferase results in sulfated TH [21].

Possible endocrine disruptions. The synergy of the different mechanisms evoked previously provides a clockwork regulation of the thyroid hormone system. However, several weak points can hinder this complex machinery:

1. TH synthesis relies heavily on iodide availability in the thyroid gland. An interruption of iodide intake, or the malfunction of the dedicated iodide transporters can then lead to severe hypothyroidism. Such effects can also result from an impaired TPO activity in the thyroid follicle [23].
2. Disruption in deiodinases activity also leads to troubles, but not necessarily as expected at first sight. Indeed, if considering only thyroid hormone levels, the absence of one of the activating deiodinase (D_1 or D_2) can be counterbalanced by the remaining activating deiodinase. However, D_2 is unable to recycle iodide efficiently. The absence of D_1 can thus lead to a iodide shortage [11]. Conversely, D_2 key role in the pituitary gland cannot be matched by D_1. An absence of D_2 thus leads to local hypothyroidism in the pituitary gland, leading to an unnecessary overproduction of TSH, finally resulting in an global hyperthyroidism [17].
3. The presence of some exogenous compounds in the organism can trigger a dramatic increase of hepatic detoxifying enzymes. This augmentation helps the organism to address the irregular presence of exogenous compounds, but also abnormally increases TH disposal, leading to global hypothyroidism [1]. The hypothyroidism is combined with excessive TSH levels, as the organism tries to counterbalance the lack of TH. In turn, the excess in TSH overstimulates the thyroid gland and can lead to the formation of tumors [8]. This sequence of events - from the presence of an exogenous compound to the apparition of a tumor in the thyroid gland - constitutes the well-defined liver-mediated thyroid toxicity pathway [13].

3   Describing Equilibrium Changes with Transformation Rules

A biological system can be abstracted as a set of biological entities interacting with each other at different concentrations. In parallel, each entity has a concentration regarded as normal in a given organism. This concentration tends to be maintained in normal conditions, and a modification of this concentration can lead to adverse effects. For instance, the normal blood concentration of glucose is about 1 g/L in an adult human, and a concentration greater than 1.3 g/L can lead to several complications.

Consequently, our domain-oriented formalism represent the evolution of the concentration of each entity as a change in its equilibrium level. In that line, we introduce four qualitative equilibrium levels depicting increasing concentrations of an entity:

- ε stands for a negligible concentration (*i.e.* a concentration too low to trigger any reaction in the biological system).
- ι stands for an abnormally low concentration (*i.e.* a relative lack of this entity, affecting some mechanisms in the biological system).
- Δ stands for a normal concentration.
- θ stands for an abnormally high concentration (*i.e.* an excess of this entity).

Notation 1 (Concentration levels). *We note* $\mathbb{L}$ *the set* $\{\varepsilon, \iota, \Delta, \theta\}$ *equipped with the total order relation such that:* $\varepsilon < \iota < \Delta < \theta$. *The elements of* $\mathbb{L}$ *are called* concentration levels.

In a given biological system and depending on the studied issue, not all levels are regarded as useful. For example, the modeler may be only interested in the normal (Δ) or excessive (θ) presence of an entity. Therefore, an entity must have at least two levels, but not necessarily more. The signature of a biological system allows the definition of the set of biological entities considered in the system and, for each entity, its admissible concentration levels.

Definition 1 (Signature). *A signature is a map* $\mathcal{E} : E \rightarrow \mathcal{P}(\mathbb{L})$ *where* E *is a finite set and for all* $e \in E$, $|\mathcal{E}(e)| \geqslant 2$. *Elements of* E *are called* entities *and for each entity* e, $\mathcal{E}(e)$ *is the set of* admissible levels *of* e.

For instance, $E = \{T_3, T_4, TPO, I\}$ can be the signature of a thyroid model, with $\mathcal{E}(T_3) = \{\varepsilon, \iota, \Delta, \theta\}$, $\mathcal{E}(T_4) = \{\varepsilon, \iota, \Delta, \theta\}$, $\mathcal{E}(TPO) = \{\varepsilon, \iota, \Delta, \theta\}$ and $\mathcal{E}(I) = \{\varepsilon, \Delta, \theta\}$.

After defining the system signature, a state of the system is defined as the qualitative level of each entity present in the system. For example, a state η_0 where T_3 is at the level Δ, noted $\eta_0(T_3) = \Delta$ and where $\eta_0(T_4) = \varepsilon$, $\eta_0(TPO) = \iota$ and $\eta_0(I) = \theta$. This state can also be written:

$$\eta_0 = (\Delta, \varepsilon, \iota, \theta) \tag{1}$$

where the entities order is (T_3, T_4, TPO, I).

Definition 2 (State). *A signature* $\mathcal{E}$ *being given, the set of states* ζ *is the set of functions* $\eta : E \rightarrow \mathbb{L}$ *such that for all* $e \in E$, $\eta(e) \in \mathcal{E}(e)$.

In our formalism, the evolution of an entity can follow two functions: the incrementation, *incr*, and the decrementation, *decr*. They return the level of this entity just above (resp. below) its current level. For instance, as $\mathcal{E}(TPO) = \{\varepsilon, \Delta, \theta\}$, $incr_{TPO}(\Delta) = \theta$ and $decr_{TPO}(\Delta) = \varepsilon$. Note that the incrementation (resp. decrementation) function is not defined on the maximal (resp. minimal) admissible levels. As such, $incr_{TPO}(\eta(TPO))$ is not defined if $\eta(TPO) = \theta$.

Besides these functions, the formalism also makes use of formulas to describe properties about the entities concentration levels.

Definition 3 (Formula). *The set $\mathcal{F}$ of formulas on a signature $\mathcal{E}$ is inductively defined by:*

- *any atomic formula of the form $a \leqslant b$ (where a and b can be any element of $E \cup \mathbb{L}$) belongs to $\mathcal{F}$.*
- *if φ and ψ are elements of $\mathcal{F}$, then $\neg\varphi$, $\varphi \wedge \psi$, $\varphi \vee \psi$, $\varphi \Rightarrow \psi$ are also elements of $\mathcal{F}$.*

Definition 4 (Satisfaction relation). *A state η and a formula $\varphi \in \mathcal{F}$ on a signature $\mathcal{E}$ being given, the satisfaction relation $\eta \vDash \varphi$ is inductively defined by:*

- *if φ is an atom of the form $a \leqslant b$, then $\eta \vDash \varphi$ if and only if $\overline{\eta}(a) \leqslant \overline{\eta}(b)$ where $\overline{\eta}$ is the extension of η to $E \cup \mathbb{L}$ by the identity on $\mathbb{L}$.*
- *if φ is of the form $\varphi_1 \wedge \varphi_2$ then $\eta \vDash (\varphi_1 \wedge \varphi_2)$ if and only if $\eta \vDash \varphi_1$ and $\eta \vDash \varphi_2$. We proceed similarly for the other connectives.*

"$\eta \vDash \varphi$" is read "η satisfies φ."

We use the abbreviation $a = b$ as a shortcut for $(a \leqslant b) \wedge (b \leqslant a)$ and we proceed similarly for $a < b$, $a > b$ and $a \geqslant b$.

Examples of formulas can be $\varphi \equiv (\mathrm{I} = \theta)$, stating an excessive presence of I or $\psi \equiv (\mathrm{T}_4 > \mathrm{TPO})$, stating that the qualitative level of T_4 is strictly greater than the one of TPO. The state η_0, previously described in Eq. 1, satisfies φ but not ψ.

To describe possible evolutions of the system, a set of rules of the following form is then used:

$$r : A_1 + \cdots + A_m \Rightarrow A_{m+1} + \cdots + A_n \quad when(\varphi) \quad boost(\psi)$$

Beside its identifier r, each rule includes two sets of entities A_i. The first one, for all i in $[1, m]$, constitutes the set of *consumables*, whose level may be reduced by the application of the rule. The other set, for all i in $[m + 1, n]$, represents the set of *produceables* whose level may be increased by the application of the rule. A rule also includes two modulating conditions $when(\varphi)$ and $boost(\psi)$ (φ and ψ being formulas). Intuitively, φ the role of the guard of the rule and ψ will relax some restriction on the increasing of produceable levels.

Definition 5 (Biological action network). *A biological action network on a signature $\mathcal{E}$, or $\mathcal{E}$-action network, is a set R of rules of the form:*

$$r : A_1 + \cdots + A_m \Rightarrow A_{m+1} + \cdots + A_n \quad when(\varphi) \quad boost(\psi)$$

where:

- *r is an identifier such that there are not two rules in N with the same r.*
- *$\forall i = 1 \ldots n, A_i \in E$.*
- *$\{A_1 \ldots A_m\} \cap \{A_{m+1} \ldots A_n\} = \varnothing$.*
- *φ and ψ are elements of $\mathcal{F}$.*

For short, we will call such rules $\mathcal{E}$-rules and we will call a "state of R" a state on the signature of R.

Let us emphasize that a rule represents *possible equilibrium changes*. Therefore, it makes no sense to have an entity being part of both consumables and produceables of a same rule.

Moreover, a rule can be devoid of any consumable or produceable: In the previous definition, the index m can be equal to zero (the rule does not need any consumable from the signature $\mathcal{E}$) or m can be equal to n (the rule has no produceable from the signature $\mathcal{E}$). A rule without consumable can be considered as a constitutive production of an entity in a given model and a rule without produceable can be interpreted as a constitutive depletion of an entity. In either cases, conventionally, the empty set of entities is denoted Ω, depicting the biological system in the broad sense, outside the signature.

Also, if no modulation is known for a given rule, *when* and *boost* regulations are not displayed in the rule representation, *i.e. when*(True) and *boost*(False) are left implicit.

It is worth mentioning that despite the obvious syntactic resemblance between a rule and a chemical reaction, a rule must *not* be interpreted as quanta of consumables converted into quanta of produceables but as a *possible evolution* of the levels of entities present in the rule, representing possible equilibrium shifts.

As a basic example of rule, the production of T_3 and T_4 from I can be represented by the following rule:

$$r_A : I \Rightarrow T_3 + T_4 \ \ when(\text{TPO} > \varepsilon)$$

In order to be applicable at a given state, a rule must meet basic criteria. First, since the level ε is interpreted as a negligible concentration, a rule is applicable only if all its consumables are present at least at the level ι. In addition, a rule cannot be applied if the formula φ of the modulating condition *when* is not satisfied.

Definition 6 (Applicable rule). *Let us consider a state η on a signature $\mathcal{E}$. An $\mathcal{E}$-rule $r \in R$ of the form:*

$$r : A_1 + \cdots + A_m \Rightarrow A_{m+1} + \cdots + A_n \ \ when(\varphi) \ \ boost(\psi)$$

is said applicable at the state η *if and only if:*

- $\forall i = 1 \ldots m, \ \eta(A_i) \neq \varepsilon.$
- $\eta \vDash \varphi.$

For instance, the rule r_A is applicable if and only if the levels of I and TPO are strictly greater than ε. By the way, note that the catalysis, namely the necessary presence of an enzyme to the proper conduct of a reaction, can be expressed using the *when* condition as in the previous example, but the catalyst cannot be present both on the left and right parts of the rule.

Definition 7 (Potential next level). *Let R be an $\mathcal{E}$-action network, let η be a state of R and r be a rule of R of the form:*

$$r : A_1 + \cdots + A_m \Rightarrow A_{m+1} + \cdots + A_n \ \ when(\varphi) \ \ boost(\psi)$$

We note $\eta_r^{\rhd} : E \to \mathbb{L}$ the partial function such that $\eta_r^{\rhd}(e)$ is defined if and only if r is applicable and one of the following conditions is satisfied:

- *$e \in \{A_1 \ldots A_m\}$ and $\eta_r^{\rhd}(e) = decr_e(\eta(e))$.*
- *$e \in \{A_{m+1} \ldots A_n\}$, $\eta(e) < max(\mathcal{E}(e))$, and:*
  - *if $\eta \not\models \psi$ and $\eta(e) < \min\limits_{i \in \{1...m\}} (\eta(A_i))$ then $\eta_r^{\rhd}(e) = incr_e(\eta(e))$*
  - *if $\eta \models \psi$ then $\eta_r^{\rhd}(e) = incr_e(\eta(e))$.*

Where conventionally, $\min\limits_{i \in \varnothing}(\eta(A_i)) = \eta(\Omega) = \Delta$.

If the entity A_i acts as a consumable, its potential next level is the one returned by the decrementation function.

If it acts as a produceable, its potential next level depends on the *boost* statement:

- if the *boost* statement ψ is not satisfied, a produceable level can increase *only if all the consumables levels are strictly greater*. In this case, the potential next level of a produceable is thus the one returned by the incrementation function applied to the produceable.
- if the *boost* statement ψ is satisfied, the previous restriction no longer applies. In such cases, the potential next level of a produceable is returned by the incrementation function applied to it, independently of the consumable levels. These levels must still be greater than ε, as the rule is applicable.

So, in the case of a rule deprived of consumables, produceables levels cannot exceed Δ unless the *boost* statement is satisfied.

Moreover, let us note that the potential next level is returned either by the incrementation or decrementation function. Therefore, when these functions are not defined, the potential next level of an entity is also not defined.

Keeping the synthesis of T_3 and T_4 as an example, we can also specify that an excess of TPO can cause trouble in T_3 and T_4 levels by adding a *boost* condition to the rule r_A:

$$r_B : I \Rightarrow T_3 + T_4 \ \ when(TPO > \varepsilon) \ \ boost(TPO > \Delta)$$

Here, assuming that the rule is applicable at the state η_0 and that $\eta_0(T_3) = \Delta$, the potential next level of T_3 by this rule can be θ only if $\eta_0(I) = \theta$ or if $\eta_0(TPO) = \theta$.

The dynamics is fully asynchronous. Among all the applicable rules at a given state, at most one is applied at a time. When a rule is applied, one and only one of its entities sees its level changing to its potential next level. Similar ideas have been firstly developed for discrete gene models by Thomas and Snoussi [18, 20]. This behavior reflects the possibility for an entity to cross a threshold without all the other entities levels doing likewise.

In brief, starting from a given state, it is possible to determine which rules of the system are applicable at that state. The application of one of these rules is not required but if so, it changes the level of one entity. It is possible to stay indefinitely at a same system state thanks to a special transition called Id (whose application does not change the levels of the system entities and that is always applicable).

It is then possible to establish a transition graph, mapping all the possible transitions between the states of a system. An infinite succession of transitions such that the output state of a transition is the input state of the next one is here called a path of the transition graph.

Definition 8 (Transition graph). *The* transition graph *of an $\mathcal{E}$-action network R is the labeled graph whose set of vertices is the set of states ζ and the set of edges T is the set of transitions of the form $\eta \xrightarrow{r} \eta'$ such that one of the following condition is satisfied:*

- *$r = Id$ and $\eta' = \eta$*
- *$r \in R$ and there exists an entity $e \in E$ such that $\eta_r^{\triangleright}(e)$ is defined and:*
  - *$\eta'(e) = \eta_r^{\triangleright}(e)$*
  - *$\forall\, e' \in E \setminus \{e\},\ \eta'(e') = \eta(e')$.*

So, the transition graph of an $\mathcal{E}$-action network R canonically defines a labeled Kripke structure $L = (\mathcal{L}, \Sigma, T)$ as follows:

- *$\mathcal{L}(\eta) = \{\alpha \in \mathcal{A} \mid \eta \vDash \alpha\}$ where $\mathcal{A} \subset \mathcal{F}$ is the set of atomic formulas.*
- *$\Sigma = R \cup \{Id\}$.*
- *T can obviously be seen as the set of triplets (η, r, η') such that $(\eta \xrightarrow{r} \eta')$ is a transition of T.*

A path $(\pi \equiv \eta_0 \xrightarrow{r_0} \eta_1 \xrightarrow{r_1} \ldots \xrightarrow{r_{i-1}} \eta_i \xrightarrow{r_i} \ldots)$ is then an infinite sequence of labeled transitions such that the input state of r_i is equal to the output state of r_{i-1} for all $i > 0$. The set of paths is called Π_R.

4   Integrating Toxicological Knowledge into Constraints

As the transition graph of a biological system includes many toxicologically improbable paths, it is necessary to filter out the irrelevant ones and to characterize the interesting paths for toxicologists. Temporal logic and model checking tools have already been successfully applied to biological systems [2,7]. Here, since we seek to filter paths in details, we need a logic able to express both state and transition properties, such as the state/event linear temporal logic (SE-LTL) developed by Chaki [4].

Since a path can be seen as an infinite alternation between states and transitions, atomic temporal formulas concern either a state or a transition. For states, atomic temporal formulas are similar to atomic formulas of Definition 3. For transitions, atomic temporal formulas involve a rule identifier or the identity transition.

Definition 9 (Temporal formula). *Given an $\mathcal{E}$-action network R, the set $\mathcal{T}_R$ of temporal formulas on R is inductively defined by:*

- $(\mathcal{A} \cup R \cup \{Id\}) \subset \mathcal{T}_R$
- *if φ and ψ are formulas of $\mathcal{T}_R$, then $\neg\varphi$, $\varphi \wedge \psi$, $\varphi \vee \psi$, $\varphi \Rightarrow \psi$, $X\varphi$, $F\varphi$, $G\varphi$, $\varphi U \psi$ are formulas of $\mathcal{T}_R$.*

Definition 10 (Temporal formula satisfaction). *Given an $\mathcal{E}$-action network R and a path $(\pi \equiv \eta_0 \xrightarrow{r_0} \eta_1 \xrightarrow{r_1} \ldots) \in \Pi_R$, the satisfaction relation $\models \subset \Pi_R \times \mathcal{T}_R$ is inductively defined by:*

- *$\pi \models \alpha$ (where the atom α belongs to $\mathcal{A}$) if and only if $\eta_0 \models \alpha$,*
- *$\pi \models r$ where $r \in R \cup \{Id\}$ if and only if $r = r_0$,*
- *$\pi \models \varphi \wedge \psi$ where $(\varphi, \psi) \in \mathcal{T}_R{}^2$ if and only if $\pi \models \varphi$ and $\pi \models \psi$, other propositional logic connectives are treated similarly,*
- *$\pi \models X\varphi$ where $\varphi \in \mathcal{T}_R$ if and only if $(\eta_1 \xrightarrow{r_1} \eta_2 \xrightarrow{r_2} \ldots) \models \varphi$,*
- *$\pi \models G\varphi$ where $\varphi \in \mathcal{T}_R$ if and only if for all $i \in \mathbb{N}$, $(\eta_i \xrightarrow{r_i} \eta_{i+1} \xrightarrow{r_{i+1}} \ldots) \models \varphi$,*
- *$\pi \models F\varphi$ where $\varphi \in \mathcal{T}_R$ if and only if there exists $i \in \mathbb{N}$, $(\eta_i \xrightarrow{r_i} \eta_{i+1} \xrightarrow{r_{i+1}} \ldots) \models \varphi$,*
- *$\pi \models \varphi U \psi$ where $(\varphi, \psi) \in \mathcal{T}_R{}^2$ if and only if there exists $j \in \mathbb{N}$, $(\eta_j \xrightarrow{r_j} \ldots) \models \psi$ and for all $0 \leqslant i < j$, $(\eta_i \xrightarrow{r_i} \ldots) \models \varphi$.*

Furthermore, for all $r \in R$ of the form $r : A_1 + \cdots + A_m \Rightarrow A_{m+1} + \cdots + A_n$ $when(\varphi)$ $boost(\psi)$, we note $app(r)$ the temporal formula $(\bigwedge_{i=1}^m A_i > \varepsilon) \wedge \neg\psi$ stating that r is applicable at the current state (see Definition 6).

In addition, for all $e \in \mathcal{E}$, we note $\downarrow e$ the temporal formula stating that the level of the entity e decreases in the next state:

$$\bigvee_{l \in \mathcal{E}(e) \setminus \{\varepsilon\}} \left(e = l \wedge X\big(e = decr_e(l)\big) \right).$$

We proceed similarly for $\uparrow e$.

For instance in our running example, the property χ characterizing paths where an excess of I leads to a future excess of T_3 can be written as: $G((I > \Delta) \Rightarrow F(T_3 > \Delta))$ and the formula ξ stating that the rule r_B is the first applied when T_4 is absent from the system can be written as: $G((T_4 = \varepsilon) \Rightarrow r_B)$. In this situation, the path beginning with $(\eta_0 \xrightarrow{r_B} \eta_1)$, where $\eta_0 = (\Delta, \varepsilon, \iota, \theta)$ and $\eta_1 = (\theta, \varepsilon, \iota, \theta)$ satisfies both χ and ξ.

Finally, the association of the transition graph of a system with a set of properties representing the relevant biological pathways is called a *constrained network*. This constrained network is actually a subset of paths from the transition graph, with each path in this subset satisfying all the SE-LTL biological properties.

Definition 11 (Constrained network). *An $\mathcal{E}$-constrained network is a couple $N = (R, Ax)$ where R is an $\mathcal{E}$-action network and Ax is a set of temporal formulas.*

Definition 12 (Dynamics of a constrained network). *Given an $\mathcal{E}$-constrained network $N = (R, Ax)$, the dynamics of N is the subset Π_N of Π_R such that $\pi \in \Pi_R$ belongs to Π_N if and only if $\pi \models Ax$.*

Since properties filter out irrelevant paths from the transition graph, it is thus possible to use them in conjunction to formal methods to insure that the final constrained network satisfies basic biological and toxicological properties as well as specific properties related to the studied issue.

5   Application to the Thyroid Hormone System

The formalism described in the previous section can be illustrated with the thyroid hormone system developed in Sect. 2. This system contains the following entities: blood iodide (I_B), thyroid iodide (I_T), thyroid peroxydase (TPO), blood triiodothyronine (T_{3B}), blood tetraiodothyronine (T_{4B}), pituitary triiodothyronine (T_{3Pit}), thyroid-stimulating hormone (TSH), type 1 to 3 deiodinases (D_1, D_2, D_3) and hepatic detoxifying enzymes (Detox).

On top of these endogenous entities, we can also introduce exogenous compounds able to disrupt the thyroid hormone system. Each compound is an endocrine disruptor triggering one of the disruptions listed in Sect. 2: X_I impacts the dedicated iodide transporters in thyroid, X_{D1} and X_{D2} respectively inactivates D_1 and D_2, and X_{Hep} increases hepatic enzymes levels.

Finally, the signature of our example corresponds to the set:

$$\mathcal{E}_{thy} = \{I_B, I_T, TPO, T_{3B}, T_{4B}, T_{3Pit}, TSH, D_1, D_2, D_3, Detox, X_I, X_{D1}, X_{D2}, X_{Hep}\}$$

The $\mathcal{E}_{thy}$-action network R_{thy} is made of 21 rules. However, for the sake of clarity, only a part of these rules is presented in this section. The complete model, including the list of rules, is available in appendix.

Central regulation. The HPT axis is modeled thanks to the following rules:

$$I_{transfer} : I_B \Rightarrow I_T \ \ when(TSH > \varepsilon \wedge X_I = \varepsilon)$$
$$TPO_{synth} : \Omega \Rightarrow TPO \ \ when(TSH > \varepsilon) \ boost(TSH = \theta)$$
$$TPO_{destr} : TPO \Rightarrow \Omega \ \ when(TSH = \varepsilon \vee (TPO = \theta \wedge TSH < \theta))$$
$$TH_{synth} : I_T \Rightarrow T_{3B} + T_{4B} \ \ when(TPO > \varepsilon) \ boost(TPO = \theta)$$
$$Pit_{synth} : T_{4B} \Rightarrow T_{3Pit} \ \ when(D_2 > \varepsilon) \ boost(D_2 = \theta)$$
$$Pit_{destr} : T_{3Pit} \Rightarrow \Omega \ \ when(T_{4B} = \varepsilon \vee D_2 = \varepsilon \vee (T_{3Pit} = \theta \wedge D_2 < \theta))$$
$$TSH_{synth} : \Omega \Rightarrow TSH \ \ when(T_{3Pit} < \theta) \ boost(T_{3Pit} = \varepsilon)$$
$$TSH_{destr} : TSH \Rightarrow \Omega \ \ when(T_{3Pit} = \theta \vee (TSH = \theta \wedge T_{3Pit} > \varepsilon))$$

Rule TPO_{synth} expresses the ability of the organism to restore normal levels of TPO only when TSH is present in the system. Conversely, TPO_{destr} conveys that levels of TPO tend to decrease when TSH is absent. Moreover, TSH is also required for the production of the dedicated iodide transporters. Note that these transporters are abstracted in this model. Consequently, actions of TSH and X_I directly apply to $I_{transfer}$.

The synthesis of TH requires the presence of both I_T and TPO. However, TPO levels are not affected by TH_{synth} since TPO is a catalyst of the reaction.

The negative feedback of T_{4B} on TSH production mediated exclusively by D_2 (as illustrated in Fig. 1) is highlighted in TSH_{synth} and TSH_{destr}. Indeed, T_{3Pit} can only be obtained through deiodination of T_{4B} by D_2 (rules Pit_{synth} and Pit_{destr}).

Activation and metabolism of TH. The activation of blood TH is handled by D_1 and D_2. Their equilibria and their impact on the system are handled thanks to the following rules:

$$D1_{synth} : \Omega \Rightarrow D_1 \quad when(X_{D1} = \varepsilon) \quad boost(T_{3B} = \varepsilon)$$
$$D1_{destr} : D_1 \Rightarrow \Omega \quad when(X_{D1} > \varepsilon \vee (D_1 = \theta \wedge T_{3B} > \varepsilon))$$
$$D2_{synth} : \Omega \Rightarrow D_2 \quad when(T_{3B} < \theta \wedge X_{D2} = \varepsilon) \quad boost(T_{3B} = \varepsilon)$$
$$D2_{destr} : D_2 \Rightarrow \Omega \quad when(T_{3B} = \theta \vee X_{D2} > \varepsilon \vee (D_2 = \theta \wedge T_{3B} > \varepsilon))$$
$$I_{recycling} : T_{4B} \Rightarrow I_B \quad when(D_1 > \varepsilon)$$
$$TH_{activation} : T_{4B} \Rightarrow T_{3B} \quad when(D_1 = \theta \vee D_2 > \varepsilon) \quad boost(D_2 = \theta)$$

Both D_1 and D_2 levels are induced by a lack of T_{3B} in the system. On the contrary, D_2 levels are reduced by an excess of T_{3B}. On top of that, the presence of exogenous disruptors such as X_{D1} or X_{D2} alters D_1 and D_2 levels.

As the vast majority of sulfated TH is composed of T_{4B}, $I_{recycling}$ models accurately the preponderant recycling role of D_1. D_1 also intervenes marginally in T_{4B} deiodation, as shown in $TH_{activation}$, where D_1 needs to be at level θ to satisfy the *when* statement. As for D_2, the enzyme acts essentially on T_{4B} deiodation in T_{3B}, as shown by both *when* and *boost* statements of $TH_{activation}$.

The metabolism of TH is mainly provided by D_3 and Detox. The rules involving these entities are:

$$D3_{synth} : \Omega \Rightarrow D_3 \quad when(T_{3B} > \varepsilon) \quad boost(T_{3B} = \theta)$$
$$D3_{destr} : D_3 \Rightarrow \Omega \quad when(T_{3B} = \varepsilon \vee (D_3 = \theta \wedge T_{3B} < \theta))$$
$$Detox_{synth} : \Omega \Rightarrow Detox \quad boost(X_{Hep} > \varepsilon)$$
$$Detox_{destr} : Detox \Rightarrow \Omega \quad when(Detox = \theta \wedge X_{Hep} = \varepsilon)$$
$$T3_{destr} : T_{3B} \Rightarrow \Omega \quad when(D_3 = \theta \vee Detox = \theta \vee (T_{3B} = \theta \wedge D_3 > \varepsilon))$$
$$T4_{destr} : T_{4B} \Rightarrow \Omega \quad when(D_3 = \theta \vee Detox = \theta \vee (T_{4B} = \theta \wedge D_3 > \varepsilon))$$

The regulation of D_3 levels is symmetrical to D_2 regulation, as T_{3B} is an inducer of D_3 levels. The case of Detox is interesting: since we are only interested in an excessive activity of the hepatic detoxifying enzymes, the set of admissible levels of this entity is $\{\Delta, \theta\}$. An excess of Detox is only triggered by the presence of X_{Hep}, as seen in $Detox_{synth}$ and $Detox_{destr}$.

Furthermore, the *when* statements of rules $T3_{destr}$ and $T4_{destr}$ reflects two important notions on the metabolism of T_{3B} and T_{4B}:

1. An excess of D_3 or Detox is enough to decrease the levels of both TH.
2. If TH are in excess, the presence of D_3 is enough to restore normal TH levels. It is capital to note that when both levels of D_3 and TH are normal, D_3 does not trigger the decrease of TH levels.

The set of all these rules allows to generate the system dynamics. It is then possible to constrain these dynamics thanks to biological observations expressed through SE-LTL properties. For instance, we express the fact that $TH_{activation}$ always primes on $I_{recycling}$ as long as there is no shortage of blood iodide:

$$\varphi_0 \equiv G((\mathrm{I_B} > \varepsilon \wedge app(TH_{activation}) \wedge app(I_{recycling})) \rightarrow \neg I_{recycling})$$

Literally, φ_0 means that if $\mathrm{I_B}$ is present in the system, and both $TH_{activation}$ and $I_{recycling}$ are applicable, then $I_{recycling}$ does not apply. The G operator indicates that the property must be satisfied at every step of the path.

It has also been observed that an excess of detoxifying enzymes quickly leads to the depletion of $\mathrm{T_{4B}}$:

$$\varphi_1 \equiv G((\mathrm{Detox} = \theta \wedge app(T4_{destr})) \rightarrow T4_{destr})$$

We can also check that the model verifies global biological properties such as the fact that without any disruption in $\mathrm{D_2}$ functionning, an hypothyroidic state (*i.e.* where $\mathrm{T_{4B}}$ is lacking) leads to a state where TSH is in excessive concentration:

$$\varphi_2 \equiv G(\mathrm{T_{4B}} = \varepsilon \wedge \mathrm{X_{D2}} = \varepsilon) \rightarrow F(\mathrm{TSH} = \theta))$$

Of course, several paths belonging to R_{thy} do not satisfy the previous properties. For instance, let π_1 and π_2 the portions of path represented in Fig. 2. In state η_{12} of π_1, the conditions of φ_0 are satisfied and $I_{recycling}$ should not be applied. Therefore, π_1 does not satisfy φ_0, contrary to π_2.

As such, if we consider $Ax_{thy} = \{\varphi_0, \varphi_1, \varphi_2\}$ and the constrained network $N_{\mathrm{thy}} = (R_{\mathrm{thy}}, Ax_{thy})$, paths including π_1 do not belong to N_{thy}.

$$\pi_1 = \underbrace{\Delta\varepsilon\varepsilon}_{\eta_{10}} \xrightarrow{TH_{synth}} \underbrace{\Delta\Delta\varepsilon}_{\eta_{11}} \xrightarrow{TH_{activation}} \underbrace{\Delta\Delta\Delta}_{\eta_{12}} \xrightarrow{I_{recycling}} \underbrace{\Delta\varepsilon\Delta}_{\eta_{13}}$$

$$\pi_2 = \underbrace{\varepsilon\varepsilon\varepsilon}_{\eta_{20}} \xrightarrow{TH_{synth}} \underbrace{\varepsilon\Delta\varepsilon}_{\eta_{21}} \xrightarrow{TH_{activation}} \underbrace{\varepsilon\Delta\Delta}_{\eta_{22}} \xrightarrow{I_{recycling}} \underbrace{\Delta\Delta\Delta}_{\eta_{12}}$$

Fig. 2. Possible path segments belonging to R_{thy}. For the sake of simplicity, states depicted here only contain the levels of respectively $\mathrm{I_B}$, $\mathrm{T_{4B}}$ and $\mathrm{T_{3B}}$.

Finally, we can use the constrained network N_{thy} to search for existing pathways of toxicity. Indeed, possible disruptions described in Sect. 2 correspond to sets of paths belonging to N_{thy}. These sets of paths can be identified thanks to temporal formulas. For example, the inactivation of $\mathrm{D_2}$ by $\mathrm{X_{D2}}$ leading to hyperthyroidism corresponds to paths satisfying φ_{D2}:

$$\varphi_{\mathrm{D2}} \equiv G(\mathrm{X_{D2}} {>} \varepsilon \;\; U \;\; (\mathrm{D_2} {=} \varepsilon \;\; U \;\; (\mathrm{T_{3Pit}} {=} \varepsilon \;\; U \;\; (\mathrm{TSH} {=} \theta \;\; U \;\; (\mathrm{T_{4B}} {=} \theta)))))$$

$$\pi_3 = \underbrace{\Delta\Delta\Delta\Delta\Delta}_{\eta_{30}} \xrightarrow{D2_{destr}} \underbrace{\Delta\Delta\Delta\varepsilon\Delta}_{\eta_{31}} \xrightarrow{Pit_{destr}} \underbrace{\Delta\varepsilon\Delta\varepsilon\Delta}_{\eta_{32}} \xrightarrow{TSH_{synth}} \underbrace{\Delta\varepsilon\theta\varepsilon\Delta}_{\eta_{33}} \xrightarrow{TH_{synth}} \underbrace{\theta\varepsilon\theta\varepsilon\Delta}_{\eta_{33}}$$

Fig. 3. Possible path segment belonging to R_{thy}. For the sake of simplicity, states depicted here only contain the levels of respectively T_{4B}, T_{3Pit}, TSH, D_2 and X_{D2}.

$$\pi_4 = \underbrace{\Delta\Delta\Delta\Delta\Delta}_{\eta_{40}} \xrightarrow{Detox_{synth}} \underbrace{\Delta\Delta\Delta\theta\Delta}_{\eta_{41}} \xrightarrow{T4_{destr}} \underbrace{\varepsilon\Delta\Delta\theta\Delta}_{\eta_{42}} \xrightarrow{Pit_{destr}} \underbrace{\varepsilon\varepsilon\Delta\theta\Delta}_{\eta_{42}} \xrightarrow{TSH_{synth}} \underbrace{\varepsilon\varepsilon\theta\theta\Delta}_{\eta_{43}}$$

Fig. 4. Possible path segment belonging to R_{thy}. For the sake of simplicity, states depicted here only contain the levels of respectively T_{4B}, T_{3Pit}, TSH, Detox and X_{Hep}.

The effect of X_{Hep}, namely the trigger of hepatic detoxifying enzymes leading to decreased levels of T_{4B} and then high levels of TSH can also be expressed thanks to φ_{Hep}:

$$\varphi_{\text{Hep}} \equiv G(X_{\text{Hep}} > \varepsilon \ U \ (\text{Detox} = \theta \ U \ (T_{4B} = \varepsilon \ U \ (\text{TSH} = \theta))))$$

Path π_3 and π_4 as depicted in Figs. 3 and 4, are examples of interesting trajectories for toxicologists. Indeed, both these paths start in the *initial state* (init) defined as follow: the biological system is considered healthy (all the endogenous entities at Δ) but contains also an exogenous compound (X_{D2} or X_{Hep} greater than ε). Then, as an exogenous compound leads the organism towards pathological states (here respectively a chronic hyperthyroidism and a thyroid cancer), we can enumerate its possible pathways of toxicity by filtering paths satisfying temporal formulas (here, init $\wedge$ $FG(T_{4B} = \theta)$ and init $\wedge$ $FG(\text{TSH} = \theta)$).

6   Conclusion

We presented a new formal framework able to handle several specificities of the toxicology domain not taken into account so far. This rule-based modeling framework allows for a direct description of equilibrium changes happening in a biological system. This description does not model the strength differences between equilibrium rules, which can affect the global behavior of the system. For this reason, we integrated biological and toxicological knowledge about equilibria kinetics through formulas expressed in SE-LTL. As demonstrated on a simple model of the thyroid hormone system, the expressive power of the formalism enable us to describe equilibrium changes in the organism as well as knowledge about equilibrium kinetics. This knowledge allows then the filtering out of irrelevant paths from the initial model and the search for toxicity pathways.

In the future, our formalism will be coupled with a SE-LTL model checker in order to list the most probable toxicity pathways present in a model. Indeed, it is possible to define pathological states and enumerate the paths leading to these states. Furthermore, filtering the resulting paths could also highlight gaps in the current toxicological knowledge and help toxicologists in their design of new experimental strategies.

Finally, as this formalism is now well-defined, it will serve as a basis to develop a software platform dedicated to toxicology. This platform is currently under development and it is already possible to run simulations on biological action networks. In the near future, the platform will also be able to integrate the temporal formulas and to use these biological constraints to filter out irrelevant paths. This will be achieved by generating all the paths allowed by a biological action network while checking these paths for their biological relevance. Finally, by defining states regarded as pathologic, the platform will then be able to compute all the paths leading to pathologic states and propose putative pathways of toxicity.

Appendix

See Tables 1 and 2

Table 1. The signature of $\mathcal{E}_{\text{thy}}$, including the different set of admissible levels.

Entity	Biological name	Admissible levels
I_B	Blood iodide	$\{\varepsilon, \Delta\}$
I_T	Thyroid iodide	$\{\varepsilon, \Delta\}$
TPO	Thyroid peroxydase	$\{\varepsilon, \Delta, \theta\}$
T_{3B}	Blood triiodothyronine	$\{\varepsilon, \Delta, \theta\}$
T_{4B}	Blood tetraiodothyronine	$\{\varepsilon, \Delta, \theta\}$
T_{3Pit}	Pituitary triiodothyronine	$\{\varepsilon, \Delta, \theta\}$
TSH	Thyroid-stimulating hormone	$\{\varepsilon, \Delta, \theta\}$
D_1	Type 1 deiodinase	$\{\varepsilon, \Delta, \theta\}$
D_2	Type 2 deiodinase	$\{\varepsilon, \Delta, \theta\}$
D_3	Type 3 deiodinase	$\{\varepsilon, \Delta, \theta\}$
Detox	Hepatic detoxifying enzymes	$\{\Delta, \theta\}$
X_I	Iodide transporter inactivator	$\{\varepsilon, \Delta\}$
X_{D1}	D_1 inactivator	$\{\varepsilon, \Delta\}$
X_{D2}	D_2 inactivator	$\{\varepsilon, \Delta\}$
X_{Hep}	Detoxifying enzymes inducer	$\{\varepsilon, \Delta\}$

Table 2. The R_{thy} action network.

I_{intake}	: Ω	$\Rightarrow$ I_B	
$I_{transfer}$	: I_B	$\Rightarrow$ I_T	$when(\text{TSH} > \varepsilon \wedge X_I = \varepsilon)$
TPO_{synth}	: Ω	$\Rightarrow$ TPO	$when(\text{TSH} > \varepsilon)$ $boost(\text{TSH} = \theta)$
TPO_{destr}	: TPO	$\Rightarrow$ Ω	$when(\text{TSH} = \varepsilon \vee (\text{TPO} = \theta \wedge \text{TSH} < \theta))$
TH_{synth}	: I_T	$\Rightarrow$ $T_{3B} + T_{4B}$	$when(\text{TPO} > \varepsilon)$ $boost(\text{TPO} = \theta)$
Pit_{synth}	: T_{4B}	$\Rightarrow$ T_{3Pit}	$when(D_2 > \varepsilon)$ $boost(D_2 = \theta)$
Pit_{destr}	: T_{3Pit}	$\Rightarrow$ Ω	$when(T_{4B} = \varepsilon \vee D_2 = \varepsilon \vee (T_{3Pit} = \theta \wedge D_2 < \theta))$
TSH_{synth}	: Ω	$\Rightarrow$ TSH	$when(T_{3Pit} < \theta)$ $boost(T_{3Pit} = \varepsilon)$
TSH_{destr}	: TSH	$\Rightarrow$ Ω	$when(T_{3Pit} = \theta \vee (\text{TSH} = \theta \wedge T_{3Pit} > \varepsilon))$
$D1_{synth}$	: Ω	$\Rightarrow$ D_1	$when(X_{D1} = \varepsilon)$ $boost(T_{3B} = \varepsilon)$
$D1_{destr}$	: D_1	$\Rightarrow$ Ω	$when(X_{D1} > \varepsilon \vee (D_1 = \theta \wedge T_{3B} > \varepsilon))$
$D2_{synth}$	: Ω	$\Rightarrow$ D_2	$when(T_{3B} < \theta \wedge X_{D2} = \varepsilon)$ $boost(T_{3B} = \varepsilon)$
$D2_{destr}$	: D_2	$\Rightarrow$ Ω	$when(T_{3B} = \theta \vee X_{D2} > \varepsilon \vee (D_2 = \theta \wedge T_{3B} > \varepsilon))$
$D3_{synth}$	: Ω	$\Rightarrow$ D_3	$when(T_{3B} > \varepsilon)$ $boost(T_{3B} = \theta)$
$D3_{destr}$	: D_3	$\Rightarrow$ Ω	$when(T_{3B} = \varepsilon \vee (D_3 = \theta \wedge T_{3B} < \theta))$
$Detox_{synth}$	: Ω	$\Rightarrow$ Detox	$boost(X_{Hep} > \varepsilon)$
$Detox_{destr}$	: Detox	$\Rightarrow$ Ω	$when(\text{Detox} = \theta \wedge X_{Hep} = \varepsilon)$
$I_{recycling}$	: T_{4B}	$\Rightarrow$ I_B	$when(D_1 > \varepsilon)$
$TH_{activation}$	: T_{4B}	$\Rightarrow$ T_{3B}	$when(D_1 = \theta \vee D_2 > \varepsilon)$ $boost(D_2 = \theta)$
$T3_{destr}$	: T_{3B}	$\Rightarrow$ Ω	$when(D_3 = \theta \vee \text{Detox} = \theta \vee (T_{3B} = \theta \wedge D_3 > \varepsilon))$
$T4_{destr}$	: T_{4B}	$\Rightarrow$ Ω	$when(D_3 = \theta \vee \text{Detox} = \theta \vee (T_{4B} = \theta \wedge D_3 > \varepsilon))$

References

1. Barter, R.A., Klaassen, C.D.: Reduction of thyroid hormone levels and alteration of thyroid function by four representative UDP-glucuronosyltransferase inducers in rats. Toxicol. Appl. Pharmacol. **128**(1), 9–17 (1994)
2. Bernot, G., Comet, J.P., Richard, A., Guespin, J.: Application of formal methods to biological regulatory networks: extending Thomas' asynchronous logical approach with temporal logic. J. Theor. Biol. **229**(3), 339–347 (2004)
3. Chabrier-Rivier, N., Fages, F., Soliman, S.: The biochemical abstract machine BIOCHAM. In: Danos, V., Schachter, V. (eds.) CMSB 2004. LNCS, vol. 3082, pp. 172–191. Springer, Heidelberg (2005). doi:10.1007/978-3-540-25974-9_14
4. Chaki, S., Clarke, E.M., Ouaknine, J., Sharygina, N., Sinha, N.: State/Event-based software model checking. In: Boiten, E.A., Derrick, J., Smith, G. (eds.) IFM 2004. LNCS, vol. 2999, pp. 128–147. Springer, Heidelberg (2004). doi:10.1007/978-3-540-24756-2_8
5. Danos, V., Feret, J., Fontana, W., Harmer, R., Hayman, J., Krivine, J., Thompson-Walsh, C., Winskel, G.: Graphs, rewriting and pathway reconstruction for rule-based models. In: LIPIcs-Leibniz International Proceedings in Informatics, vol 18. Schloss Dagstuhl-Leibniz-Zentrum fuer Informatik (2012)
6. Dentice, M., Salvatore, D.: Local impact of thyroid hormone inactivation deiodinases: the balance of thyroid hormone. J. Endocrinol. **209**(3), 273–282 (2011)
7. Fages, F., Soliman, S.: Formal cell biology in biocham. In: Bernardo, M., Degano, P., Zavattaro, G. (eds.) SFM 2008. LNCS, vol. 5016, pp. 54–80. Springer, Heidelberg (2008). doi:10.1007/978-3-540-68894-5_3

8. Isler, H., Leblond, C., Axelrad, A.: Influence of age and of iodine intake on the production of thyroid tumors in the rat 2. J. Natl. Cancer Inst. **21**(6), 1065–1081 (1958)

9. Kavlock, R.J., Ankley, G., Blancato, J., Breen, M., Conolly, R., Dix, D., Houck, K., Hubal, E., Judson, R., Rabinowitz, J., et al.: Computational toxicologya state of the science mini review. Toxicol. Sci. **103**(1), 14–27 (2008)

10. Kopp, P.: Thyroid Hormone Synthesis. Werner and Ingbars The Thyroid. A Fundamental and Clinical Text, 10th edn., pp. 48–74 (2012)

11. Maia, A.L., Goemann, I.M., Meyer, E.L.S., Wajner, S.M.: Type 1 iodothyronine deiodinase in human physiology and disease deiodinases: the balance of thyroid hormone. J. Endocrinol. **209**(3), 283–297 (2011)

12. Marians, R., Ng, L., Blair, H., Unger, P., Graves, P., Davies, T.: Defining thyrotropin-dependent and-independent steps of thyroid hormone synthesis by using thyrotropin receptor-null mice. Proc. Natl. Acad. Sci. **99**(24), 15776–15781 (2002)

13. McClain, R.M.: Thyroid gland neoplasia: non-genotoxic mechanisms. Toxicol. Lett. **64**, 397–408 (1992)

14. Moeller, L.C., Cao, X., Dumitrescu, A.M., Seo, H., Refetoff, S.: Thyroid hormone mediated changes in gene expression can be initiated by cytosolic action of the thyroid hormone receptor beta through the phosphatidylinositol 3-kinase pathway. Nucl. Recept. Sig. **4**, e020 (2006)

15. Nagasaki, M., Onami, S., Miyano, S., Kitano, H.: Bio-calculus: Its concept and molecular interaction. Genome Inf. **10**, 133–143 (1999)

16. Regev, A., Panina, E.M., Silverman, W., Cardelli, L., Shapiro, E.: Bioambients: an abstraction for biological compartments. Theoret. Comput. Sci. **325**(1), 141–167 (2004)

17. Rosene, M.L., Wittmann, G., Arrojo e Drigo, R., Singru, P.S., Lechan, R.M., Bianco, A.C.: Inhibition of the type 2 iodothyronine deiodinase underlies the elevated plasma TSH associated with amiodarone treatment. Endocrinology **151**(12), 5961–5970 (2010)

18. Snoussi, E.H.: Qualitative dynamics of piecewise-linear differential equations: a discrete mapping approach. Dyn. Stab. Syst. **4**(3–4), 565–583 (1989)

19. Talcott, C.: Pathway logic. In: Bernardo, M., Degano, P., Zavattaro, G. (eds.) SFM 2008. LNCS, vol. 5016, pp. 21–53. Springer, Heidelberg (2008). doi:10.1007/978-3-540-68894-5_2

20. Thomas, R.: Regulatory networks seen as asynchronous automata: a logical description. J. Theor. Biol. **153**(1), 1–23 (1991)

21. Visser, T.J.: Role of sulfation in thyroid hormone metabolism. Chem. Biol. Interact. **92**(1–3), 293–303 (1994)

22. Williams, G.R., Bassett, J.D.: Local control of thyroid hormone action: role of type 2 deiodinase deiodinases: the balance of thyroid hormone. J. Endocrinol. **209**(3), 261–272 (2011)

23. Zoeller, R.T., Tan, S.W., Tyl, R.W.: General background on the hypothalamic-pituitary-thyroid (HPT) axis. Crit. Rev. Toxicol. **37**(1–2), 11–53 (2007)

Data-Driven Robust Control for Type 1 Diabetes Under Meal and Exercise Uncertainties

Nicola Paoletti[1(✉)], Kin Sum Liu[1], Scott A. Smolka[1], and Shan Lin[2]

[1] Department of Computer Science, Stony Brook University, Stony Brook, USA
nclpltt@gmail.com
[2] Department of Electrical and Computer Engineering, Stony Brook University, Stony Brook, USA

Abstract. We present a fully closed-loop design for an artificial pancreas (AP) which regulates the delivery of insulin for the control of Type I diabetes. Our AP controller operates in a fully automated fashion, without requiring any manual interaction (e.g. in the form of meal announcements) with the patient. A major obstacle to achieving closed-loop insulin control is the uncertainty in those aspects of a patient's daily behavior that significantly affect blood glucose, especially in relation to meals and physical activity. To handle such uncertainties, we develop a data-driven robust model-predictive control framework, where we capture a wide range of individual meal and exercise patterns using uncertainty sets learned from historical data. These sets are then used in the controller and state estimator to achieve automated, precise, and personalized insulin therapy. We provide an extensive *in silico* evaluation of our robust AP design, demonstrating the potential of this approach, without explicit meal announcements, to support high carbohydrate disturbances and to regulate glucose levels in large clusters of virtual patients learned from population-wide survey data.

1 Introduction

Type 1 diabetes (T1D) is an autoimmune disease where the pancreas is not able to autonomously produce a sufficient amount of insulin to regulate blood glucose (BG) levels, thereby inhibiting glucose uptake in muscle and adipose (fatty) tissue. In healthy subjects, pancreatic β cells are responsible for the release of insulin in amounts commensurate with current BG levels. This regulation maintains healthy BG values within tight ranges, normally between 70–200 mg/dL. In T1D, T cell–mediated destruction of insulin-producing β cells occurs, leading to high BG levels.

In the U.S. alone, more than 29 million people suffer from diabetes, among which about 5% have T1D [2]. T1D patients need to wear an insulin pump for the injection of *basal* and *bolus* insulin. Basal insulin is a low and continuous dose that covers insulin needs outside meals. Bolus insulin is a single high dose for covering meals.

© Springer International Publishing AG 2017
J. Feret and H. Koeppl (Eds.): CMSB 2017, LNBI 10545, pp. 214–232, 2017.
DOI: 10.1007/978-3-319-67471-1_13

The concept of closed-loop control of insulin, a.k.a. the artificial pancreas (AP), involves a continuous glucose monitor (CGM) that provides glucose measurements (with a typical period of 5 min) to a control algorithm running inside the insulin pump or on a peripheral device (e.g. smartphone or tablet) connected to the pump [38]. The controller adjusts the insulin therapy to maintain healthy BG levels and to avoid *hyperglycemia* (BG above the healthy range) as well as *hypoglycemia* (BG below the healthy range). AP systems have been extensively studied in the last 20 years [10], but only lately cleared for clinical trials [17,22] and commercialization.

The recently FDA-approved MINIMED 670G by Medtronic[1] is the first commercial AP system, and can regulate the basal insulin rate automatically. It is referred to as a "hybrid closed-loop" device as patients need to manually announce the amount of carbohydrate (CHO) and time of each meal to receive the appropriate bolus insulin dose. This manual procedure is a burden to the patient and inherently dangerous as incorrect information can lead to incorrect insulin dosage and, in turn, harmful BG levels.

While meals are the major source of uncertainty in BG control, another important factor is physical activity, which accelerates glucose absorption and thus requires a reduced insulin dosage. To build fully automated *closed-loop* AP systems, it is essential to design insulin control algorithms that are *robust* to the patient's behavior and activities.

In this paper, we propose a *data-driven, robust model-predictive control* (robust MPC) framework for the closed-loop control of insulin administration, both basal and bolus, for T1D patients under uncertain meal and exercise events. Such a framework seeks to eliminate the need for meal announcements by the patient, to fully automate insulin regulation. We capture the wide range of individual meal and exercise patterns using *uncertainty sets* learned from historical data.

Following [1], we construct uncertainty sets from data so that they cover the underlying (unknown) distribution with prescribed probabilistic guarantees. Leveraging such information, our robust MPC system computes the insulin administration profile that minimizes the worst-case performance with respect to these uncertainty sets, so providing a principled way to deal with uncertainty.

Besides uncertainty, another challenging aspect of closed-loop control is *state estimation*, which is needed to recover the full state of the model (used within MPC) from CGM measurements. Not only are these measurements noisy and delayed with respect to BG (the CGM detects glucose in the interstitial fluid), but we also need to estimate, along with the state, current meal and exercise uncertainties.

For this purpose, we designed a moving-horizon state estimator (MHE) [6,20,27] that, similar to MPC, exploits a prediction model to find the most likely state estimate given the observations. Crucially, data-driven uncertainty sets improve the estimation by constraining the admissible meal and exercise uncertainties.

---

[1] https://www.medtronicdiabetes.com/products/minimed-670g-insulin-pump-system.

To the best of our knowledge, our robust MPC design for an AP is the first approach to leverage data-driven techniques to enhance robust insulin control and state estimation, supporting at the same time both meal and exercise uncertainties. In summary, our main contributions are the following.

- We formulate a closed-loop AP design based on robust MPC to optimize BG levels under meal and exercise uncertainties.
- We apply data-driven techniques to construct uncertainty sets that provide probabilistic guarantees on the robust MPC solution.
- We design an MHE that leverages data to make informed estimates for BG and uncertainty parameters.
- We provide an extensive *in-silico* evaluation of our design, including one-meal simulations, one-day high carbohydrate intake scenarios, and one-day simulations of large clusters of virtual patients learned from population-wide survey data sets (CDC NHANES).
- Overall, our robust closed-loop AP is able to keep BG within safe levels between 84% and 100% of the time, outperforming an implementation of a hybrid closed-loop AP and state-of-the art robust control algorithms [31].

2   System Overview

The design of our proposed data-driven robust artificial pancreas is illustrated in Fig. 1. The *robust MPC* component (described in Sect. 4) is responsible for computing the insulin administration strategy (both basal and bolus) that optimizes, over a finite time horizon, the predicted BG profile against worst-case realizations of the uncertainty parameters, used to capture unknown meal and exercise information.

Uncertainty sets describe the domains of the uncertainty parameters and are derived by the *data-driven learning* component (see Sect. 4.2), starting from a dataset about the patient's meal and exercise schedules. Uncertainty sets can be also updated online as new data (estimated or announced) comes along, in this way enabling the continuous learning of the patient's behavior.

At this stage, we analyze our robust artificial pancreas design *in silico*. Thus, the *plant* is given by a system of differential equations (see Sect. 3) describing the gluco-regulatory dynamics of a virtual T1D patient, as well as the effects of insulin and random disturbances (i.e. unknown realizations of the uncertainty parameters).

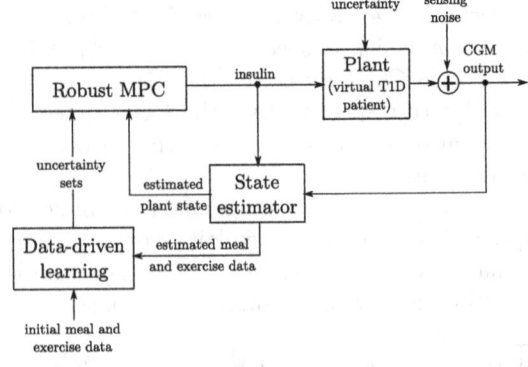

Fig. 1. Robust artificial pancreas design.

In order to faithfully reproduce real-life settings, we assume that the state of the plant (BG) cannot be observed by the controller, but that we can only access (noisy) CGM measurements. We designed a *moving-horizon state estimator* (described in Sect. 4.1) that, based on a bounded history of CGM measurements and estimations, computes the most likely plant state. Importantly, this component also provides estimates for the uncertainty parameters, which can be used to update the uncertainty sets.

3   Plant Model

3.1   Uncertainty Parameters

To account for uncertainty in meal consumption, we consider the parameter D_G^t, which describes the *rate of CHO ingestion* at time t. As in the exercise model of [9,13,21,28], physical activity is represented by parameters MM^t, the *percentage of active muscular mass* at time t, and $O2^t$, the *percentage of maximum oxygen consumption* which can be combined to reproduce arbitrary kinds of physical activity.

MM^t corresponds to the ratio between the active muscular mass and the total muscular mass, with typical values being $MM^t = 0\%$ at rest and $MM^t = 25\%$ for a two-legged exercise. $O2^t$ describes the oxygen consumed relative to the maximum oxygen consumption of the subject, and thus, represents a subject-independent measure of exercise workload. As in [9,21], typical values are 8% at rest, 30% for light activity, 60% for moderate activity, and 90% for intense activity. In our scenario, these meal and exercise parameters are not observed or measured, and are thus represented by an uncertainty parameter vector $\mathbf{u}^t = (D_G^t, MM^t, O2^t)$. The effects of these parameters on blood glucose are described in Sect. 3.2, in which the patient's gluco-regulatory model is presented.

3.2   Patient Model

We consider the nonlinear ODE gluco-regulatory model of Jacobs et al. [13,28], which extends Hovorka's well-established model [11,36,37] to capture the effect of exercise on BG. The model describes the dynamics of glucose and insulin in the human body, i.e., their absorption, metabolism, excretion and transport between compartments (tissues and organs). In addition to insulin, Jacobs' model also allows for the automated control of glucagon, i.e. the hormone antagonistic to insulin that protects against hypoglycemia. In our work, however, we leave aside glucagon. Model parameters (available in the technical report [24]) are deterministic and represent the physiological characteristics (e.g. transport or consumption rates) of a single virtual subject.

At time t, the inputs to the system are the subcutaneous insulin infusion rate, ι^t (mU/min), and the uncertainty parameter values, $\mathbf{u}^t = (D_G^t, MM^t, O2^t)$. The output corresponds to the CGM measurement. The state-space representation of the system is as follows:

$$\dot{\mathbf{x}}(t) = \mathbf{F}\left(\mathbf{x}(t), \iota^t, \mathbf{u}^t\right) \tag{1}$$

$$y(t) = h\left(\mathbf{x}(t)\right) + v^t \tag{2}$$

where $\mathbf{x}$ is the 14-dimensional state vector that evolves according to the ODE system $\mathbf{F}$, which is given below (see the technical report [24] for the full set of equations). Equation 2 describes the CGM measurement y, which is derived from $\mathbf{x}$ with the measurement model h and subject to an additive measurement noise $v^t \in \mathcal{N}(0, q^t)$, where q^t is the noise variance. We fix $q^t = 0.1521 \text{ mmol}^2/\text{L}^2$ constant for all t, corresponding to a standard deviation equal to 5% of the ideal glucose value.

Figure 2 illustrates a high-level schema of the ODE system $\mathbf{F}$. The *gut absorption* subsystem [37] uses a chain of two compartments, G_1 and G_2 (mmol), to describe digestion of ingested CHO, given by the uncertainty parameter D_G^t.

The *glucose kinetics* subsystem describes the glucose masses in the accessible (where BG measurements are made) and non-accessible compartments, respectively through variables Q_1 and Q_2 (mmol). BG concentration, G (mmol/L), is the main variable we aim to control, and is derived from Q_1 as $G(t) = Q_1(t)/V_G$, where V_G is the glucose distribution volume. Variable C is the glucose concentration in the interstitial fluid, which has a delayed response w.r.t. the concentration in the blood G. C corresponds to the glucose detected by the CGM sensor and thus, the measurement function h of Eq. 2 maps the state vector $\mathbf{x}(t)$ to $C(t)$.

The *insulin kinetics* subsystem models the absorption of the fast-acting insulin ι^t, i.e. our control input (in mU/min), and its transport through compartments Q_{1a}, Q_{1b}, Q_{2i} and Q_3 (in mU) [36]. This model assumes a slow insulin absorption pathway consisting of compartments Q_{1a} (subcutaneous insulin mass) and Q_{2i} (non-accessible insulin), and a fast pathway that includes only Q_{1b} (subcutaneous). K represents the proportion in which the input insulin ι^t is distributed into the two pathways. Q_3 is the plasma insulin mass, from which we derive the plasma insulin concentration I (mU/L) as $I(t) = Q_3(t)/V_I$, where V_I is the insulin distribution volume.

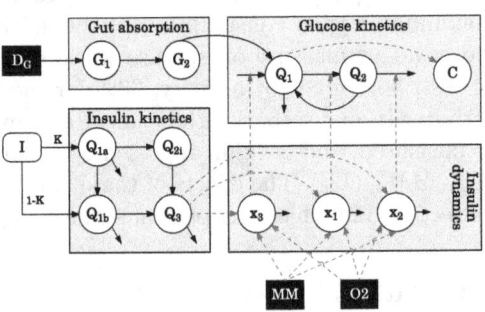

Fig. 2. Schema of the gluco-regulatory ODE system and its four main subsystems. White circles: ODE variables; black boxes: uncertainty parameters; white rounded box: insulin input; solid black arrows: flows of glucose or insulin; dashed green/red arrows: positive/negative interactions between variables. (Color figure online)

The *insulin dynamics* subsystem defines the effects of insulin on blood glucose through variables x_1, x_2, x_3. Variable x_1 (min^{-1}) promotes glucose distribution; x_2 (min^{-1}) promotes glucose disposal; and x_3 (unitless) inhibits endogenous glucose production. The overall subsystem decrease blood glucose masses Q_1 and

Q_2 and in turn, BG concentration G. Plasma insulin levels I directly increase x_1, x_2, x_3. Uncertainty parameters MM^t (active muscular mass) and $O2^t$ (target workload in terms of oxygen consumption) increase x_1, x_2, x_3 indirectly, through state variables UA (mg/min) and $O2_m$ (unitless), not shown in the figure. They characterize physical activity and describe, respectively, the glucose uptake due to active muscular tissue and the actual percentage of maximum oxygen consumption.

Initial Conditions: The initial state of the system is derived at a steady-state BG level of 7.8 mmol/L [31], assuming no meal and exercise. We use a non-linear equation solver (MATLAB's `fsolve`) to find $\mathbf{x}(0)$ and the basal insulin level $\bar{\iota}$ such that $\dot{\mathbf{x}}(0) = F\left(\mathbf{x}(0), \bar{\iota}, \mathbf{u}^0\right) = \mathbf{0}$ (see Eq. 1), where the uncertainty parameters $\mathbf{u}^0$ are given by $D_G^0 = 0$, $MM^0 = 0$ and $O2^0 = 8$ (oxygen consumption at rest). Following [13], we further assess the physiologic feasibility of the initial conditions by checking that: (1) in absence of insulin, steady-state BG is above 300 mg/dL, and (2) delivery of high-dose insulin (15 U/h) results in a steady-state BG below 100 mg/dL.

4   Robust MPC

Since we want to optimize the BG profile against worst-case realizations of the uncertainty parameters, at each time step t, the robust MPC computes the insulin infusion ι^t as the solution of the following non-linear minimax optimization problem:

$$\min_{\iota^t,\ldots,\iota^{t+N_c-1}} \max_{\mathbf{u}^t,\ldots,\mathbf{u}^{t+N_p-1}} \sum_{k=1}^{N_p} d(\tilde{x}(t+k)) + \beta \cdot \sum_{k=0}^{N_c-1} (\Delta\iota^{t+k})^2 \quad (3)$$

subject to: $\iota^{t+k} \in D_\iota$ $(k = 0, \ldots, N_c - 1)$ (4)

$\iota^{t+k} = \bar{\iota}$ $(k = N_c, \ldots, N_p - 1)$ (5)

$\mathbf{u}^{t+k} \in \mathcal{U}^{t+k}$ $(k = 0, \ldots, N_p - 1)$ (6)

$\tilde{\mathbf{x}}(t) = \hat{\mathbf{x}}(t)$ (7)

$\dot{\tilde{\mathbf{x}}}(t+k) = F(\tilde{\mathbf{x}}(t+k), \iota^{t+k}, \mathbf{u}^{t+k})$ $(k = 0, \ldots, N_p - 1)$ (8)

where N_c and N_p are the control and prediction horizon (in minutes), respectively; constraint (4) states that the control input ι must belong to some set D_ι of admissible insulin infusion rates; through (5), we impose that ι is fixed to the basal insulin rate $\bar{\iota}$ outside the control horizon; (6) states that, at any time point $t + k$ in the prediction horizon, uncertainty parameters $\mathbf{u}^{t+k}$ must belong to the corresponding uncertainty sets $\mathcal{U}^{t+k}$; constraint (7) and (8) restrict how the robust MPC computes the predicted state vector $\tilde{\mathbf{x}}$: for the initial state, it uses the estimated plant state at time t, $\hat{\mathbf{x}}(t)$, while following states are predicted using the same plant model (see Eq. 1). We set control and prediction horizons to $N_c = 100$ min and $N_p = 150$ min, respectively, as opposed to [28] where

$N_c = 20$ and $N_p = 200$: preliminary experiments suggested that large N_p values and small N_c values cause excessive insulin therapy and hypoglycemia.

We design the cost function so as to optimize the following two objectives:

1. Minimize the sum of squared distances between the predicted BG level $\tilde{x}_G(t+k)$ and a target trajectory $R(t + k)$:

$$d(\tilde{x}(t + k)) = \gamma(t + k) \cdot (\tilde{x}_G(t + k) - R(t + k))^2 \qquad (9)$$

where $\gamma(t + k) = \gamma$ if $\tilde{x}_G(t + k) < R(t + k)$ and 1 otherwise. (Remind that $x_G(t) = G(t) = Q_1(t)/V_G$ in the glucose kinetics subsystem) Parameter $\gamma \geq 1$ allows defining asymmetric cost functions where predicted BG values below the target are penalized more than those above the target. Glucose control is naturally asymmetric given that hypoglycemia leads to more severe consequences than (temporary) hyperglycemia, and, as shown in [7], asymmetric costs effectively contribute avoiding hypoglycemia.

2. Minimize step-wise changes in the control input $(\Delta \iota^{t+k})^2$, where $\Delta \iota^{t+k} = \iota^{t+k} - \iota^{t+k-1}$, and ι^{t-1} corresponds to the control input in the previous iteration, or to the basal insulin rate $\bar{\iota}$ if $t = 0$.

In our setup, we fix the target trajectory to $R(t + k) = 7.8$ mmol/L for all time instants and set penalty β to $1/50$. We set the asymmetric cost penalty to $\gamma = 2$, after experimenting with different values (reported in [24]).

Optimization Algorithm: We solve problem (3) using non-linear optimization techniques, where, for a fixed control strategy $\iota^t, \ldots, \iota^{t+N_c-1}$, the objective function value is given in turn as the result of maximizing the objective function over the uncertainty parameters (and with fixed $\iota^t, \ldots, \iota^{t+N_c-1}$). To solve both minimization and maximization problems, we use MATLAB's `fmincon`. To reduce the computational cost of this optimization method, we decrease the number of decision variables by assuming that, in the prediction model, control inputs change with period 10 min, and uncertainty parameters with period 30 min.

Hybrid Closed-Loop (HCL) Variant: To compare with our robust MPC approach, we develop a hybrid closed-loop insulin pumps where only basal insulin is automatically regulated and the patient is responsible for bolus insulin. This reduces to a MPC that has no knowledge of meals and exercise, and thus, approximates the behavior of a current state-the-art approved device that requires explicit meal announcement. In our settings, this is equivalent to fixing the uncertainty parameters to their default values at rest.

Then the optimization problem of the HCL controller reduces to:

$$\min_{\iota^t, \ldots, \iota^{t+N_c-1}} \sum_{k=1}^{N_p} d(\tilde{x}(t + k)) + \beta \cdot \sum_{k=0}^{N_c-1} (\Delta \iota^{t+k})^2 \qquad (10)$$

subject to $(4, 5, 7, 8)$ and $\mathbf{u}^{t+k} = (0, 0, 8)$ $(k = 0, \ldots, N_p - 1)$.

Note that the constraints on the insulin therapy are the same of the robust controller (4-5) meaning that the HCL controller is free to synthesize bolus-like therapy profiles too. This will also serve as the baseline controller in the evaluation part of Sect. 5.

4.1   State Estimation

This component allows to recover an estimate of the current state, which is used in the following iteration by the robust MPC as the initial state for its predictions (see Eq. 7). Following [8,27], we designed a moving-horizon state estimator (MHE) that works in a finite-horizon fashion similar to an MPC problem, and allows estimating the current state starting from previous estimations and a bounded history of observed CGM measurements.

For an estimation window of size N, MHE is based on simulating a model of the plant from time $t - N$ to t and aims at finding the model trajectory $\mathbf{x}(t - N), \ldots \mathbf{x}(t)$ that minimizes the discrepancies between simulated and estimated states, and between simulated and measured outputs (CGM). Then, $\hat{\mathbf{x}}(t)$ is chosen as the final state of the optimal trajectory.

Crucially, our estimator also works as a meal and physical activity detector [3,19,34]: in addition to the plant state, we compute the most likely sequence of uncertainty parameters $\mathbf{u}^{t-N}, \ldots, \mathbf{u}^t$, corresponding to decision variables in our optimization problem as they are inputs of the model. The MHE problem boils down to the following non-linear optimization problem:

$$\min_{\mathbf{x}(t-N),\ldots\mathbf{x}(t),\mathbf{u}^{t-N},\ldots,\mathbf{u}^t} \mu \cdot \|\mathbf{x}(t - N) - \hat{\mathbf{x}}(t - N)\|^2 + \sum_{k=0}^{N-1} \frac{\|v^{t-k}\|^2}{q^{t-k}} \tag{11}$$

$$\text{subject to: } v^{t-k} = y(t - k) - h(\mathbf{x}(t - k)) \qquad (k = N - 1, \ldots, 0) \tag{12}$$

$$\dot{\mathbf{x}}(t - k) = F(\mathbf{x}(t - k), \iota^{t-k}, \mathbf{u}^{t-k}) \qquad (k = N, \ldots, 0) \tag{13}$$

$$\mathbf{u}^{t-k} \in \mathcal{U}^{t-k} \qquad (k = N, \ldots, 0) \tag{14}$$

where (12) defines the measurement discrepancy v^{t-k} at time $t - k$ as the difference between the measured and simulated output, $y(t - k)$ and $h(\mathbf{x}(t - k))$, respectively (see also Eq. 2); and (13) states that $\mathbf{x}$ evolves according to the same ODE model of the plant, with ι^{t-k} being the insulin input previously computed by the robust MPC. We remark that data-driven uncertainty sets play an important role also in state estimation, since they constrain the domain of the corresponding estimated uncertainty parameters, as per (14). The problem is solved using MATLAB's `fmincon` non-linear solver.

The first addend of the cost function penalizes the discrepancy between the initial state of the simulated trajectory and the corresponding state estimation, where $\mu > 0$ is a weighting factor. The second addend penalizes measurement discrepancies, weighted by the inverse of the measurement noise variance q^{t-k} (see Eq. 2). In the original formulation of the MHE [8,27], the cost function includes discrepancies for all the states in the trajectory. Our simplification comes from the fact that we do not consider random noise in the model (but only in the measurements), and thus, the trajectory $\mathbf{x}(t - N), \ldots, \mathbf{x}(t)$ is fully determined by the initial state $\mathbf{x}(t - N)$ and by the uncertainty parameters $\mathbf{u}^{t-N}, \ldots, \mathbf{u}^t$. Further, this greatly improves computational efficiency because variables $\mathbf{x}(t-N+1), \ldots, \mathbf{x}(t)$ are strictly constrained by the ODE in Eq. (13). In practice, this means that the decision variables reduce to $\mathbf{x}(t-N), \mathbf{u}^{t-N}, \ldots, \mathbf{u}^t$.

The MHE has an important probabilistic interpretation: when $N = t$ (unbounded horizon), the MHE problem corresponds to maximizing the joint probability for the trajectory of states $\mathbf{x}(t - N), \ldots, \mathbf{x}(t)$ given the measurements $y(t - N), \ldots, y(t)$ [27].

4.2   Building Data-Driven Uncertainty Sets

In this section, we describe how to build the uncertainty sets used within the robust MPC and the state estimator to restrict the domain of the admissible meal and exercise parameters. We apply the approach of [1] where the authors present a general schema for designing uncertainty sets from data for robust optimization (of which robust MPC is an instance). The key idea is to define an uncertainty set that captures possible realizations of the uncertain parameters and then optimize against worst-case realizations within this set. Importantly, this method requires no information about the underlying distribution of the parameters and provides a probabilistic guarantee (an upper bound) on the likelihood that the true realized cost is higher than the optimal 'worst-case' cost computed by the robust controller.

Let us characterize an uncertainty set $\mathcal{U}$ by means of a so-called robust constraint $f(\mathbf{u}, \mathbf{x}) \leq 0$, where $\mathbf{u}$ is the uncertainty parameter and $\mathbf{x}$ is the optimization variable, corresponding in our case to the state vector plus insulin input. Recall that the true distribution $\mathbb{P}^*$ of $\mathbf{u}$ is unknown. Given confidence level $\epsilon > 0$, $\mathcal{U}$ should satisfy two conditions: (1) the robust constraint f is computationally tractable. (2) $\mathcal{U}$ *implies a probabilistic guarantee for* $\mathbb{P}^*$ *at level* ϵ, that is, for any solution $\mathbf{x}^* \in \mathbb{R}^k$ and for any function $f(\mathbf{u}, \mathbf{x})$ concave in $\mathbf{u}$ for all $\mathbf{x}$,

$$\text{if } f(\mathbf{u}, \mathbf{x}^*) \leq 0 \, \forall \mathbf{u} \in \mathcal{U}, \text{then } \mathbb{P}^*(f(\mathbf{u}, \mathbf{x}^*) \leq 0) \geq 1 - \epsilon.$$

The data-driven schema we follow is based on sampling a set of data points $\mathcal{S}$ i.i.d. from the true distribution $\mathbb{P}^*$ and uses hypothesis testing to construct the uncertainty sets with such guarantees. In particular, for confidence level $\alpha < 1$, the schema employs the corresponding $(1 - \alpha)$ confidence region to build $\mathcal{U}$. With the proper construction, the following theorem from [1, Sect. 3.2] holds.

Theorem 1. *With probability at least* $1 - \alpha$ *with respect to the sampling, the resulting set* $\mathcal{U}(\mathcal{S}, \epsilon, \alpha)$ *implies a probabilistic guarantee at least* ϵ *for* $\mathbb{P}^*$.

In [1], the authors show how different uncertainty sets are built depending on the assumptions about $\mathbb{P}^*$, and, in turn, on the suitable statistical test. In this work we consider box sets (i.e. multi-dimensional intervals), which make no assumptions on $\mathbb{P}^*$ and are suitable for data with missing values (see the technical report [24] for further details on assumptions and set construction). The application of other types of uncertainty sets, able for instance to capture temporal dependencies and correlation between meals and exercise, is in our future plans.

To shrink the size of uncertainty set, we employ the following two strategies: (1) prior to set construction, we classify the input data and partition it into a number of clusters so as to obtain tighter sets and more customized, patient-specific control strategies. (2) based on Algorithm 1 of [1], we use bootstrapping [5] to approximate the threshold of the test statistics, by estimating the sampling distribution of the statistics through re-sampling with replacement.

We remark that the construction of uncertainty sets is performed off-line and thus has no computational footprint on the robust controller.

5 Results and Discussion

We evaluate our robust control algorithm through a number of experiments for simulating: intake of a single meal (Sect. 5.1), exercise (Sect. 5.2), one-day meal intake scenario with patient behavior learned from population-wide survey data (Sect. 5.3), and two-day scenario with irregular meal timing and unusually high CHO intake (Sect. 5.4). Section 5.5 is dedicated to the analysis of state estimation. For each experiment, we compare the robust controller with the non-robust, hybrid closed-loop (HCL) variant introduced in Sect. 4. We also report the ideal performance by running a so-called *perfect controller*, that can access both the full plant state (i.e. does not need state estimation) and the exact values of the uncertainty parameters in the plant.

Hardware and Performance: We ran the experiments on a Windows 8 machine with an Intel Core i7 processor and 32 GB of DDR3 memory. We used MATLAB version 2016b. With this configuration, the average time to compute the insulin therapy over all the experiments ranged from 4 to 18 s, which is well within the CGM measurement period of 5 min. This means that the controller works faster than real-time. Given the significant performance improvement of modern embedded and mobile devices, we expect our algorithm to perform similarly as well once deployed on such hardware platforms.

Performance Indicators: To measure the efficacy of our robust controller design over multiple runs, we consider the following indicators:

- $t_{<3.9}$, $t_{3.9-11.1}$, $t_{>11.1}$: mean percentage of time spent in, respectively, hypoglycemia (BG < 3.9 mmol/L), normal ranges (BG between 3.9 and 11.1), and hyperglycemia (BG > 11.1). Clearly, we wish to maximize $t_{3.9-11.1}$ and minimize the other two indicators, keeping in mind that we can tolerate some temporary postpandrial hyperglycemia while hypoglycemia should be avoided as much as possible.
- $BG_{\min}$, $BG_{\max}$: average low BG level and peak BG level, respectively, in mmol/L. An effective robust controller should keep $BG_{\min}$ and $BG_{\max}$ as close as possible to the target BG level.
- $\sum \iota$: mean total non-basal insulin (in U). This indicator measures the amount of insulin injected by the controller in order to cover meals, and thus excludes the contribution of basal insulin.

To evaluate state estimation, we further consider indicators E_{D_G}, E_{MM}, E_{O2}, i.e. the mean absolute error between plant and estimated uncertain variable values, and E_{BG}, the mean absolute error between plant BG and estimated BG.

5.1   One-Meal Experiments

We consider 300-minute simulations comprising a single meal, and three different synthetic scenarios (illustrated in Fig. 3(a–c)), i.e. where meals are sampled from arbitrary distributions. For each scenario and controller, we collect results for 50 repetitions. Further details on the construction of uncertain sets from arbitrary distributions are available in the technical report [24].

Scenario 1, Meals as Expected: in the uncertain plant, we assume a uniformly distributed meal with start time $t_s = \text{unif}(30, 90)$, total amount of CHO (grams) $CHO = \text{unif}(42, 78)$ and meal duration fixed to 20 min, during which CHO ingestion happens at a constant rate. Given that uniform distributions have bounded support, we can build tight box-type uncertainty sets (i.e. intervals) that contain all possible realizations. This scenario allows us evaluating the adequacy of the controller when the plant behaves according to a known distribution, in other words, when we have accurate information for building uncertainty sets.

Scenario 2, Outliers: in this case, random meals behave as statistical outliers, i.e. they are constantly distant from the expected value of the underlying distribution. To this purpose, we build the uncertainty sets under the assumption that meals are normally distributed with parameters $t_s = \mathcal{N}(60, 15)$ and $CHO = \mathcal{N}(60, 9)$. The uncertainty sets are built so as to cover all possible realizations with z-score between -3 and 3 (i.e. between -3 S.D. and $+3$ S.D. around the mean). However, to reproduce outliers, meals in the uncertain plant are sampled from the tails of the distributions (z-scores in $[-4, -3]$ and $[3, 4]$).

Scenario 3, Late Meals: here we consider the same settings as in Scenario 1, but with each random meal delayed of one hour. This models the situation where the controller has wrong information about the meal schedule, since it expects the meal to start, on the average, one hour earlier.

Results in Fig. 3 show that our robust controller attains very good performance, closely following the ideal behavior of the perfect controller in the first and third scenarios, where the virtual patient stays in normal ranges for >97% of the time. In the outliers scenario, we register some postprandial hyperglycemia, because this scenario is characterized by frequent high CHO intake. Overall, the robust controller is able to limit the time spent in hypoglycemia below 1% and consistently outperforms the HCL controller, staying in normal BG ranges for 3% to 31% more (full statistics are available in the report [24]).

5.2   Regulation During Exercise

We evaluate the behavior of the robust controller when the virtual patient is involved in physical activity, which, contrarily to meals, contributes to decreasing BG levels. We simulate a two-legged exercise consisting of two phases:

1. **Moderate activity**, with start time $t_s = \text{unif}(40, 80)$, duration $d = \text{unif}(24, 36)$, active muscular mass $MM = \text{unif}(0.15, 0.35)$, and oxygen consumption $O2 = \text{unif}(45, 75)$; followed by

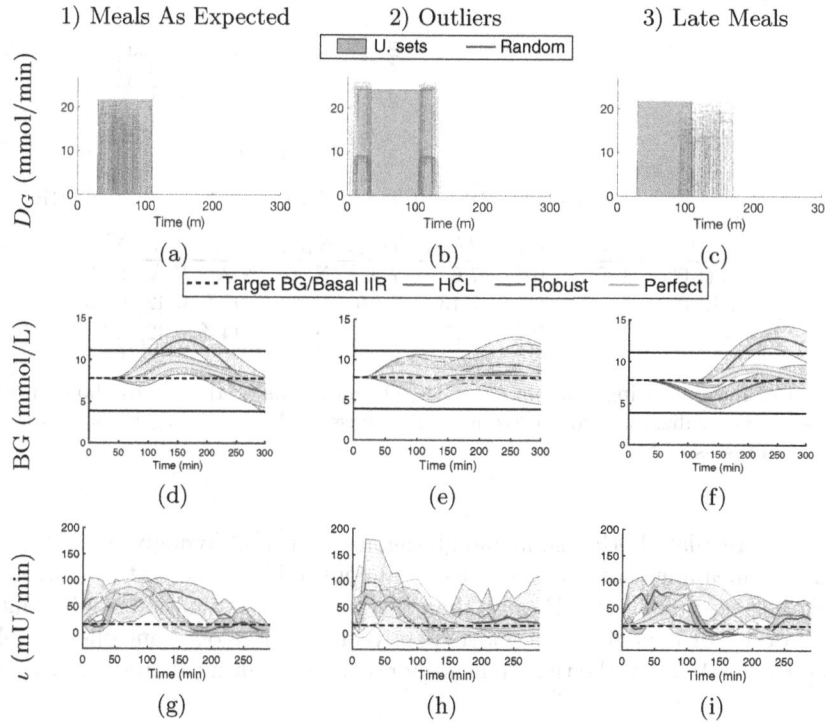

Fig. 3. One-meal, 300-minute experiments (50 repetitions). **Top:** uncertainty sets and random realizations of parameter D_G (rate of CHO ingestion). **Middle:** BG profiles (with solid black lines indicating the normal BG range). **Bottom:** synthetized insulin therapies. Thick solid lines indicate average BG/insulin values, and are surrounded by an area spanning $\pm$ 1 S.D. In the table, we highlight in bold the best value of each index between the robust and the HCL controllers.

2. **Light activity**, where parameters stay as in the previous phase except for $02 = \text{unif}(15, 45)$.

Results, reported in Fig. 4, evidence that both the robust and the HCL controller can maintain BG within very tight ranges, as confirmed by the $BG_{\min}$ and $BG_{\max}$ indicators. BG profiles are almost indistinguishable from the ideal ones (i.e. those of the perfect controller) and for 100% of the times within healthy ranges. Note that both controllers correctly reduce the insulin therapy below the basal level to counteract the decrease of BG due to exercise. Hence, the negative values of $\sum \iota$. The main difference is that the robust controller, due to the superior predictive capabilities, is more timely in cutting insulin therapy than the HCL controller, leading to a smaller excursion from the target BG value.

Resalat et al. [28] realized a similar scenario to test their dual-hormone MPC (300-minute simulation with a 45-minute exercise at fixed $02 = 60$ and $MM = 0.8$). While we use their same plant model, their MPC design is different in two

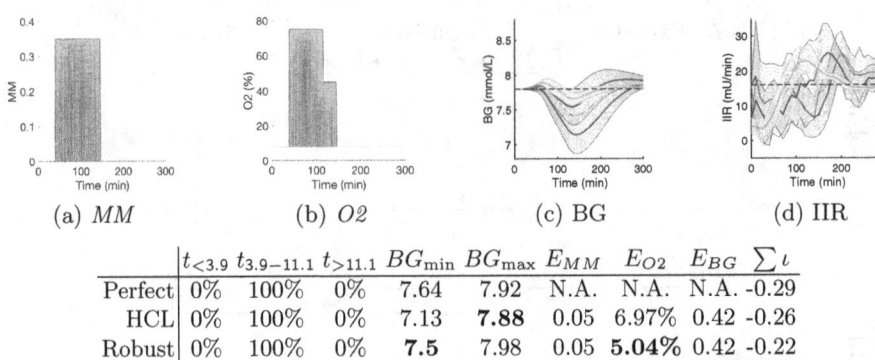

	$t_{<3.9}$	$t_{3.9-11.1}$	$t_{>11.1}$	BG_{min}	BG_{max}	E_{MM}	E_{O2}	E_{BG}	$\sum \iota$
Perfect	0%	100%	0%	7.64	7.92	N.A.	N.A.	N.A.	-0.29
HCL	0%	100%	0%	7.13	**7.88**	0.05	6.97%	0.42	-0.26
Robust	0%	100%	0%	**7.5**	7.98	0.05	**5.04%**	0.42	-0.22

Fig. 4. Regulation during random exercise (50 repetitions). (a) and (b) show uncertainty sets and realizations for active muscular mass (MM) and oxygen consumption ($O2$). Legend is as in Fig. 3.

ways: it can regulate both insulin and glucagon (to prevent hypoglycemia) and is not robust, meaning that exercise must be announced in order for the controller to make correct predictions. Despite that, however, their evaluation resulted into some episodes of hypoglycemia and hyperglycemia, while our controller is able to keep BG for 100% of the time in healthy ranges without meal announcements.

5.3   One-Day Experiments Using NHANES Survey Data

We test our robust controller with real population data from the CDC's National Health and Nutrition Examination Survey (NHANES) database.[2] We consider the 2013 survey, comprising 8,611 participants, and classify the participants into 10 groups using k-means clustering. In this experiment, we selected the cluster whose meal patterns are characterized by a CHO-rich breakfast at around 9am, as visible in the uncertainty set of Fig. 5(a). From this cluster, we extract meal information to parameterize the virtual patient and build the uncertainty sets as explained in Sect. 4.2 (choosing $\alpha = 0.2$ and $\epsilon = 0.2$). Due to the poor quality of physical activity data in NHANES, we generated one random exercise event for each patient. Details on the other clusters and on extraction and processing of data are reported in [24].

Results were obtained with 20 repetitions and are reported in Fig. 5. In this experiment, our robust controller has a close-to-ideal performance, with >93% of time spent in normal BG ranges. It outperforms the HCL controller, which fails to predict the correct BG levels during sleep (time < 500 min), leading to excessive insulin therapy and to dangerous overnight hypoglycemia.

---

[2] https://www.cdc.gov/nchs/nhanes/.

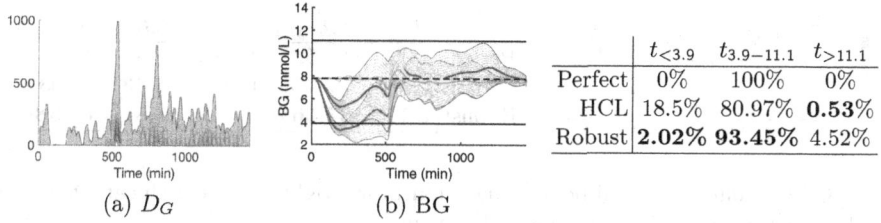

	$t_{<3.9}$	$t_{3.9-11.1}$	$t_{>11.1}$
Perfect	0%	100%	0%
HCL	18.5%	80.97%	**0.53%**
Robust	**2.02%**	**93.45%**	4.52%

(a) D_G (b) BG

Fig. 5. BG regulation for virtual patient learned from NHANES database (20 repetitions). Legend is as in Fig. 3.

5.4 High Carbohydrate Intake Scenario

We assess the behavior of the controller under irregular meal timing and unusually high CHO intake, following the protocol of [31], reported in Table 1. In this protocol, no physical activity is considered. Uncertainty sets were derived following the same construction of the one-meal experiments. Results, obtained with 50 repetitions, are shown in Fig. 6.

Our robust controller resulted in 87.56% of time within healthy BG ranges, against the 80.6% of the HCL controller. Despite hypoglycemia amounts to 3.11% of the total time, it corresponds only to minor episodes, as visible by the standard deviation intervals in the plot and by the average minimum BG

Table 1. High carbohydrate intake simulation parameters of [31]. Meals in the plant are sampled uniformly based on the above intervals and probabilities.

	Chance of occurrence	CHO (g)	Time of day (h)
Breakfast	100%	40–60	6:00–10:00
Snack 1	50%	5–25	8:00–11:00
Lunch	100%	70–110	11:00–15:00
Snack 2	50%	5–25	15:00–18:00
Dinner	100%	55–75	18:00–22:00
Snack 3	50%	5–15	22:00–00:00

$(BG_{min} = 3.84 \, \text{mmol/L})$ that falls only slightly below the hypoglycemic level (3.9 mmol/L).

We also report that our approach outperforms the robust LPV approach of Jacobs et al. [31], discussed in the related work (Sect. 6). With the same plant model and scenario, they obtain $t_{<3.9} = 0\%$, $t_{3.9-11.1} = 83.08\%$ and $t_{>11.1} = 16.92\%$, meaning that our robust controller stays >4% of the time longer in healthy ranges. We remark that the results of Jacobs et al. are as reported in [31], and were not obtained by running their controller on our machine.

5.5 Evaluation of State Estimator

We chose an MHE scheme for state estimation (see Sect. 4.1) after having evaluated *extended Kalman filters (EKF)* [35], which are commonly employed for the state estimation of non-linear systems. MHE overcomes some of the typical problems of Kalman filtering, namely, the inability to accurately incorporate state constraints (e.g. non-negative concentrations); poor use of the nonlinear

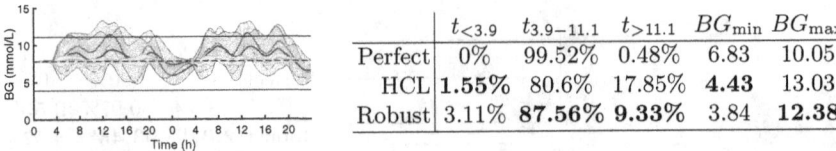

	$t_{<3.9}$	$t_{3.9-11.1}$	$t_{>11.1}$	$BG_{\min}$	$BG_{\max}$
Perfect	0%	99.52%	0.48%	6.83	10.05
HCL	**1.55%**	80.6%	17.85%	**4.43**	13.03
Robust	3.11%	**87.56%**	**9.33%**	3.84	**12.38**

Fig. 6. BG profile (left) and performance indicators (right) for the high carbohydrate intake scenario (50 repetitions). Legend is as in Fig. 3.

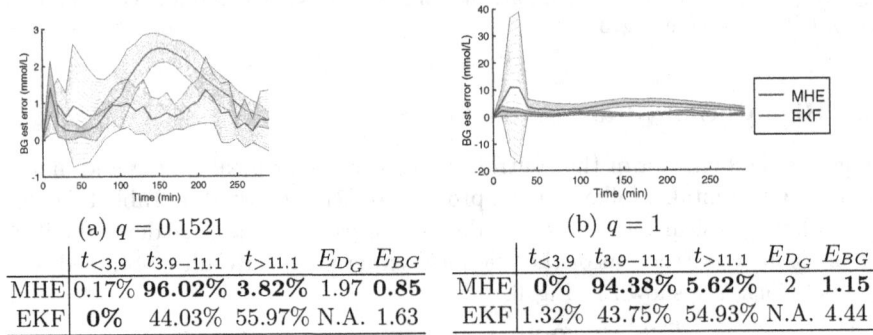

(a) $q = 0.1521$

	$t_{<3.9}$	$t_{3.9-11.1}$	$t_{>11.1}$	E_{D_G}	E_{BG}
MHE	0.17%	**96.02%**	**3.82%**	1.97	**0.85**
EKF	**0%**	44.03%	55.97%	N.A.	1.63

(b) $q = 1$

	$t_{<3.9}$	$t_{3.9-11.1}$	$t_{>11.1}$	E_{D_G}	E_{BG}
MHE	**0%**	**94.38%**	**5.62%**	2	**1.15**
EKF	1.32%	43.75%	54.93%	N.A.	4.44

Fig. 7. BG estimation error of Moving Horizon Estimator (MHE) and Extended Kalman Filter (EKF), at different sensing noise variances q (20 repetitions).

model [8]; and estimations that often diverge, or converge to wrong state predictions [26,32]. Moreover, "off-the-shelf" Kalman filters only support zero-mean disturbances (white Gaussian noise), thus preventing the estimation of random meal and exercise episodes.

We compare the state estimation accuracy between our MHE design and an EKF scheme, according to the *meals as expected* scenario (see Sect. 5.1). In the EKF, to predict the state estimate at time t, $\hat{\mathbf{x}}(t)$, we use the model of Sect. 3 as follows: $\dot{\hat{\mathbf{x}}}(t) = F(\hat{\mathbf{x}}(t), \iota^t, \mathrm{E}[\mathbf{u}^t])$, where ι^t is the (known) insulin input and uncertainty parameters $\mathbf{u}^t$ are replaced with their expected value $\mathrm{E}[\mathbf{u}^t]$[3].

To evaluate if the estimators are robust with respect to sensing noise, we tested two different variance values for the sensing noise: $q = 0.1521$ (default) and $q = 1$ (increased noise). As visible in Fig. 7, the MHE outperforms the EKF, with a consistently lower state estimation error. The imprecise state predictions of the EKF lead to a wrong behavior of the overall closed-loop system, with only $\sim 44\%$ of time spent within normal BG ranges, against $>94\%$ of the MHE. Unlike the EKF, the MHE is robust to sensing noise, with an average estimation error (column E_{BG}) that stays relatively constant from $q = 0.1521$ to $q = 1$.

---

[3] The real expected value of $\mathbf{u}^t$ is known because here we work with arbitrary distributions.

6   Related Work

Robust control methods are able to minimize the impact of input disturbances on the plant, and thus have the potential to enable fully closed-loop insulin delivery. Earlier approaches [14,25,29] are based on the theory of H_∞ control [30], a technique where the robust controller is synthesized offline as the result of an optimization problem that minimizes the worst-case closed-loop performance of the controlled system. However, H_∞ control only supports linear systems, thus requiring linearization of physiological, non-linear gluco-regulatory models, with inevitable loss of accuracy.

Kovacs et al. [15,16,31] introduce robust linear parameter varying (LPV) control, a technique that consists on deriving a piecewise-linear approximation of the non-linear plant and synthesizing a robust H_∞ controller for each linear region, and thus, improves on previous H_∞ approaches. In Sect. 5.4, we have compared our robust controller to [31], showing that our algorithm is able keep glucose levels within normal ranges for a longer time.

In contrast to the above techniques, our data-driven robust MPC supports not just meal disturbances, but also physical activity, and is based on non-linear optimization, meaning that it does not require to approximate the system dynamics, leading to more precise predictions. Further, MPC is known to be superior for individualized control strategies [4,23,33], even though is computationally more demanding than offline techniques like H_∞ or LPV control, but still feasible within the update periods typical of the artificial pancreas (5–10 min). Finally, our data-driven scheme supports continuous learning of the patient's behavior, thus enabling the synthesis of robust and adaptive insulin therapies. On the other hand, H_∞ and LPV controllers are offline and need to be synthesized from scratch in order to adapt to changing patient conditions.

A simpler strategy employed in a number of AP studies, see e.g. [12,18], is that of PID control, where the control input results from applying tunable gains to the error between the system output and a desired setpoint. Synthesizing these gains to obtain robustness guarantees, however, becomes difficult for systems with nonlinear and probabilistic dynamics.

7   Conclusions

Thanks to modern wearable sensing devices, patient-specific data about meals and physical activity is becoming more readily available, making it possible to offer significantly enhanced personalized medical therapy for type 1 diabetes. Accordingly, we presented a data-driven robust MPC framework for T1D that leverages meal and exercise data to provide enhanced control and state estimation. Our results show that learning a patient's behavior from data is key to achieving fully closed-loop therapy that does not require meal and exercise announcements.

Acknowledgments. Research supported in part by AFOSR Grant FA9550-14-1-0261 and NSF Grants IIS-1447549, CNS-1446832, CNS-1445770, CNS-1445770, CNS-1553273, CNS-1536086, and IIS-1460370.

References

1. Bertsimas, D., Gupta, V., Kallus, N.: Data-driven robust optimization. arXiv preprint arXiv:1401.0212 (2013)
2. Centers for Disease Control and Prevention. National Diabetes Statistics Report: Estimates of Diabetes and its Burden in the United States. US Department of Health and Human Services, Atlanta (2014)
3. Dassau, E., Bequette, B.W., Buckingham, B.A., Doyle, F.J.: Detection of a meal using continuous glucose monitoring. Diab. Care **31**(2), 295–300 (2008)
4. De Nicolao, G., Magni, L., Dalla Man, C., Cobelli, C.: Modeling and control of diabetes: towards the artificial pancreas. IFAC Proc. Vols. **44**(1), 7092–7101 (2011)
5. Efron, B., Tibshirani, R.J.: An Introduction to the Bootstrap. CRC Press, Boca Raton (1994)
6. Gondhalekar, R., Dassau, E., Doyle, F.J.: Moving-horizon-like state estimation via continuous glucose monitor feedback in MPC of an artificial pancreas for type 1 diabetes. In: 2014 IEEE 53rd Annual Conference on Decision and Control (CDC), pp. 310–315. IEEE (2014)
7. Gondhalekar, R., Dassau, E., Doyle, F.J.: Periodic zone-MPC with asymmetric costs for outpatient-ready safety of an artificial pancreas to treat type 1 diabetes. Automatica **71**, 237–246 (2016)
8. Haseltine, E.L., Rawlings, J.B.: Critical evaluation of extended Kalman filtering and moving-horizon estimation. Ind. Eng. Chem. Res. **44**(8), 2451–2460 (2005)
9. Hernandez-Ordonez, M., Campos-Delgado, D.: An extension to the compartmental model of type 1 diabetic patients to reproduce exercise periods with glycogen depletion and replenishment. J. Biomech. **41**(4), 744–752 (2008)
10. Hovorka, R.: Closed-loop insulin delivery: from bench to clinical practice. Nat. Rev. Endocrinol. **7**(7), 385–395 (2011)
11. Hovorka, R., et al.: Nonlinear model predictive control of glucose concentration in subjects with type 1 diabetes. Physiol. Meas. **25**(4), 905 (2004)
12. Huyett, L.M., Dassau, E., Zisser, H.C., Doyle III, F.J.: Design and evaluation of a robust PID controller for a fully implantable artificial pancreas. Ind. Eng. Chem. Res. **54**(42), 10311–10321 (2015)
13. Jacobs, P.G., et al.: Incorporating an exercise detection, grading, and hormone dosing algorithm into the artificial pancreas using accelerometry and heart rate. J. Diab. Sci. Technol., 1932296815609371 (2015)
14. Kienitz, K.H., Yoneyama, T.: A robust controller for insulin pumps based on H-infinity theory. IEEE Trans. Biomed. Eng. **40**(11), 1133–1137 (1993)
15. Kovács, L., Benyó, B., Bokor, J., Benyó, Z.: Induced L2-norm minimization of glucose-insulin system for Type I diabetic patients. Comput. Methods Programs Biomed. **102**(2), 105–118 (2011)
16. Kovács, L., Szalay, P., Almássy, Z., Barkai, L.: Applicability results of a nonlinear model-based robust blood glucose control algorithm. J. Diab. Sci. Technol. **7**(3), 708–716 (2013)
17. Kovatchev, B., et al.: Feasibility of long-term closed-loop control: a multicenter 6-month trial of 24/7 automated insulin delivery. Diab. Technol. Ther. (2017)

18. Laxminarayan, S., Reifman, J., Steil, G.M.: Use of a food and drug administration-approved type 1 diabetes mellitus simulator to evaluate and optimize a proportional-integral-derivative controller. J. Diab. Sci. Technol. **6**(6), 1401–1412 (2012)
19. Lee, H., Buckingham, B.A., Wilson, D.M., Bequette, B.W.: A closed-loop artificial pancreas using model predictive control and a sliding meal size estimator. J. Diab. Sci. Technol. **3**(5), 1082–1090 (2009)
20. Lee, J.J., Gondhalekar, R., Doyle, F.J.: Design of an artificial pancreas using zone model predictive control with a moving horizon state estimator. In: 2014 IEEE 53rd Annual Conference on Decision and Control (CDC), pp. 6975–6980. IEEE (2014)
21. Lenart, P.J., Parker, R.S.: Modeling exercise effects in type i diabetic patients. IFAC Proc. Vols. **35**(1), 247–252 (2002)
22. Ly, T.T., et al.: Day and night closed-loop control using the integrated Medtronic hybrid closed-loop system in type 1 diabetes at diabetes camp. Diab. Care **38**(7), 1205–1211 (2015)
23. Magni, L., et al.: Model predictive control of glucose concentration in type I diabetic patients: an in silico trial. Biomed. Signal Process. Control **4**(4), 338–346 (2009)
24. Paoletti, N., Liu, K.S., Smolka, S.A., Lin, S.: Data-driven robust control for type 1 diabetes under meal and exercise uncertainties. CoRR, 1707.02246 (2017)
25. Parker, R.S., Doyle, F.J., Ward, J.H., Peppas, N.A.: Robust H∞ glucose control in diabetes using a physiological model. AIChE J. **46**(12), 2537–2549 (2000)
26. Perea, L., How, J., Breger, L., Elosegui, P.: Nonlinearity in sensor fusion: divergence issues in EKF, modified truncated GSF, and UKF. In: AIAA Guidance, Navigation and Control Conference and Exhibit, p. 6514 (2007)
27. Rao, C.V., Rawlings, J.B., Mayne, D.Q.: Constrained state estimation for nonlinear discrete-time systems: stability and moving horizon approximations. IEEE Trans. Autom. Control **48**(2), 246–258 (2003)
28. Resalat, N., El Youssef, J., Reddy, R., Jacobs, P.G.: Design of a dual-hormone model predictive control for artificial pancreas with exercise model. In: 2016 IEEE 38th Annual International Conference of the Engineering in Medicine and Biology Society (EMBC), pp. 2270–2273. IEEE (2016)
29. Ruiz-Velázquez, E., Femat, R., Campos-Delgado, D.: Blood glucose control for type I diabetes mellitus: a robust tracking H∞ problem. Control Eng. Pract. **12**(9), 1179–1195 (2004)
30. Stoorvogel, A.A.: The H∞ Control Problem: A State Space Approach. Prentice Hall, Upper Saddle River (1992)
31. Szalay, P., Eigner, G., Kovács, L.A.: Linear matrix inequality-based robust controller design for type-1 diabetes model. IFAC Proc. Vols. **47**(3), 9247–9252 (2014)
32. Van Der Merwe, R.: Sigma-point Kalman filters for probabilistic inference in dynamic state-space models. Ph.D. thesis, Oregon Health & Science University (2004)
33. Wang, Y., Zisser, H., Dassau, E., Jovanovič, L., Doyle, F.J.: Model predictive control with learning-type set-point: Application to artificial pancreatic β-cell. AIChE J. **56**(6), 1510–1518 (2010)
34. Weimer, J., Chen, S., Peleckis, A., Rickels, M.R., Lee, I.: Physiology-invariant meal detection for type 1 diabetes. Diab. Technol. Ther. **18**(10), 616–624 (2016)
35. Welch, G., Bishop, G.: An Introduction to the Kalman Filter. Technical report, University of North Carolina at Chapel Hill, Chapel Hill, NC, USA (1995)

36. Wilinska, M.E., et al.: Insulin kinetics in type-1 diabetes: continuous and bolus delivery of rapid acting insulin. IEEE Trans. Biomed. Eng. **52**(1), 3–12 (2005)
37. Wilinska, M.E., et al.: Simulation environment to evaluate closed-loop insulin delivery systems in type 1 diabetes. J. Diab. Sci. Technol. **4**(1), 132–144 (2010)
38. Zavitsanou, S., Chakrabarty, A., Dassau, E., Doyle, F.J.: Embedded control in wearable medical devices: application to the artificial pancreas. Processes **4**(4), 35 (2016)

Graph Representations of Monotonic Boolean Model Pools

Robert Schwieger(✉) and Heike Siebert

Freie Universität Berlin, Berlin, Germany
rschwieger@zedat.fu-berlin.de

Abstract. In the face of incomplete data on a system of interest, constraint-based Boolean modeling still allows for elucidating system characteristics by analyzing sets of models consistent with the available information. In this setting, methods not depending on consideration of every single model in the set are necessary for efficient analysis. Drawing from ideas developed in qualitative differential equation theory, we present an approach to analyze sets of monotonic Boolean models consistent with given signed interactions between systems components. We show that for each such model constraints on its behavior can be derived from a universally constructed state transition graph essentially capturing possible sign changes of the derivative. Reachability results of the modeled system, e.g., concerning trap or no-return sets, can then be derived without enumerating and analyzing all models in the set. The close correspondence of the graph to similar objects for differential equations furthermore opens up ways to relate Boolean and continuous models.

1 Introduction

Mathematical modeling in systems biology is often hampered by lack of information on mechanistic detail and parameters. Constraint-based Boolean modeling still allows investigations based on restricted knowledge, e.g., on component dependencies and impact of certain interactions, by considering sets of models consistent with such constraints. However, analysis of every single model in the set is costly. Exploiting formal verification techniques still allows to investigate the behavior of large numbers of models in this context [10,14]. A different approach aims at avoiding enumeration and explicit analysis of every model in the set by deriving properties directly from the given constraints, e.g., inferring dynamical information from coinciding structural characteristics of all models [9,11–13]. Here, we adopt the latter approach for sets of Boolean networks consistent with a given signed interaction graph Σ capturing dependencies between system components and the type of influence exerted, activating or inhibiting. This constitutes a scenario of particular interest in application, where interaction information is usually more readily available than details on the processing logic of multiple influences on a target component.

© Springer International Publishing AG 2017
J. Feret and H. Koeppl (Eds.): CMSB 2017, LNBI 10545, pp. 233–248, 2017.
DOI: 10.1007/978-3-319-67471-1_14

A similar scenario motivates the theory of qualitative differential equations (QDE) [4]. Here, a signed interaction graph is interpreted as the sign structure of a Jacobi matrix of an ordinary differential equation (ODE) system. All ODE-systems, which are consistent with a given interaction graph are collected in a so-called monotonic ensemble $\mathscr{M}(\Sigma)$. For this ensemble, a qualitative state transition graph (QSTG) can be constructed whose nodes represent derivative signs of the system components and edges indicate possible changes in the derivative over time. It can then be used to describe the behavior of the ensemble. Similar ideas have been exploited successfully for piecewise linear differential equations by de Jong and colleagues for systems biology modeling [2,3].

Motivated by the results in the QDE setting, we show that a similar graph G carries meaning in the Boolean framework as well. Here, for each Boolean function f consistent with a given interaction graph, we are interested in the asynchronous state transition graph (ASTG) capturing the dynamics of the model. We show that while the ASTG cannot be related directly to G this becomes possible for a quotient graph derived from the ASTG by identifying system states with the same image under f. This quotient graph needs to be a subgraph of G. Consequently, analysis of G allows to infer reachability constraints valid for all models consistent with the interaction graph. In particular, universal statements about trap sets and attractors become possible. The close correspondence of G to the QSTG of a family of ODEs furthermore allows to relate discrete and continuous model ensembles, which can facilitate the preprocessing of continuous data for Boolean models as well as prove useful in model validation.

Our paper is structured in the following way: In the first section we state definitions and notions about Boolean regulatory networks. Afterwards, we review existing results for monotonic ensembles in the continuous setting and transfer these ideas to the Boolean framework in Sect. 3. Subsequently, we exploit the results by investigating how information about trap sets and no-return sets can be obtained for sets of models consistent with a given interaction graph without enumeration. Section 5 then touches upon some aspects relating Boolean and ODE models. A short discussion concludes the paper. To allow for easy reproduction of our results, our Python implementation is publicly accessible in the following git-repository: https://github.com/RSchwieger/QDE.

2   Preliminaries

Throughout the paper we consider a system of components $1, \ldots, n$, $n \in \mathbb{N}$. As general notation for different and coinciding entries or arbitrary vectors, we use for $v, w \in S^n$, S any set:

$$\operatorname{diff}(v, w) := \{i \in \{1, \ldots, n\} | v_i \neq w_i\},$$
$$\operatorname{comm}(v, w) := \{1, \ldots, n\} \backslash \operatorname{diff}(v, w).$$

2.1   Boolean Networks

We consider an arbitrary Boolean function $f : \{0,1\}^n \to \{0,1\}^n$, $n \in \mathbb{N}$ capturing the dynamics of n interacting components represented by 0/1 variables.

Definition 1. *The* discrete derivative *of the function $f : \{0,1\}^n \to \{0,1\}^n$ is defined by*

$$(\partial_j f_i)(x) := \frac{f_i(x \oplus e_j) - f_i(x)}{(x_j \oplus 1) - x_j} \in \{-1,0,1\},$$

where $\oplus$ is the addition modulo 2. Furthermore, we denote with ∇f_i the vector $(\partial_1 f_i, \ldots, \partial_n f_i)^t$.

As is standard, we then derive an interaction graph from the derivatives that captures dependencies between the components, either locally, i.e., in a given state, or globally summarizing all possible interactions between components.

Definition 2. *The* local interaction graph $IG_f(x) := (V, E)$, $x \in \{0,1\}^n$ *of a Boolean function $f : \{0,1\}^n \to \{0,1\}^n$ consists of n vertices $V := \{1,\ldots,n\}$ and a signed edge-set $E(IG_f(x))$, which is defined as*

$$(i, j, \epsilon) \in E(IG_f(x)) \Leftrightarrow (\partial_i f_j)(x) = \epsilon$$

with $\epsilon \in \{-1,1\}$. We denote with $IG^{global}(f)$ the global interaction graph defined as the union of all local interaction graphs, i.e.,

$$IG^{global}(f) = \bigcup_{x \in \{0,1\}^n} IG_f(x)$$

For convenience, we often identify an interaction graph with its signed adjacency matrix.

In general, the global interaction graph can contain two edges with opposite signs between two components. However, here we consider only functions f, which lead to interaction graphs with maximally one edge between two components. That is, in the following we only consider Boolean functions $f : \{0,1\}^n \to \{0,1\}^n$, where

$$\forall x, y \in \{0,1\}^n \forall \epsilon \in \{1,-1\} : (s, t, \epsilon) \in E(IG_f(x)) \Rightarrow (s, t, -\epsilon) \notin E(IG_f(y))$$

holds. We call such functions *monotonic*. This should not be confused with the notion of monotone functions as defined for example in [8], which is more restrictive. The assumption poses no severe restriction for application since most models of bioregulatory systems lack parallel edges of different signs (cf. model repositories as e.g. for PyBoolNet [7][1]).

---

[1] https://github.com/hklarner/PyBoolNet/tree/master/PyBoolNet/Repository.

In addition to the (local) interaction graph(s), we attribute to f a second graph, which describes the dynamics of the components. Here, we use an asynchronous update scheme attributing to f an *asynchronous state transition* graph $G_{\mathrm{async}}(f) = (V_{\mathrm{async}}(f), E_{\mathrm{async}}(f))$ with $V_{\mathrm{async}}(f) := \{0,1\}^n$ and

$$E_{\mathrm{async}}(f) = \{(s,t) \in V_{\mathrm{async}}(f) \times V_{\mathrm{async}}(f) \mid (\mathrm{diff}(s,t) = \{i\} \text{ and } f_i(s) = t_i)$$
$$\text{or } s = t = f(s)\}.$$

Example 1. Let $f(x_1, x_2, x_3, x_4) = (1 - x_4, x_1, x_2 \cdot (1 - x_4), 1 - x_3)$ be a Boolean function. Its global interaction graph and asynchronous state transition graph is depicted in Fig. 1. The global interaction graph is given by the adjacency matrix

$$\Sigma = \begin{pmatrix} 0 & 0 & 0 & - \\ + & 0 & 0 & 0 \\ 0 & + & 0 & - \\ 0 & 0 & - & 0 \end{pmatrix},$$

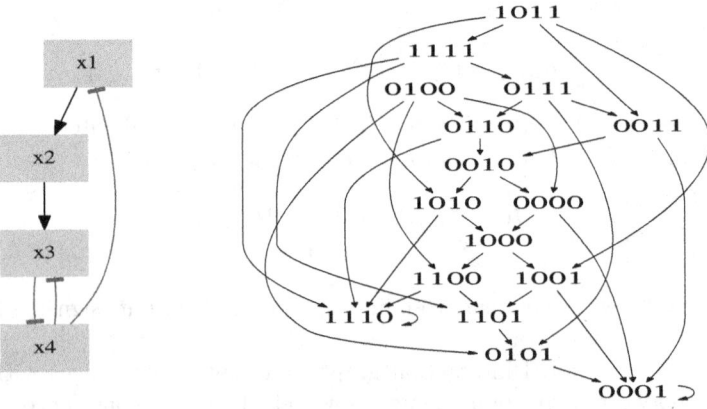

Fig. 1. Left: Global interaction graph of the running example. Right: ASTG of the running example.

2.2   Monotonic Ensembles and Qualitative Differential Equations

The motivation of our results on sets of Boolean models comes from approaches to analyze families of ODE models $\dot{x} = f(x)$ which share some qualitative properties. Mainly, these are sign constraints on the Jacobi matrix of the right hand sides of the ODE-System. Instead of the solutions $x(\cdot)$ of the ODE-systems, so-called "abstractions" are considered. In the context of this paper, these abstractions are sequences of sign vectors of the derivatives of the solutions. A state transition graph on the sign vectors can be constructed based on the sign matrix, which captures restrictions on the behavior of the solutions.

This graph will be our focus in the subsequent sections. Notions and definitions in this section are taken from [4].

We define an ensemble of ODE systems whose corresponding Jacobi matrices share a sign structure. The usual *sign operator* is denoted $[\cdot] := sign(\cdot)$, taking values in $\{-1, 0, 1\}$ and extended componentwise to vectors and matrices.

Definition 3 ([4, p. 22]). *For a given $n \times n$ matrix of signs $\Sigma = (\sigma_{i,j})_{i,j=1,\ldots,n}$, $\sigma_{i,j} \in \{-1, 0, 1\}$ and a state space $X \subseteq \mathbb{R}^n$ we define the* monotonic ensemble

$$\mathcal{M}(\Sigma, X) = \{f \in C^1(X, \mathbb{R}^n) \mid \forall x \in X : [J(f)(x)] = \Sigma\}, where$$

$J(f)$ denotes the Jacobian of f.

We call a function $x \in C^1([0, T], \mathbb{R}^n)$, $T \in [0, \infty]$, reasonable, if there is only a finite set of points t with $\dot{x}(t) = 0$ in any bounded interval. We define the space *of admissible trajectories by*

$$\mathcal{E} = \{x \in C^1([0, T], X) | x \text{ is reasonable}\}.$$

The solution set *$S_{\mathcal{M}(\Sigma,X)}$ for an initial value problem with $x(0) = x_0 \in X$ contains all reasonable solutions of corresponding ODE systems whose right hand side function is contained in the monotonic ensemble $\mathcal{M}(\Sigma, X)$, i.e.,*

$$S_{\mathcal{M}(\Sigma,X)} := \{x \in \mathcal{E} | \exists f \in \mathcal{M}(\Sigma, X), x_0 \in X \text{ s.t. } \dot{x} = f(x), x(0) = x_0\}.$$

The restriction to reasonable solutions is a technical detail needed to allow for a discretization which tracks the sign vectors $[\dot{x}(t)]$ for each solution and can be deduced directly from Σ [4, p. 22]. For conciseness, we will identify an ODE with its right hand side function, talking about ODEs as elements of the monotonic ensemble

We illustrate the notions on our running example.

Example 2. Consider all solutions of ODE-systems $\dot{x} = f(x)$ with $f \in C^1([0, 1]^4, \mathbb{R}^4)$ having a Jacobi matrix with the sign structure

$$\Sigma = \begin{pmatrix} 0 & 0 & 0 & - \\ + & 0 & 0 & 0 \\ 0 & + & 0 & - \\ 0 & 0 & - & 0 \end{pmatrix},$$

corresponding to the signed adjacency matrix of the interaction graph of the Boolean function given in Example 1. They constitute a monotonic ensemble denoted by $\mathcal{M}(\Sigma)$.

As an example for elements of this monotonic ensemble, we construct now a function $f \in \mathcal{M}(\Sigma)$. To connect to our Boolean example, we construct an

ODE-System from the function f given in Example 1 by using one of the methods explained in [15][2]. We obtain:

$$\dot{x} = \tilde{f}(x) - x$$
$$x(0) = x_0$$

with $\tilde{f} : [0,1]^4 \to [0,1]^4$ given by

$$\tilde{f}(x) = \left(1 - \frac{x_4}{x_4+0.5}\, \frac{x_1}{x_1+0.5}\, \frac{x_2}{x_2+0.5} \cdot \left(1 - \frac{x_4}{x_4+0.5}\right)\, 1 - \frac{x_3}{x_3+0.5}\right)^t .$$

It can easily be checked that the map $\tilde{f}$ is in the monotonic model ensemble $\mathscr{M}(\Sigma)$. Figure 2 shows the solution of the ODE-System, if we choose $x_0 = \left(0.6\ 0.6\ 0.6\ 0.6\right)^t$, and illustrates that it is a reasonable function, thus belonging to the solution set.

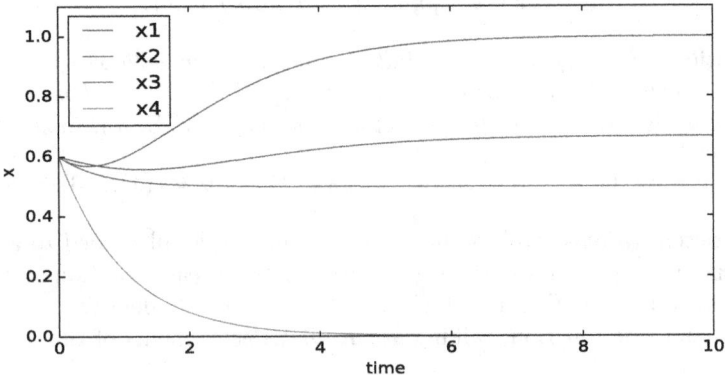

Fig. 2. Trajectories of a solution of an ODE in the model ensemble. Its abstraction is given by $(-1,-1,-1,-1) \to (1,-1,-1,-1) \to (1,1,-1,-1)$. The ODE was parametrized with $x_0 = (0.6, 0.6, 0.6, 0.6)$, $d = (1,1,1,1)$, $\theta = (0.5, 0.5, 0.5, 0.5)$ and $k = (1,1,1,1)$.

We are now looking for qualitative features of solutions in order to find properties common to all ODEs in the ensemble. The idea is to obtain a rough description of solution trajectories by keeping track of the sign changes in the derivative. In general, to each reasonable solution $x : [0,T] \to X$, T finite, we can assign a unique ordered, maximal sequence $(t_j)_j \in \{0, \ldots, M\}$, $t_j \in [0,T]$, with $t_0 = 0$, $t_M = T$ and $t_j \in (0,T)$ with the vector $\dot{x}(t_j)$ having a zero entry for $j \in \{1, \ldots, M-1\}$ indicating sign jumps of the trajectory. If T is infinite we can define a similar sequence not ending in T. In the interval between the

---

[2] More specifically speaking, we use a multivariate interpolation and a subsequent concatenation with Hill Cubes to obtain an ODE-System, which is guaranteed to have a Jacobi matrix, whose abstraction coincides with the matrix Σ on the off diagonals. Then we choose arbitrarily Hill coefficients and thresholds.

points in the sequence the sign of the solution derivative is then constant. We are now interested in the sequence of those signs. Formally, we get the following definition.

Definition 4 ([4, p. 23]). *For a solution* $x \in \mathcal{E}$, *consider the ordered sequence* (t_j) *in* $[0, T]$ *consisting of 0 and all boundary points of the closure of all sets* $\{t \in [0, T] | [\dot{x}(t)] = v\}$ *with* $v \in \{-1, 1\}^n$. *A sequence* (τ_j) *s.t.* $\tau_j \in (t_j, t_{j+1})$ *gives rise to the sequence* $\tilde{x} = (\tilde{x}_j) := ([\dot{x}(\tau_j)])$ *which is called* abstraction of $x(\cdot)$. *We denote the* set of abstractions *of the solutions of a monotonic ensemble* $\mathcal{M}(\Sigma)$ *by*

$$\tilde{S}_{\mathcal{M}(\Sigma)} := \{\tilde{x} \mid \tilde{x} \text{ is the abstraction of } x(\cdot) \text{ for some } x \in S_{\mathcal{M}(\Sigma)}\}.$$

To illustrate the notion, we extract the abstraction for a solution of the running example.

Example 3. Consider the solution of the ODE depicted in Fig. 2. Its abstraction is given by $((-1, -1, -1, -1), (1, -1, -1, -1), (1, 1, -1, -1))$, since on the beginning of the trajectory all components are decreasing. Then the first component starts increasing followed by the second one.

Based on the abstractions a state transition graph can now be constructed. The states correspond to the signs of $\dot{x}(\cdot)$ and edges indicate subsequent sign vectors in some abstraction.

Definition 5 ([4, p. 23]). *The* directed state transition graph $G_{\mathrm{QDE}}(\Sigma)$ *of the monotonic ensemble is defined by the vertex set*

$$V_{QDE}(\Sigma) := \{-1, 1\}^n,$$

called qualitative states, *and the edge set*

$$E_{QDE}(\Sigma) := \{(v, w) | \exists \tilde{x} \in \tilde{S}_{\mathcal{M}(\Sigma)}, j \in \mathbb{N} : \tilde{x}_j = v \text{ and } \tilde{x}_{j+1} = w\},$$

called qualitative transitions.

In the following, we indicate the edge relation with $\rightarrow$.

Example 4. The state transition graph of our running example is depicted in Fig. 3. We see that we can find the trajectory $(-1, -1, -1, -1) \rightarrow (1, -1, -1, -1) \rightarrow (1, 1, -1, -1)$ of Example 3 in this graph.

Naturally, this graph would not be very helpful in application if we would need to solve all ODEs in the ensemble to construct it. The following proposition constitutes a different approach. It basically says that a change in sign of a component i must be caused by a consistent dependency on a component j in the right hand side function f, as captured in the i, j-th entry σ_{ij} of the sign matrix Σ.

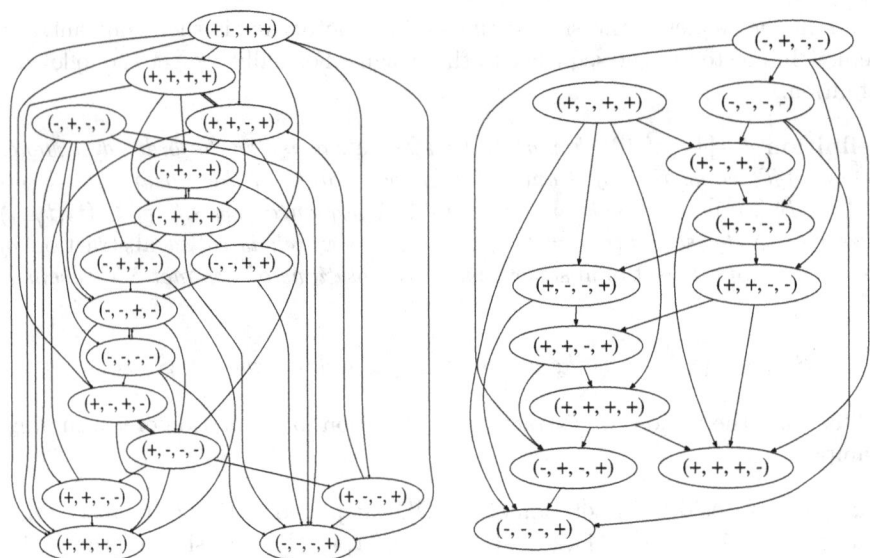

Fig. 3. Left: State transition graph of the monotonic ensemble in the running example. Right: The quotient graph $G_{\text{async}}(f)/\phi_f$ from Example 5. Self-loops are not shown.

Proposition 1 ([4, p. 25]). *Let* $v, w \in \{-1, 1\}^n = V_{QDE}(\Sigma)$, $v \neq w$. *Then,* $(v, w) \in E_{QDE}(\Sigma)$ *iff*

$$\forall i \in \text{diff}(v, w) \exists j \in \text{comm}(v, w) : w_i \cdot v_j = \sigma_{i,j} \qquad (1)$$

The proof of Proposition 1 is given in [4, p. 25]. In the following section, we will transfer the statement into the Boolean setting.

To conclude this section, we note that there is no one-to-one correspondence between the qualitative state transition graph and the corresponding sign matrix Σ. It is possible to change elements on the diagonal of Σ without changing the graph $G_{QDE}(\Sigma)$. This is due to the fact that the sets $\text{diff}(v, w)$ and $\text{comm}(v, w)$ are disjoint and thus the diagonal elements do not play a role in Eq. (1). Consequently, the edge set does not change when changing the diagonal of Σ.

Remark 1. Let Σ be a sign matrix. Then

$$G_{QDE}(\Sigma) = G_{QDE}(\Sigma - D),$$

where D is any diagonal matrix with entries in $\{-1, 0, 1\}$.

In the following we will examine analogues to the graph $G_{QDE}(\Sigma)$ and the set $\mathcal{M}(\Sigma)$ in the Boolean setting.

3   Quotient graphs for Boolean networks

In the Boolean setting, paths in the asynchronous state transition graph can be seen as trajectories of the system. To transfer the results from the previous section, we first need to adapt the notion of derivative sign vectors and abstractions. A first approach could be to consider the difference vectors of a state and its image under the Boolean function f to capture the tendencies to increase or decrease. It turns out, however, that for transferring the results of Sect. 2.2 the better choice is to just assign values 1, if the image value under f is 1, and -1, if the image value is 0. This choice becomes more meaningful when relating Boolean to corresponding ODE systems as we will discuss later.

To formalize this value assignment, we introduce the function

$$\phi_f : \{0,1\}^n \to \{-1,1\}^n, \quad \phi_f(s) \mapsto \left((-1)^{1-f_i(s)}\right)_{i=1,\ldots,n}. \tag{2}$$

Using ϕ_f we can now assign a sign vector to every state in the state transition graph of f. Naturally, several states correspond to the same sign vector. To associate the set of these states to the corresponding sign vector, we consider the equivalence relation induced by ϕ_f on the vertices $V_{\text{async}}(f)$:

$$s \approx t :\Leftrightarrow \phi_f(s) = \phi_f(t).$$

Since two states are in the same equivalence class if and only if their images under ϕ_f are the same we identify the equivalence classes with the states of ϕ_f. In a next step, we want to obtain abstractions. The trajectories of the Boolean system are the paths in the asynchronous state transition graph, so we infer edges between sign vectors (representing sets of nodes of the state transition graph) from the edges in the state transition graph. For this, we simply consider the quotient graph $G_{\text{async}}(f)/\phi_f$, which consists of the node set induced by the above equivalence class, identified with the images of ϕ_f, and an edge between two equivalence classes if there is at least one edge between two nodes in the preimages of ϕ_f in $G_{\text{async}}(f)$.

Since $f_i(s) = 0 \Leftrightarrow \phi_f(s)_i = -1$ and $f_i(s) = 1 \Leftrightarrow \phi_f(s)_i = 1$, we see that f and ϕ_f induce the same equivalence relation, and thus $G_{\text{async}}(f)/\phi_f = G_{\text{async}}(f)/f$.

Example 5. We illustrate the notions using the Boolean function f from Example 1 again. Its ASTG $G_{\text{async}}(f)$ is depicted in Fig. 1. In Fig. 3 the quotient graph $G_{\text{async}}(f)/\phi_f$ is depicted. Each vertex represents an equivalence class on the vertices of $G_{\text{async}}(f)$ represented using ϕ_f. The state $(-1,1,-1,-1)$ for example represents the equivalence class $\phi_f^{-1}(-1,1,-1,-1) = \{(1,0,1,1),(1,1,1,1)\}$.

Looking at the example, we can detect some similarity between the quotient graph and the state transition graph of the monotonic ensemble in Fig. 3. To understand this similarity better we will derive a Boolean version of Proposition 1. We obtain a nearly identical statement. One adjustment concerns negative self-loops, which in the time-discrete other than in the continuous setting potentially instigate oscillation in one component. We need to exclude such an immediate sign switch in the following statement.

Theorem 1. *If there is an edge $(s,t) \in E_{async}(f)$ for a Boolean function $f : \{0,1\}^n \to \{0,1\}^n$ then*

$$\left(\forall i \in \text{diff}(f(s), f(t)) \ \exists j \in \text{comm}(f(s), f(t)): \quad \phi_f(t)_i \cdot \phi_f(s)_j = \partial_j f_i(s)\right) \tag{3}$$

$$\text{or} \quad \left(\text{diff}(s,t) \cap \text{diff}(f(s), f(t)) = \{j\} \text{ and } \partial_j f_j(s) = -1\right).$$

Proof. Let $(s,t) \in E_{\text{async}}(f)$. By definition of the asynchronous update $\text{diff}(s,t)$ is either empty, and the statement holds trivially, or contains one element. In the following let $j \in \text{diff}(s,t)$ be this element. Consider the following two cases:

Case 1: $j \in \textbf{comm} \, f(s)f(t)$. For $i \in \text{diff}(f(s), f(t))$, we can build the following table by considering all possible values of $f_i(s)$ and listing in the corresponding row some inferred values.

$f_i(s) = 0$	$\phi_f(s)_i = -1$	$f_i(t) = 1$	$\phi_f(t)_i = 1$	$f_i(t) - f_i(s) = 1$
$f_i(s) = 1$	$\phi_f(s)_i = 1$	$f_i(t) = 0$	$\phi_f(t)_i = -1$	$f_i(t) - f_i(s) = -1$

Since $j \in \text{comm}(f(s), f(t))$, $j \in \text{diff}(s,t)$ and $(s,t) \in E_{\text{async}}(f)$, we obtain:

$s_j = 1$	$f_j(s) = 0$	$t_j = 0$	$f_j(t) = 0$	$\phi_f(s)_j = -1$	$t_j - s_j = -1$
$s_j = 0$	$f_j(s) = 1$	$t_j = 1$	$f_j(t) = 1$	$\phi_f(s)_j = 1$	$t_j - s_j = 1$

Taking the last two columns of these tables, we obtain by substitution

$$\phi_f(t)_i \cdot \phi_f(s)_j = \left[f_i(t) - f_i(s)\right] \cdot \left[t_j - s_j\right] = \frac{f_i(t) - f_i(s)}{t_j - s_j} = \frac{f_i(s \oplus e_j) - f_i(s)}{s_j \oplus 1 - s_j} = \partial_j f_i(s).$$

Case 2: $j \in \textbf{diff} \, f(s)f(t)$. Due to $t_j \neq s_j$, $t_j = f_j(s)$ and $f_j(t) \neq f_j(s)$ we get:

s_j	t_j	$f_j(s)$	$f_j(t)$	$t_j - s_j$	$f_j(t) - f_j(s)$
0	1	1	0	1	-1
1	0	0	1	-1	1

From the last two columns we can deduce that there is a negative self-loop:

$$\partial_j f_j(s) = \frac{f_j(s \oplus e_j) - f_j(s)}{s_j \oplus 1 - s_j} = \frac{f_j(t) - f_j(s)}{t_j - s_j} = -1.$$

$\square$

Before we continue, we want to emphasize the importance of Theorem 1 with respect to two aspects. First, if we restrict ourselves to Boolean functions without self-loops, it gives us a condition for an edge in $G_{\text{async}}(f)$ purely in terms of f. Indeed, we will see later that in principle it is enough to know the image of f and $IG_f(\cdot)$ This means, we can deduce information about $G_{\text{async}}(f)/\phi_f = G_{\text{async}}(f)/f$ from $IG_f(\cdot)$ without having full information about f. Second, if we compare Theorem 1 with Proposition 1, we notice that they are identical, if we consider monotonic functions without negative self-loops. This suggests that it is reasonable to interpret the states of the quotient graph as signs of changes in the continuous setting as well. This gives us another set of predictions Boolean models are capable to make. We will discuss this aspect further in Sect. 5.

To explain the first point better, we state the following corollary.

Corollary 1. *Consider a Boolean monotonic function* $f : \{0,1\}^n \rightarrow \{0,1\}^n$ *with an interaction graph* $IG^{global}(f) = (\sigma_{i,j})_{i,j \in \{1,...,n\}}$ *without negative self-loops. Assume there is an edge* (v,w) *in* $G_{async}(f)/\phi_f$ *then*

$$\forall i \in \text{diff}(v,w) \exists j \in \text{comm}(v,w) : w_i \cdot v_j = \sigma_{i,j}, \tag{4}$$

where we identified the equivalence classes v, w *with the corresponding elements in the image of* ϕ_f.

Proof. Assume there is an edge (v,w) in $G_{async}(f)/\phi_f$. According to the definition of the quotient graph there exists an edge $(s,t) \in E_{async}(f)$ such that $\phi_f(s) = v$ and $\phi_f(t) = w$. According to Theorem 1 and the assumption that the interaction graph has no negative self loops Condition 3 is satisfied in the local interaction graph $IG_f(s)$ and consequently also in $IG^{global}(f)$.   □

Example 6. Consider the states $s = (1,0,1,1), t = (1,0,1,0)$ for our running example. Their images under ϕ_f are $v = (-1,1,-1,-1), w = (1,1,-1,-1)$. As can be seen in Fig. 3 there is an edge (v,w) in $G_{async}(f)/\phi_f$ and Condition 4 should be satisfied for these nodes. Indeed $w_1 \cdot v_4 = 1 \cdot (-1) = \sigma_{1,4}$.

4   Boolean Monotonic Ensembles

Motivated by the treatment of families of ODEs using QDEs, we now discuss families of Boolean functions defined by a $n \times n$ sign matrix $\Sigma = (\sigma_{i,j})_{i,j \in \{1,...,n\}}$ taking values in $\{-1,0,1\}$. Here, Σ fixes the structure of Boolean model. In addition to the constraints captured by Σ we assume throughout the section that the considered functions do not have negative self loops.

Definition 6. *For a given $n \times n$ matrix of signs* $\Sigma = (\sigma_{i,j})_{i,j = 1,...,n}$ *we define the* Boolean monotonic ensemble

$$\mathcal{M}_{\mathbb{B}}(\Sigma) = \{f : \{0,1\}^n \rightarrow \{0,1\}^n | IG^{global}(f) = \Sigma\}.$$

As in the QDE setting, we are now looking for a compact representation of possible trajectories for all models in the ensemble. The QDE graph $G_{QDE}(\Sigma)$ and Theorem 1 motivate the following definition.

Definition 7. *We call the graph* $G_{QDE}^{Boolean}(\Sigma) = (V_{QDE}^{Boolean}, E_{QDE}^{Boolean}(\Sigma))$ *with*

$$V_{QDE}^{Boolean} = \{-1,1\}^n,$$
$$E_{QDE}^{Boolean}(\Sigma) = \{(v,w)|\forall i \in \text{diff}(v,w) \exists j \in \text{comm}(v,w) : w_i \cdot v_j = \sigma_{i,j}\}$$

the Ensemble state transition graph (ESTG) of Σ.

This graph, which is due to Proposition 1 the same object as $G_{QDE}(\Sigma)$, can be constructed without any consideration of specific functions in the ensemble. Nevertheless, Theorem 1 enables us to extract information on specific state transition graphs using the quotient graph.

Theorem 2. *Let $f \in \mathscr{M}_{\mathbb{B}}(\Sigma)$. The map ϕ_f is a graph homomorphism from $G_{async}(f)$ into $G_{QDE}^{Boolean}(\Sigma)$.*

Proof. By definition, ϕ_f maps the vertex set $\{0,1\}^n$ of $G_{async}(f)$ into the vertex set of $G_{QDE}^{Boolean}(\Sigma)$. Now, we show that ϕ_f conserves the edges. Let $(s,t) \in E_{async}(f)$ and set $v := \phi_f(s)$, $w := \phi_f(t)$.

By definition and as discussed in Sect. 3, ϕ_f is a graph epimorphism from $G_{async}(f)$ onto $G_{async}(f)/\phi_f$. Therefore, we only need to show that $G_{async}(f)/\phi_f$ is a subgraph of $G_{QDE}^{Boolean}(\Sigma)$. But this is clear, since the criterion for an edge in $G_{async}(f)/\phi_f$ is according to Theorem 1 just Condition (4), which in turn defines the edges in $G_{QDE}^{Boolean}(\Sigma)$. $\square$

We can utilize this theorem to obtain reachability constraints for every Boolean function f in the monotonic ensemble since the lack of an edge in $G_{QDE}^{Boolean}(\Sigma)$ implies the lack of the corresponding edge in the ASTG of f. The following corollary illustrates this point for trap sets, i.e., sets of states no trajectory can leave, and no-return sets, i.e., sets of states that no trajectory enters.

Corollary 2. *Assume the set $T \subseteq V_{QDE}^{Boolean}$ is a trap set (or a no-return set) in $G_{QDE}^{Boolean}(\Sigma)$. Then for any $f \in \mathscr{M}_{\mathbb{B}}(\Sigma)$ the set $\phi_f^{-1}(T)$ is a trap set (no-return set) in $G_{async}(f)$. More generally speaking: If $T_1, T_2 \subseteq V_{QDE}^{Boolean}$ and there is no path between T_1 and T_2 in $G_{QDE}^{Boolean}(\Sigma)$ then for any $f \in \mathscr{M}_{\mathbb{B}}(\Sigma)$ there is no path from $\phi_f^{-1}(T_1)$ to $\phi_f^{-1}(T)$ in $G_{async}(f)$.*

Trap sets are of particular interest in application since they always contain at least one attractor. However, not every trap set in $G_{QDE}^{Boolean}(\Sigma)$ gives rise to a trap set in $G_{async}(f)$ since the preimage might be empty. Conversely, since $G_{QDE}^{Boolean}(\Sigma)$ can be interpreted as a supergraph of a corresponding ASTG, there might be trap sets in $G_{async}(f)$ that cannot be identified in $G_{QDE}^{Boolean}(\Sigma)$. In any case, to obtain explicit information on a function f in the ensemble, we need to calculate preimages, which in general is a hard problem [6]. However, in some situations we can test very cheaply if a preimage is empty or if it is worth computing it with respect to the long term behavior of the system as we illustrate with the following statements.

For easier notation, we now identify the nodes $G_{QDE}^{Boolean}(\Sigma)$ with the elements in the image of any $f \in \mathscr{M}_{\mathbb{B}}(\Sigma)$ (and thus Boolean states), as previously done for ϕ_f.

Corollary 3. *Assume the set $\{t\} \subseteq V_{QDE}^{Boolean}$ is a trap set of cardinality 1 in $G_{QDE}^{Boolean}(\Sigma)$. Then for any $f \in \mathscr{M}_{\mathbb{B}}(\Sigma)$ the set $f^{-1}(t)$ is either empty or $f(t) = t$. If $f(t) = t$ every trajectory in $f^{-1}(t) \subseteq G_{async}(f)$ ends in t.*

Proof. Since $\{t\} \subseteq V_{\text{QDE}}^{\text{Boolean}}$ is a trap set in $G_{\text{QDE}}^{\text{Boolean}}(\Sigma)$, $f^{-1}(t)$ is a trap set in $G_{\text{async}}(f)$. For each $s \in f^{-1}(t)$ it holds $f(s) = t$, so there must exists a path to t in the trap set. In particular, t is in the trap set, so $f(t) = t$ $\qquad\square$

We illustrate the last two corollaries by finding trap-sets and no-return sets of the Boolean monotonic ensemble in our running example.

Example 7. Consider the sign matrix from our running Example 1, depicted in Fig. 1. The graph $G_{\text{QDE}}^{\text{Boolean}}$ with its strongly connected components is depicted in Fig. 4. They are:

$$A_0 = \{(0,0,0,1)\}$$
$$A_1 = \{(1,1,1,0)\}$$
$$A_2 = \{0,1\}^4 \backslash (A_0 \cup A_1 \cup A_3 \cup A_4)$$
$$A_3 = \{(1,0,1,1)\}$$
$$A_4 = \{(0,1,0,0)\}.$$

We can now infer that for any $f \in \mathscr{M}_{\mathbb{B}}(\Sigma)$ the sets $f^{-1}(A_1), f^{-1}(A_0)$ are trap-sets while the sets $f^{-1}(A_3), f^{-1}(A_4)$ are no-return sets.

Both the function f introduced in Example 1 as well as $g \in \mathscr{M}_{\mathbb{B}}(\Sigma)$ given by $g(x) = (1 - x_4, x_1, x_2 + (1 - x_4) - x_2 \cdot (1 - x_4))$ are elements of the ensemble.

For function f we see its asynchronous state transition graph $G_{\text{async}}(f)$ in Fig. 1 and its quotient graph $G_{\text{async}}(f)/f$ in Fig. 3. We easily compute that $f^{-1}(A_1), f^{-1}(A_0)$ are non-empty by checking that $(0,0,0,1), (1,1,1,0)$ are fixed points of f.

5 The Quotient Graph $G_{\text{async}}(f)/\phi_F$ as Discretization of Continuous Data

In application, the choice of modeling formalism is not always straight forward, since continuous and discrete approaches have complementing strengths. Bridging the formalisms is therefore of high interest. The QDE formalism is a suitable way to do so, as illustrated by the results for piecewise linear systems utilizing similar ideas with successful applications in systems biology [3].

The results we presented here give a precise link between a Boolean model f and a QDE system, since the ESTG $G_{\text{QDE}}^{\text{Boolean}}(\Sigma)$ of a Boolean monotonic ensemble coincides with the QDE state transition graph corresponding to Σ, and the quotient graph $G_{\text{async}}(f)/\phi_f$ is a subgraph of this ESTG. Trajectories in $G_{\text{async}}(f)/\phi_f$ thus capture the qualitative behavioral patterns encoded in abstractions of ODEs. The -1 and 1 values of ϕ_f reflect the decreasing or increasing tendencies of a continuous trajectory, taking into account that for example a negative value of $(\phi_f)_i$ for some $i \in \{1, \ldots, n\}$ is either caused by a component i in the Boolean system switching from 1 to 0 or by remaining 0, which is normally in a continuous system realized by an asymptotically decreasing behavior.

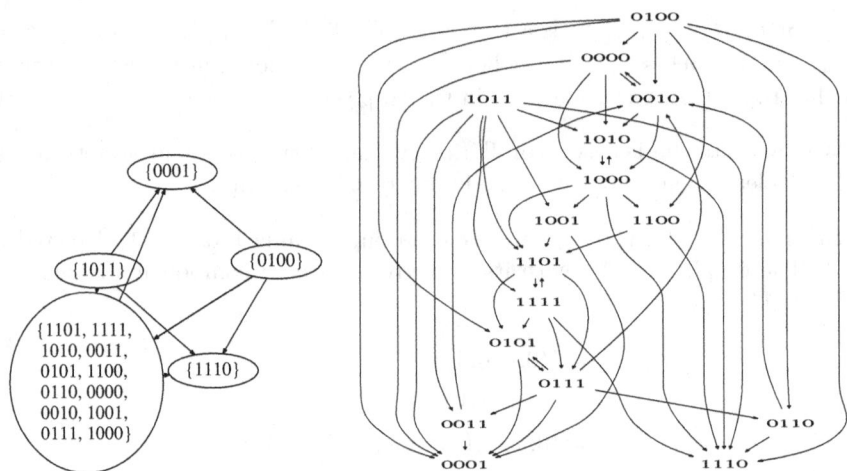

Fig. 4. This example illustrates how statements about $\mathscr{M}_{\mathbb{B}}(\Sigma)$ can be derived from Σ using its strongly connected components (see Example 7). Left: Strongly connected components of $G_{\mathrm{QDE}}^{\mathrm{Boolean}}$. Right: The graph $G_{\mathrm{QDE}}^{\mathrm{Boolean}}$. Self-loops are not shown.

Since $G_{\mathrm{QDE}}(\Sigma)$ summarizes the behavior of the corresponding ODE systems, one could argue that the more realistic trajectories of the Boolean system are to be found in $G_{\mathrm{async}}(f)/\phi_f$ rather than the ASTG.

The nodes of the graph $G_{\mathrm{async}}(f)/\phi_f$ can be interpreted as the slope signs of the component trajectories in the regulatory networks directly without using the ASTG at all. This addresses one of the difficulties often occurring when informing a Boolean model with experimental data, where it is often not at all obvious what discretization thresholds and basal values to assign when processing quantitative data. Here, considering just changes in the measurements, e.g., when exploiting time series data, allows the direct comparison with the ESTG resp. $G_{\mathrm{async}}(f)/\phi_f$.

6   Conclusion

Motivated by the theory of QDEs, we presented the notion of an ensemble state transition graph $G_{\mathrm{QDE}}^{\mathrm{Boolean}}(\Sigma)$ for a family of monotonic Boolean functions with coinciding global interaction graphs without negative loops. Every asynchronous state transitions graph for a function f in the ensemble can be mapped by a graph homomorphism via their natural quotient graph $G_{\mathrm{async}}(f)/f$ into the graph $G_{\mathrm{QDE}}^{\mathrm{Boolean}}(\Sigma)$. Consequently, analysis of the ensemble graph yields information valid across all ASTGs in the ensemble, in particular, on trap sets, no-return sets and reachability properties in general. This may be exploitable when looking for control strategies for model ensembles, e.g., for identifying knock-out candidates in the interaction graph that assure certain reachability properties in the ASTGs of the corresponding functions as has been done for differently defined ensembles [5].

Construction of the graph $G_{\mathrm{QDE}}^{\mathrm{Boolean}}(\Sigma)$ is easily done based on information encoded in Σ. At first glance the computational cost for the construction using Condition (4) is not cheap. However, in application interaction graphs are often rather sparse and the condition need only be tested for non-zero entries of Σ.

Interpretation of the information encoded in the ESTG for a particular function in the ensemble is hampered by the fact that the nodes represent sets of preimages of the node state. Since the possibility of these sets being empty is not excluded, information can not always be transferred in a straight-forward manner. For this problem, we want to explore two aspects in future research. First, we would like to clarify whether the existence of an edge in the ESTG implies the existence of an ensemble function f with corresponding edges in its ASTG. In the QDE setting this statement is true. Second, we want to investigate which subgraphs of $G_{\mathrm{QDE}}^{\mathrm{Boolean}}(\Sigma)$ arise as quotient graphs of Boolean functions $f \in \mathscr{M}_{\mathbb{B}}(\Sigma)$ and, related, derive constraints from ensemble graph edges for the models that exhibit them.

In the last section, we shortly touched upon the connection the QDE framework offers between ODE and Boolean models. In the constraint-based view adopted here, we saw that the ensemble graphs $G_{\mathrm{QDE}}(\Sigma)$ and $G_{\mathrm{QDE}}^{\mathrm{Boolean}}(\Sigma)$ are the same. The interpretation of the nodes as signs of change gives us an alternative way of interpreting the Boolean states of the graph $G_{\mathrm{async}}(f)/f$ in a continuous setting, based on differences rather than absolute values, as is often more natural when processing experimental data. The fact that graphs $G_{\mathrm{QDE}}(\Sigma)$ and $G_{\mathrm{QDE}}^{\mathrm{Boolean}}(\Sigma)$ are identical, but carry different meaning, could in future lead to new network inference algorithms for predicting edges in Σ, based on network inference algorithms for Boolean functions. Since QDEs and ODEs are intrinsically related, such predictions could be very robust. Beyond this aspect, we would like to investigate ways to assign a family of ODE systems to a Boolean function $f \in \mathscr{M}_{\mathbb{B}}(\Sigma)$ such that the subgraph of $G_{\mathrm{QDE}}(\Sigma)$ induced by this family via their set of abstractions $\tilde{S}_{\mathscr{M}(\Sigma)}$ resembles the quotient graph $G_{\mathrm{async}}(f)/f$ more closely. This could provide approaches for validating Boolean models taking dynamics on a higher resolution level consistent with the qualitative model properties into account.

Lastly, our results raise the question in how far logic constrains obtained from different abstractions than Proposition 1 lead to similar results [1]. For many biological applications it could make sense to consider a more restricted class of ODE-systems. For example in [5] ODE-systems (models of reaction networks) stemming from different kinetic rate laws are abstracted. It would be interesting to see in how far such results are comparable to QDEs in general and in how far they are related to the results in this paper.

References

1. Abou-Jaoudé, W., Thieffry, D., Feret, J.: Formal derivation of qualitative dynamical models from biochemical networks. Biosystems **149**, 70–112 (2016)

2. De Jong, H., Gouzé, J.-L., Hernandez, C., Page, M., Sari, T., Geiselmann, J.: Qualitative simulation of genetic regulatory networks using piecewise-linear models. Bull. Math. Biol. **66**(2), 301–340 (2004)

3. De Jong, H., Page, M., Hernandez, C., Geiselmann, J.: Qualitative simulation of genetic regulatory networks: method and application. In: IJCAI, pp. 67–73 (2001)

4. Eisenack, K.: Model ensembles for natural resource management: extensions of qualitative differential equations using graph theory and viability theory. Unpublished doctoral thesis, Free University Berlin, Germany (2006). Accessed 3 Feb 2008

5. John, M., Nebut, M., Niehren, J.: Knockout prediction for reaction networks with partial kinetic information. In: Giacobazzi, R., Berdine, J., Mastroeni, I. (eds.) VMCAI 2013. LNCS, vol. 7737, pp. 355–374. Springer, Heidelberg (2013). doi:10. 1007/978-3-642-35873-9_22

6. Keinänen, M.: Techniques for solving Boolean equation systems. Ph.D. thesis, Helsinki University of Technology (2006)

7. Klarner, H., Streck, A., Siebert, H.: PyBoolNet-a python package for the generation, analysis and visualisation of boolean networks. Bioinformatics **33**, 770–772 (2016)

8. Melliti, T., Regnault, D., Richard, A., Sené, S.: Asynchronous simulation of boolean networks by monotone boolean networks. In: El Yacoubi, S., Wąs, J., Bandini, S. (eds.) ACRI 2016. LNCS, vol. 9863, pp. 182–191. Springer, Cham (2016). doi:10. 1007/978-3-319-44365-2_18

9. Remy, É., Ruet, P., Thieffry, D.: Graphic requirements for multistability and attractive cycles in a boolean dynamical framework. Adv. Appl. Math. **41**(3), 335–350 (2008)

10. Streck, A., Thobe, K., Siebert, H.: Data-driven optimizations for model checking of multi-valued regulatory networks. Biosystems **149**, 125–138 (2016)

11. Thomas, R.: On the relation between the logical structure of systems and their ability to generate multiple steady states or sustained oscillations. In: Della, D.J., Demongeot, J., Lacolle, B. (eds.) Numerical Methods in the Study of Critical Phenomena, pp. 180–193. Springer, Heidelberg (1981)

12. Thomas, R., Kaufman, M.: Multistationarity, the basis of cell differentiation and memory. I. Structural conditions of multistationarity and other nontrivial behavior. Chaos **11**(1), 170–179 (2001)

13. Thomas, R., Kaufman, M.: Multistationarity, the basis of cell differentiation and memory. II. Logical analysis of regulatory networks in terms of feedback circuits. Chaos Interdisc. J. Nonlinear Sci. **11**(1), 180–195 (2001)

14. Videla, S., Saez-Rodriguez, J., Guziolowski, C., Siegel, A.: caspo: a toolbox for automated reasoning on the response of logical signaling networks families. Bioinformatics **33**, 947–950 (2017)

15. Wittmann, D.M., Krumsiek, J., Saez-Rodriguez, J., Lauffenburger, D.A., Klamt, S., Theis, F.J.: Transforming boolean models to continuous models: methodology and application to T-cell receptor signaling. BMC Syst. Biol. **3**(1), 98 (2009)

Explaining Response to Drugs
Using Pathway Logic

Carolyn Talcott$^{(\boxtimes)}$ and Merrill Knapp

SRI International, Menlo Park, CA 94025, USA
{carolyn.talcott,merrill.knapp}@sri.com

Abstract. Pathway Logic (PL) is a general system for modeling signal transduction and other cellular processes with the objective of understanding how cells work. Each specific model system builds on a knowledge base of rules formalizing local process steps such as post translational modification. The Pathway Logic Assistant (PLA) is a collection of visualization and reasoning tools that allow users to derive specific executable models by specifying of an initial state. The resulting network of rule instances describes possible behaviors of the modelled system. Subnets and pathways can then be computed (they are not hard wired) by specifying states to reach and/or to avoid. The STM knowledge base is a curated collection of signal transduction rules supported by experimental evidence. In this paper we describe methods for using the PL STM knowledge base and the PLA tools to explain observed perturbations of signaling pathways when cells are treated with drugs targeting specific activities or protein states. We also explore ideas for conjecturing targets of unknown drugs. We illustrate the methods on phosphoproteomics data (RPPA) from SKMEL133 melanoma cancer cells treated with different drugs targeting components of cancer signaling pathways. Existing curated knowledge allowed to us explain many of the responses. Conflicts between the STM model predictions and the data suggest missing requirements for rules to apply.

1 Introduction

Understanding how cells work is a fundamental question in Biology. It is important for basic science, as well as for practical applications including understanding disease, drug discovery, and synthetic biology. There are many aspects, including the different processes within a cell (metabolism, signaling, transcription/translation, ...), how these processes interact, what are the normal states, and what happens in response to some perturbation.

Executable mechanistic models [7] play an important role in understanding cellular processes, as they support in silico experiments, hypothesis generation, and feedback between laboratory experiments and model development. In the

The work was partially supported by funding from the DARPA Big Mechansim program. The authors would like to thank the PL team for their many contributions, and the anonymous reviewers for helpful criticisms.

© Springer International Publishing AG 2017
J. Feret and H. Koeppl (Eds.): CMSB 2017, LNBI 10545, pp. 249–264, 2017.
DOI: 10.1007/978-3-319-67471-1_15

case of drug discovery such models help to determine details of the mechanism of action (MOA) and dually, drugs with a known MOA are used to learn details about how cells work.

The work reported here was done as part of a DARPA Big Mechanism project. The challenge was to use our Pathway Logic Signal Transduction model (STM) to explain how drugs with a known mechanism of action caused the changes in protein expression and/or phosphorylation measured by Reverse Phase Protein Array (RPPA) using data from [10].

The contributions of this paper are

- methods to explain effects of drugs on exponentially growing cells as measured by high throughput phosphoproteomics assays.
- a method to build a model of exponentially growing cells from a knowledge base of rules describing cellular events.
- methods to derive the mechanism (network of events) underlying response to treatment by drugs with known specific targets
- methods to hypothesize targets of unknown drugs, i.e. perturbations of the network that could explain measured responses.

Using these methods we were able to explain many of the observed changes in expression and phosphorylation in SKMEL133 cells when treated with drugs with known targets, and to make some conjectures regarding possible targets of two of the unknown drugs.

The SKMEL133 model is available at pl.csl.sri.com/online.html as part of the Pathway Logic suite of models. The accompanying guided tour is available as a link from the Online launcher, or directly from pl.csl.sri.com/ along with a techreport version of this paper.

Plan. We provide a brief introduction to Pathway Logic and describe the general method for explaining drug study data in Sect. 2. In Sect. 3 we describe the data set and how it was processed in order to map the data to a PL model. The model of exponentially growing SKMEL133 cells is presented in Sect. 4. In Sect. 5 we use the model to explain the data for drugs with known, experimentally validated, targets. In Sect. 6 we analyze the data for two of the unknown drugs, with consistent results in one case and many mysteries in the other case. Some related work is discussed in Sect. 7, and we conclude with a summary and discussion of future work in Sect. 8.

2   Pathway Logic Models and Their Use to Analyze Data

The objective of Pathway Logic (PL) is to understand how cells work. A recent overview of PL can be found in [16]. The PL collection of models, knowledge bases, software, documentation, papers, and tutorials are available from the PL website [13]. The PL model collection includes models of metabolism, protease signaling in bacteria, protein glycosylation, and fragments of the human immune system. The most highly developed model is STM (Signal Transduction Model). This will be our starting point for modeling response to drugs.

2.1   PL Concepts and Reasoning Tools

As shown in Fig. 1, the STM Pathway Logic models are founded on two formal knowledge bases: a curated datum knowledge base (DKB), and a rules knowledge base (RKB), that share a controlled vocabulary formalized in Maude [4].

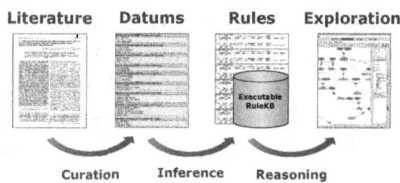

Fig. 1. From data to models in PL.

A datum formalizes an experimental observation of the state or location of protein or other biomolecule (RNA, Lipids, ...) either in some well-defined experimental condition, or a change in response to some signal or perturbation [12].

Signaling events are formalized as rewrite rules. They are generally inferred from datums, although rule sets can also be curated from review articles and text books, or simply hypothesized. A rule contains terms representing the change (before and after state) as well as terms representing the biological context required for the change to take place. A rule may be parametric, containing variables that can be instantiated in multiple ways to give different rule instances usable in different contexts. Rules in PL do not have rates.

The RKB can be thought of as a global model. Executable models of specific situations are generated by specifying initial conditions and constraints, formalized using a notion of *dish* (as in Petri dish). A dish is a term representing the initial state of the modeled system. It can be thought of as representing an experimental setup: cell type, growth conditions, and treatments or other perturbations. The cell type and growth conditions are represented by specifying which proteins and other biomolecules are present, their location, and their modification and/or activity state. The PL STM consists of rules concerning response to over 35 different stimuli as well as *common rules* that formalize local changes independent of a particular stimulus.

In PL, model elements and state are represented using a controlled vocabulary that is specified as a functional module in Maude. There is a core vocabulary shared by all PL knowledge bases/models and a model specific vocabulary that declares specific model elements (proteins, chemicals, modifications, locations, ...). The PL controlled vocabulary has several roles: organizing concepts via a sort/type hierarchy; determining legal/well-formed/meaningful terms by specifying constants and typed term constructors, and giving meaning to constants by providing metadata linking constant symbols to external references (Uniprot, HMDB, ...).

A PL executable model state is multi-set of *occurrences* of entities (proteins, chemicals, genes, ...). An occurrence specifies an entity, its modifications and/or activity state, and its location. For example `Braf-act@CLc` is an occurrence of active Braf in the cytoplasm (`CLc`), `PIP3@CLm` is an occurrence of the lipid `PIP3` in the cell membrane (`CLm`), `S6k1-phos!T412@CLc` is an occurrence in the cytoplasm of `S6k1` phosphorylated on threonine 412.

The STM model uses the term *family* for groups of proteins that cannot be differentiated by antibodies. For example, the anti-Akt antibody (CST#4691) used in [10] detects Akt1, Akt2, and Akt3. We cannot determine whether the increase in the level of protein expression is due to one and/or two and/or three of the Akts so we use the constant `Akts` to refer to some or all members of this family. Similarly, the antibody used to detect `Akt1-phos!S473` (CST#9271) also recognizes `Akt2-phos!S474` and `Akt3-phos!S472`. We use a site code (symbolic name) to represent the corresponding residues in all three proteins. The families and site codes used in the current work are shown in the table below.

Site code	Refers to	and/or	and/or
Akts-phos!FSY	Akt1-phos!S473	Akt2-phos!S474	Akt3-phos!S472
Akts-phos!KTF	Akt1-phos!T308	Akt2-phos!T309	Akt3-phos!S307
Gsk3s-phos!SFAE	Gsk3a-phos!S21	Gsk3b-phos!S9	
Mek12s-phos!SMANS	Mek1-phos!S218-phos!S222	Mek2-phos!S222-phos!S226	
Erks-phos!TEY	Erk1-phos!T202-phos!Y204	Erk2-phos!T185-phos!Y187	

An important part of the PL system is the Pathway Logic Assistant (PLA), which is a tool to generate, visualize, browse, and analyse executable PL models. Given a dish and an RKB, PLA uses a symbolic reasoning and abstraction technique called *forward collection* to infer a minimal set of rule instances that cover all situations reachable from the initial state. The resulting concrete rule set naturally forms a network, linking rules by shared output/input elements. The initial state together with the collected rules forms an executable model. A theory transformation is used to convert the model to a Petri Net to be able to use reasoning tools for Petri Nets. PLA can now be used to specify goals and/or knockouts, derive the subnet of all pathways satisfying the goals (omitting the knockouts), invoke a model checker [15] to find specific pathways, and export nets as images or data structures for use by other tools.[1] Within a subnet one can ask for all the execution pathways leading to the goal, using an inference algorithm described in [6]. Knowing all the pathways one can compute properties such as single and double knockout occurrences or essential rules. If a single knockout occurrence is removed from the model, the goal will no longer be reachable. Similarly for double knockouts and essential rules.

2.2 Use of PL to Explain Data: Generating a Model

The first step in explaining experimental results is to define a model of the unperturbed cell system being studied. For the drug studies we want a snapshot of an exponentially growing cell system that is perturbed by addition of one or more drugs. Ideally, a model is built by defining an initial state (using expert knowledge, literature, the datum KB, and the COSMIC database (for

---

[1] One can knockout an occurrence, either from the initial state or a potentially reachable occurrence, or a rule. Each choice corresponds to a different experimental perturbation.

mutations). Then, using PLA, we do a forward collection from this initial state, to collect all reachable rules in the STM RKB.

However, the world is not ideal, and the above steps may not work without some refinement. One problem is finding information about protein expression levels of a given cell line under different growth conditions, and the other is that, a priori, the rules in the RKB may capture different levels of detail (say Yphos vs phos!Y123) due to different experimental methods, and the rules may be more specific than necessary, or a rule may represents a set of more specific rules, for example by referring to a family of proteins rather than specific members.

To address the first problem, we only attempt to include in the model the measured entities and any relevant up/down stream entities. We do this by a combination of "fuzzy" backward and forward collection (currently implemented by hand). The idea is (i) identify rules that would cause the changes seen in the data; (ii) identify rules that would meet the requirements of the first set of rules; and (iii) iterate until there are no more requirements to be met. Now we prepare an initial state: for each entity in the collected rules, determine the locations and modifications that cannot be produced by any rules. Modify the result using any available information about mutations and deletions for the cell line being studied. The unperturbed network is generated from the rule set and the resulting initial state using 'fuzzy' forward collection. The idea here is that some rules may need to be generalized in order to apply to generated states. For example a rule may require `Mek1-act@CLc` but the state may contain `Mek1-act-phos!SMANS@CLi`. Adding a variable to the modification set of the occurrences of `Mek1` in the rule solves the problem. After these adaptations, the PLA forward collection process can be used to generate a model of the unperturbed system.

2.3 Use of PL to Explain Data: Using the Model

In PL, explanations for measured changes in response to treatment of a cell system with a given drug can be found in several ways. One way is to knock out the drug target and use model checking to see if increases/decreases observed in the data agree with reachability results. We can also find all the paths (in the network model) to different observed significant changes and combine this information to suggest targets if the drug or its mechanism of action is unknown.

Here we focus on direct comparison of models of untreated and treated systems. Given a drug that is known to inhibit some occurrence in the model, we generate a model of the treated system by removing that occurrence from the network and use PLA to do a forwards collection to determine the remaining reachable subnet. Now we can compare the unperturbed (untreated) and perturbed (treated) model networks to obtain a qualitative prediction of increase/decrease in levels of some of the network occurrences. Three principles for inferring expected change are illustrated in Fig. 2.

Note that some of the drugs inhibit activity by direct allosteric inhibition. The conformational change caused by the drug should not be interpreted as the inhibition or enhancement of an upstream kinase. Some of the changes cannot

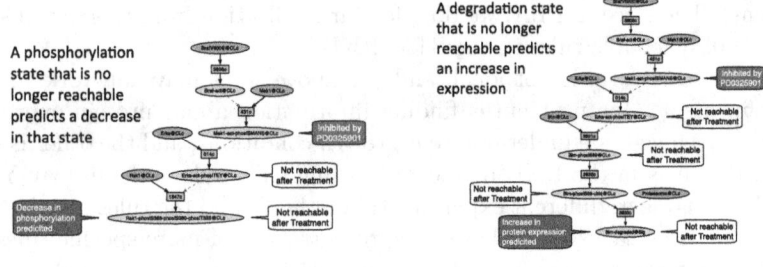

(a) Blocked Modification   (b) Blocked degradation

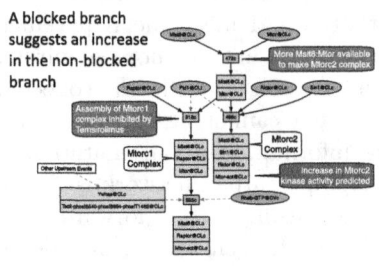

(c) Blocked branch

Fig. 2. Three principles

be explained by a PL model because they are caused by things other than signal transduction.

3   The Experiment and Data

To correctly interpret data, it is important to understand how it is generated and the criteria for interpreting measurements.

Primer on Interpreting the Results of Cell Based Assays

- An *experiment* starts with seeding cells into the containers (petri dish, flask, test tube) where they will be treated.
- The number of *biological replicates* is the number of containers used for each treatment. This detects differences in results caused by the seeding and treatment procedures.
- The number of *technical replicates* is the number of measurements made for each biological replicate. This gives you the probability that your detection method will give you the same value for the same sample.

- The number of *experimental replicate*s is the number of times the procedure is performed from different cell seedings. This gives you the probability that the change observed will occur in another experiment.
- The convention for publication in a cell biology data paper is to perform at least three independent experiments using three biological replicates for each treatment and control.
- The number of technical replicates required depends on the detection method used. The noisier the detection method, the more technical replicates required.

For the data set to be analyzed here, exponentially growing SKMEL133 cells were treated with 12 drugs at two concentrations. Change in protein expression/phosphorylation was measured for 138 entities at 24 h using Reverse Phase PhosphoProteomics Analysis (RPPA) [3].

The data to be explained was available in two formats: (i) fold-change measurements using 3 biological replicates from one experiment based on an unreported number of technical replicates; (ii) relative concentration values for each of the 3 biological replicates from one experiment and from 1 to 4 technical replicates. Variance analysis showed that the noise from the provided technical replicates was larger than that of the biological replicates. This tells us that one technical replicate is not sufficient for realistic quantitation. Without quantitative information we resorted to using the fold-change measurements with a cutoff of 1.2 fold change (up or down) based on the number of changes that we would expect to see in response to what is known about the mechanism of action of the drugs.

Only the highest drug concentration was considered. Changes in the phosphorylation of a protein were normalized to the total expression of that protein. If the total expression was not measured, the phosphorylation change could not be reliably determined, so we didn't attempt to explain those results. The one exception is the change in the Erks TEY site because the protein concentration of Erks rarely changes over 24 h perturbations.

To map the data onto a PL model it is necessary to determine what each antibody actually detects and map this to PL terms. The antibodies used in the RPPA analysis were obtained from commercial suppliers and validated by the MD Anderson Cancer Center RPPA Core Facility. Information about the validation status and source of the antibodies was obtained from the Standard Antibody List downloaded from [2]. We determined the antibody targets by mapping the antibody name reported in the data set to the Official Antibody Name used in the Standard Antibody List. Specificity and site information was obtained from the supplier. The protein or family names of the target proteins were converted into Pathway Logic names and the sites were adjusted to agree with the canonical sequence of each protein in UniProt. In the case of protein families, letter codes were used to match all members, as described in Sect. 2.

To explain the response to a drug treatment it is useful to know what the drug is, i.e. its chemical structure, to have clear experimental evidence of the target and its action on the target, and to know whether there are off-target effects. We were able to identify (find a PubChem identifier for) 8 of the 12 drugs used in the experiment. Subsequent literature search revealed solid evidence for proposed mechanisms of action for 5 of the 8. This is summarized in Sect. 5 as part of the explanation of the data.

4   Inferring the SKMEL133 Model

As discussed in Sect. 2 our idea is to build the minimal model needed to explain the data, rather than attempting a full model of SKMEL133 cells. Thus we include as a minimum the proteins such that the change in protein expression or phosphorylation passed the 1.2 fold cutoff. We carried out (by hand) the fuzzy backwards collection starting from the changed occurrences, adding occurrences with a degradation modification to represent a possible cause of change in protein expression. For example rule 3823c

```
rl[3823c.Irs1.degraded]:
Irs1-ubiq-phos!S270-phos!S307-phos!S636-phos!S1101@CLc
=>
Irs1-degraded@Sig
if Cul7@CLc
```

is collected to account for changes in `Irs1` expression level. This also introduces the protein `Cul7` into the model. Here we use informal rule notation where following the *if* are the controls (the required biological context) of the reaction.

```
rl[109c.Akts.by.Pdpk1]: Akts@CLc => Akts-phos!KTF@CLc if Pdpk1-act@CLc
```

Rule 109c is collected to produce `Akts-phos!KTF`, which then introduces a requirement for `Pdpk1-act`. This can be satisfied by rule 3818c

```
rl[3818c.Pdpk1.by.PIP3]: Pdpk1@CLc => Pdpk1-act@CLc if PIP3@CLm
```

which leads to collecting rule 3820c

```
rl[3820c.PIP3.from.PIP2]: PIP2@CLm => PIP3@CLm if  Pi3k@CLi
```

to produce *PIP3*. This chain stops here, as `PIP2` is a common component and there are no rules producing the protein `Pi3k` so we assume it is expressed by SKMEL133 cells normally.

Collecting the occurrences that can not be produced by a rule we have a preliminary version of the initial state. SKMEL133 cells contain the constitutively active mutation `BrafV600E` so we replaced wild-type `Braf` with `BrafV600E`. They also have a homozygous deletion of `Pten`, so we eliminated `Pten`. The result, called the *SKMEL133dish*, contains 31 occurrences (listed in Appendix 1 of the techreport version).

As discussed in Sect. 2 some iteration is required to achieve a connected set of rules because the curated rules reflect what experiments measured and may have different levels of detail, or need generalization. Also, the following rule was added to model the `BrafV600E` activity.

```
rl[3808c.BrafV600E.act]: BrafV600E@CLc => Braf@act@CLc
```

Unperturbed Network

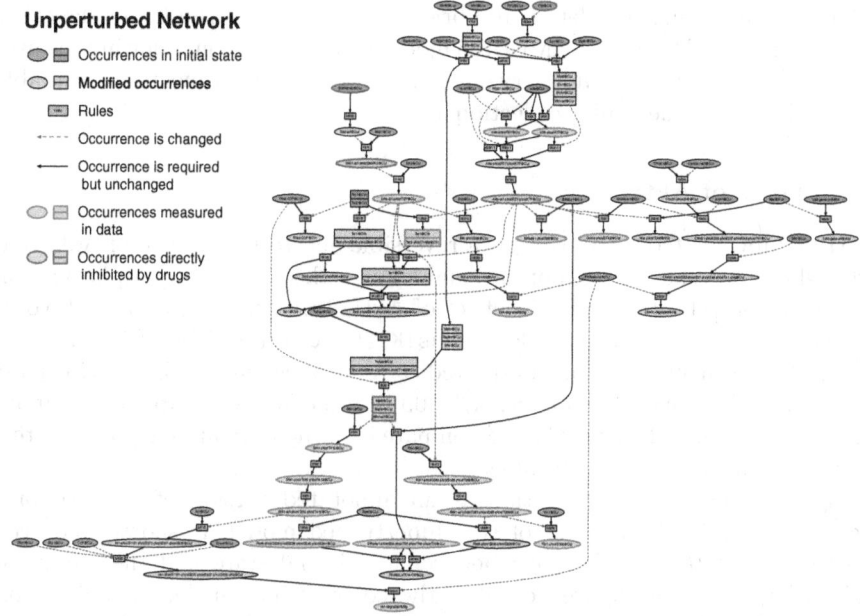

Fig. 3. The unperturbed SKMEL133 model.

This rule reflects the observation that the mutated form of Braf behaves like the active form of wild type Braf. This is a simplification which is adequate in the context of the current model, although it would fail if there were rules to deactivate Braf, since the mutated form can not be deactivated. After adding the above rule and generalizing some rules by hand, PLA is used to assemble the executable model, called the SKMEL133dishnet, shown in Fig. 3.[2]

5   Explaining Response to Known Drugs

As discussed in Sect. 3, we selected 5 drugs for which we could determine a well-defined chemical id (PUBCHEM), and for which there is reasonable evidence for the proposed mechanism of action (determined by literature search): AktI12, PD0325901, PLX4720, Temsirolimus, and ZSTK474 (described in more detail below). For each of these drugs we determined occurrences that changed significantly using the fold change table from [10] and a fold change cutoff of 1.2 for increase and 0.8 for decrease as described in Sect. 3. A table summarizing these changes is included in Appendix 1. Using the methods described in Sect. 2 we could explain 42 out of 107 changes in response to the 5 drugs. Many of the unexplained changes are in protein expression levels, which was generally not the focus of our curation efforts in the past. In the following we illustrate the analysis for AktI12

---

[2] Although in printed form the node labels are not readable, zooming in with a pdf reader reveals all the details.

and Temsirolimus in some detail, and briefly summarize the results for the other three drugs. Recall that the SKMEL133 model and a guided tour allowing the user to reproduce these results and carry out other gedanken experiments are available for download or in the Online collection at [13].

5.1   Effects of AktI12

AktI12 (PubChemCID 10196499) is a reversible allosteric inhibitor of Akt1 and Akt2 which prevents the conformational change that permits phosphorylation and activation [11]. To model the effect of AktI12 we use PLA to block (avoid) the occurrence `Akts-act-phos!FSY-phos!KTF@CLc` in the SKMEL133 dishnet. Recall, this occurrence can be interpreted as Akt1 phosphorylated at S473/T308 and/or Akt2 phosphorylated at S474/T309 in the cytoplasm. Now we compute the resulting reachable network, and compare it to the untreated model to determine what has become unreachable.

Figure 4 shows the explanation as an annotated version of network produced by PLA in the context of the unperturbed model. It shows how drug perturbations interrupt the path between the initial state and the measured goals. The key in the figure describes the color coding in detail. Yellow coloring highlights the unreachable part of the SKMEL133 dishnet. Occurrences outlined in red are directly inhibited by the drug. Occurrences outlined in green decrease in response to the drug. In particular the measured decrease in

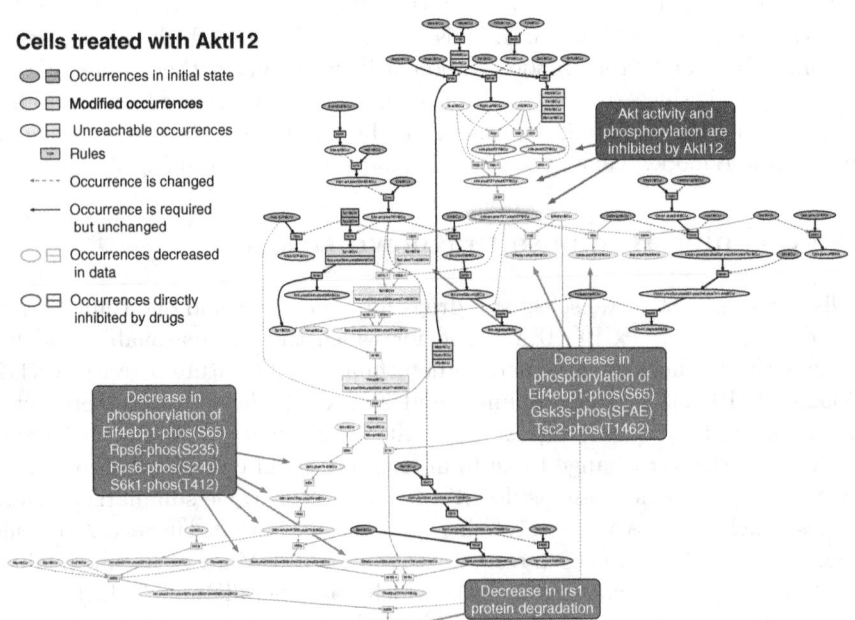

Fig. 4. The SKMEL133 model treated with AktI12.

Eif4ebp1-phos!S65, Eif4ebp1-phos!T37, Gsk3b-phos!S9, Gsk3s-phos!SFAE, Rps6-phos!S235, Rps6-phos!S240, S6k1-phos!T412, and Tsc2-phos!T1462 in response to AktI12 is explained by the unreachability of the corresponding occurrences. The increase in Irs1 protein expression is explained by the inhibition of the degradation of Irs1 by ubiquitination and degradation in the proteasome. The remaining changes are increases in protein expression of Cav1, Fn1, Pai1, and Tp53 and a decrease in Cox2 and CyclinB1, which are not represented in our model.

5.2   Effects of Temsirolimus

Temsirolimus (PubChemCID 23724530) inhibits Mtorc1 activity (a complex of Mtor, Mlst8, and Raptor) but enhances Mtorc2 activity (a complex of Mtor, Mlst8, Sin1, and Rictor) [5]. Figure 5 shows the annotated model of Temsirolimus response.

The model explains measured decrease in events downstream of Mtorc1: Eif4ebp1-phos!T37, Rps6-phos!S235, Rps6-phos!S240, S6k1-phos!T412, and Irs1-degradation. It also explains measured increase in events that are downstream of Mtorc2: Akts-phos!FSY, Akts-phos!KTF.

The model also predicts increases in Eif4ebp1-phos!S65@CLc (the data shows a decrease) and Gsk3s-phos!SFAE@CLc (the data shows no change). What might cause this discrepancy? A common cause of such discrepancy is a missing control on the phosphorylation rule, either because there are no published

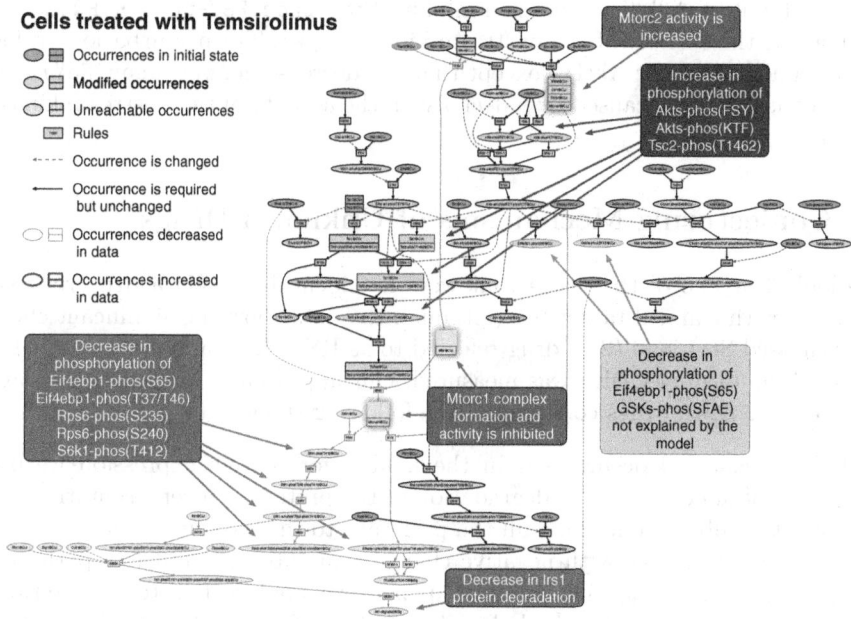

Fig. 5. The SKMEL133 model treated with Temsirolimus.

experiments giving evidence, or because they have not yet been curated. It is also possible that there are alternative activities of the Akts. Note that the RPPA experiments do not measure activity directly. Unraveling this mystery is a topic of ongoing/future work.

5.3   Effects of PD0325901, PLX4720 and ZSTK474

PD0325901 (PubChemCID 9826528) is an allosteric inhibitor of Mek1 and Mek2 kinase activity [14]. To represent the effects of PD0325901, the SKMEL133 model can be blocked at the occurrence `Mek1-act-phos!SMANS@CLc` which can be interpreted as Mek1 phosphorylated at S218 and S222. Although the antibody used in generating the data identifies both phospho-Mek1 and phospho-Mek2, the STM DKB lacks sufficient datums to include Mek2 in the rules. The resulting unreachable set explains decreases in `Erks-phos!TEY`, `Rps6-phos!S235`, `Rps6-phos!S240`, `Rsk1-phos!T359`, `S6k1-phos!T412`, and `Ybx1-phos!S102`. Using the decrease in `Bim-degraded@Sig`, it also explains the increase in Bim protein expression.

PLX4720 (PubChemCID 24180719) binds to the ATP binding site of active Braf and Raf1. It is 10 times more effective towards BrafV600E than wild-type Braf or Raf1. At the concentration used to produce the dataset (120 nM) it should be more effective on BrafV600E than Raf1 [17]. As expected, the perturbation profile PLX4720 is almost identical to that of PD0325901, since `Braf` is responsible for phosphorylation of `Meks`.

ZSTK474 (PubChemCID 11647372) inhibits all four isoforms of the catalytic subunit of Pi3k [5]. This then inhibits `Akts-phos!FSY-phos!KTF@CLc` via decrease in the activity of the upstream kinase Pdpk1. The perturbation profile is the same as that for AktI12 except that the decrease in `Akts-phos(FSY)` and `Akts-phos(KTF)` are caused by a decrease in the activity of the upstream kinase Pdpk1.

6   Conjecturing Mechanisms of Unknown Drugs

We looked at two of the drugs that were not identifiable: (1) a drug referred to as SR with claimed target Src (although the data shows no significant effect on measured Src), and (2) a drug referred to as RY, with claimed target CDK4 although no form of CDK4 was measured. Our approach to analyzing the data for these unknown drugs consisted of the following steps.

1. Identify changed occurrences in the model (for protein expression we use change of opposite sign in degradation of the protein as a representative).
2. Form the subnet containing all the pathways to these occurrences
3. For each occurrence with negative change, compute the subnet of pathways leading to that occurrence and use the pathway analysis tool to list the rules and occurrence that are single knock outs (i.e., if removed from the network the goal occurrence is no longer reachable).

4. Make a table with columns corresponding to the negatively changed occurrences and rows labeled by the knockouts. The entry in a cell is 1 if the knockout labeling the row is in knockout list of the occurrence labeling the column and 0 otherwise.

Now we want minimal subsets of rows that add to 1 for each column. Then inhibiting each of the row labels in such a subset will explain all the negative changes. Of these minimal sets, we prefer those that are furtherest down stream, since otherwise there are likely to be off-target effects.

Given a candidate drug target list, we need to check if this predicts changes consistent with the data. This can be done as for the drugs with known action. Namely, starting with the unperturbed model (the SKMEL133 dishnet), knock out the hypothesized drug target(s), compute the subnet, compare to the unperturbed net to see what is missing. Clearly, the set of occurrences used to generate the knockout lists will be unreachable and thus consistent with the hypothesized targets. Are the other unreachables plausible? We also need to look for explanations for occurrences that increased, such as blocked or diverted branches. As for the case of drugs with known targets we use the 1.2/.8 fold cutoff to determine the list of changed occurrences, and require phosphorylation change relative to protein expression change to meet the cut off criteria. In the following we discuss the for SR. The results for RY can be found in the techreport version of the paper.

6.1   Analysis of the Effects of SR

From the data for the drug SR we determined 2 instances of increase in protein expression (1 is in the model), 3 instances of decrease in protein expression (none in the model), 2 instances of increase in phosphorylation (none in the model) and 8 instances of decrease in phosphorylation (6 in the model). Converting the one increase in protein expression to a decrease in degradation, the decreases represented in the model to consider are: Bim-degraded@Sig, Eif4ebp1-phos!S65-phos!T37-phos!T46-phos!T70@CLc, Eif4ebp1-phos!S65@CLc, Erks-phos!TEY@CLc, Gsk3s-phos!SFAE@CLc, Rps6-phos!S235@CLc, and S6k1-phos!T412CLc.

After computing the subnet containing these changed occurrences and computing the knockouts for each of these occurrences, we find that no single knockout can explain the observed decreases. There are many double knockouts that can explain the decreases. They all involve blocking Mek1 activity and Akts activity, either directly or by an upstream effect. Thus the minimal pair is

[Akts-phos!FSY-phos!KTF@CLc, Mek1-act-phos!SMANS@CLc]

Although these occurrences are not decreased in response to SR, it is quite possible that the drug blocks their action and hence causes the observed downstream effects. Choosing targets upstream of this pair, say [Braf-act@CLc, Pi3k@CLi] would be inconsistent with the observed data as in this case one should observe a decrease in the phosphorylation of Akts and Mek1.

Now we check whether blocking this pair of occurrences is consistent with the measured response to SR. We start with the unperturbed model, knockout (avoid) the conjectured pair of occurrences, compute the resulting reachable subnet, and the unreachable set. The following occurrences that are predicted by the model to decrease are measured:

- `Irs1-degraded@Sig`: protein expression did not change.
- Occurrences involving `Rsk1-phos!T359`: neither `Rsk1` protein expression or `Rsk1-phos!T359` changed. Note that the antibody for `Rsk1` is labeled "use with caution" and the antibody for `Rsk1-phos!T359` is not validated.
- `Ybx1-phos!S102@CLc`: This decreased, which is consistent. The total protein for `Ybx1` was not measured, so it was not included in the list of changes to explain.

7   Related Work

We focus on the use of RPPA data to analyze cellular systems. Existing work generally focuses on inferring network models that fit the data in order to identify interactions and possible causal relations among responding proteins and/or to use the resulting models to predict response to new perturbations. To the best of our knowledge our approach of using an existing curated model to explain the mechanisms underlying cellular response to drugs, and consequently validate or find gaps or problems with the parts of the model, or to hypothesize alternative actions of a drug is unique.

The work presented in [10] is the source of the data explained in the present paper. The work was motivated by the problem of drug resistance, particularly in cancers. The paper describes a combined experimental/computational perturbation biology method to look for anti-resistant target combinations. The experiment was described in Sect. 3, with cells being treated by pair-wise combinations of drugs as well as the single drug treatments. A space of executable ODE models corresponding to influence network topologies with weighted edges are derived from the data using belief propagation techniques. The process is seeded with a prior network extracted from Pathway Commons using the PERA tool [1]. The 4000 best models were selected to make predictions of phenotypic effects of thousands of combinations of perturbations. As a result they propose cMyc as a co-target of Mek or Braf.

The results of the HPN-DREAM network inference challenge are summarized in [9]. This challenge focused on learning causal influences in signaling networks. The objective here was to train models capable of predicting context-specific phosphoprotein time courses, in contrast to the Big Mechanism objective to provide mechanistic explanations for the effects of perturbations. Participants were provided with RPPA phosphoprotein data from four breast cancer cell lines under eight ligand stimulus conditions combined with three kinase inhibitors and a vehicle control (dimethyl sulfoxide). Data for each biological context (cell line, stimulus combination) comprised time courses for approximately 45 phosphoproteins. Models were assessed using context-specific test data that were obtained under a different intervention (inhibition of the kinase mTOR). While some of the models succeeded in reasonable predictive power, more work is needed to obtain more detailed mechanistic explanations.

Reverse Phase Protein Arrays (RPPAs or RPLAs) were used in [8] to profile signaling proteins in 56 breast cancers and matched normal tissue as a method to discover phosphorylation-mediated signal transduction patterns in human tumor samples. The paper discusses the process of validating antibodies (100 antibodies validated of 400 screened), and methods for quantitation of data in some detail. Unsupervised hierarchical clustering was used as a first step in discovering patterns of co-regulation. The hierarchy was cut to yield twelve clusters, which were mapped onto pathways derived from Gene Network Central Pro. This revealed a cluster involving increased abundance of the Axl receptor tyrosine kinase (RTK) and the cMet RTK pathway. Structured Bayesian inference was then used to further analyze this cluster to find the interaction network topology with good generalization properties and that best classified cancer vs non-cancer data. The results suggested two cancerous categories: (1) where MET is highly phosphorylated and cRAF is always highly phosphorylated and (2) where MET phosphorylation is low and cRAF phosphorylation is low at sites consistent with cRaf inactivation.

8   Conclusions and Future Directions

We have shown how the Pathway Logic STM model, capturing what we know about intracellular signal transduction, can be used to explain experimental results. The rules used in the model are derived from experimental results, so if the model were complete we should be able to use the network derived from exponentially growing cultured cells to trace the paths from a known perturbation to the measured effects. In some of the cases, we were successful. Our successes were predominantly in the phosphorylation cascades and protein degradation events used in growing cells. We were less effective in explaining the decreases in expression of proteins due to inhibition of translation or transduction, or changes in the cell cycle. There is still a lot of experimental evidence in the literature to collect and make into rules. There are still a lot of experiments that need to be performed and published. Work is in progress to automate this fuzzy backwards and forwards collection carried out by hand to generate the SKMEL133 model. We are also investigating representation of executable models, network perturbations, and experimental observations as constraints and using abductive reasoning to generate potential explanations. This would unify the treatment of various aspects and help automatic the end to end reasoning process.

One caveat, not all of the unexplained results are due to an incomplete model. Only one experiment was performed so the probability that the results could be reproduced cannot be measured. Although 3 biological replicates were used - no information about the variance were provided. In addition, we obtained the mechanism of action of the drugs from a small sampling of the literature. Any of the drugs could have additional effects that we did not find.

Learning about how a cell works is still a work in progress. The Pathway Logic STM model is a tool designed to help. Hopefully it does.

References

1. Aksoy, B.A.: BioPax prior information provider (PERA) (2017). http://bit.ly/bp_prior. Accessed 26 Apr 2017
2. M.D. Anderson Cancer Center: Antibody information and protocols (2017). https://www.mdanderson.org/research/research-resources/core-facilities/functional-proteomics-rppa-core/antibody-information-and-protocols.html. Accessed 25 Apr 2017
3. M.D. Anderson Cancer Center: RPPA core facility (2017). https://www.mdanderson.org/research/research-resources/core-facilities/functional-proteomics-rppa-core.html. Accessed 25 Apr 2017
4. Clavel, M., Durán, F., Eker, S., Lincoln, P., Martí-Oliet, N., Meseguer, J., Talcott, C.: All About Maude - A High-Performance Logical Framework. LNCS, vol. 4350. Springer, Heidelberg (2007). doi:10.1007/978-3-540-71999-1
5. Dienstmann, R., Rodon, J., Serra, V., Tabernero, J.: Picking the point of inhibition: a comparative review of PI3K/AKT/mTOR pathway inhibitors. Mol. Cancer Ther. **13**(5), 1021–1031 (2014)
6. Donaldson, R., Talcott, C., Knapp, M., Calder, M.: Understanding signalling networks as collections of signal transduction pathways. In: Computational Methods in Systems Biology (2010)
7. Fisher, J., Henzinger, T.A.: Executable cell biology. Nat. Biotechnol. **25**(11), 1239–1249 (2007)
8. Gujral, T.S., Karp, R.L., Finski, A., Chan, M., Schwartz, P.E., MacBeath, G., Sorger, P.: Profiling phospho-signaling networks in breast cancer using reverse-phase protein arrays. Oncogene **32**(29), 3470–3476 (2013)
9. Hill, S.M., et al.: Inferring causal molecular networks: empirical assessment through a community-based effort. Nat. Methods **13**(4), 310–318 (2016)
10. Korkut, A., Wang, W., Demir, E., Aksoy, B.A., Jing, X., Molinelli, E.J., Babur, O., Bemis, D.L., Onur, S.S., Solit, D.B., Pratilas, C.A., Sander, C.: Perturbation biol- ogy nominates upstream-downstream drug combinations in RAF inhibitor resistant melanoma cells. Elife **18**(4) (2015). Article id 26284497
11. Logie, L., Ruiz-Alcaraz, A.J., Keane, M., Woods, Y.L., Bain, J., Marquez, R., Alessi, D.R., Sutherland, C.: Characterization of a protein kinase B inhibitor in vitro and in insulin-treated liver cells. Diabetes **56**(9), 2218–2227 (2007)
12. Nigam, V., Donaldson, R., Knapp, M., McCarthy, T., Talcott, C.: Inferring executable models from formalized experimental evidence. In: Roux, O., Bourdon, J. (eds.) CMSB 2015. LNCS, vol. 9308, pp. 90–103. Springer, Cham (2015). doi:10.1007/978-3-319-23401-4_9
13. Pathway Logic (2016). http://pl.csl.sri.com. Accessed 10 Jan 2016
14. Samatar, A.A., Poulikakos, P.I.: Targeting RAS-ERK signalling in cancer: promises and challenges. Nat. Rev. Drug Discov. **13**(12), 928–942 (2014)
15. Schmidt, K.: LoLA a low level analyser. In: Nielsen, M., Simpson, D. (eds.) ICATPN 2000. LNCS, vol. 1825, pp. 465–474. Springer, Heidelberg (2000). doi:10.1007/3-540-44988-4_27
16. Talcott, C.: The pathway logic formal modeling system: diverse views of a formal representation of signal transduction. In: Wang, Q. (ed.) Workshop on Formal Methods in Bioinformatics and Biomedicine (2016)
17. Tsai, J., et al.: Discovery of a selective inhibitor of oncogenic B-Raf kinase with potent antimelanoma activity. Proc. Natl. Acad. Sci. USA **105**(8), 3041–3046 (2008)

Automated Property Synthesis of ODEs Based Bio-pathways Models

Jun Zhou[1]([✉]), R. Ramanathan[1], Weng-Fai Wong[1], and P.S. Thiagarajan[2]

[1] Department of Computer Science, National University of Singapore,
Singapore, Singapore
{zhoujun,ramanathan,wongwf}@comp.nus.edu.sg
[2] Laboratory of Systems Pharmacology, Harvard Medical School, Boston, USA
thiagu@hms.harvard.edu

Abstract. Identifying non-trivial requirements for large complex dynamical systems is a challenging but fruitful task. Once identified such requirements can be used to validate updated versions of the system and verify functionally similar systems. Here we present a technique for discovering behavioural properties of bio-pathway models whose dynamics is modelled as a system of ordinary differential equations (ODEs). These models are usually accompanied at best by high level functional requirements while undergoing many revisions as new experimental data becomes available. In this setting we first specify a set of property templates using bounded linear-time temporal logic (BLTL). A template will have the skeletal structure of a BLTL formula but the time bounds associated with the temporal operator as well as the value bounds associated with the system variables encoded as atomic propositions will be unknown parameters. We classify a given model's behavior as corresponding to one of these templates using a convolutional neural network. We then synthesize a concrete property from this template by estimating its parameters via a standard search procedure combined with statistical model checking (SMC). We have synthesized and validated properties of a number of pathway models of varying complexity using our method.

Keywords: Property synthesis · Statistical model checking · Bounded linear-time temporal logic · ODEs models of bio-pathways

1 Introduction

Synthesizing specifications of system models is a useful but challenging task. This is especially so for bio-pathway models. These models are rarely come with concrete temporal specifications. Instead, they are accompanied by functionalities such as "EGF-NGF stimulation of PC12 cells discriminates between proliferation and differentiation". Synthesizing more concrete temporal specifications

---

This research was supported by the Singapore grant MOE2013-T2-2-033 (Ministry of Education) and the US grants P50GM107681 (Laboratory of Systems Pharmacology, Harvard Medical School), W911NF-14-1-0397, W911NF-15-1-0544 (DARPA).

J. Feret and H. Koeppl (Eds.): CMSB 2017, LNBI 10545, pp. 265–282, 2017.
DOI: 10.1007/978-3-319-67471-1_16

from these models is appealing for at least two reasons. First, the synthesized specifications can point to mechanistic chains of events that determine the overall functionality such as "transient activation of Erk1/2 leads to proliferation while its sustained activation results in differentiation". (We hasten to add that this is merely an illustration using the functional specification and the concrete mechanistic property presented in [5]). Second, the construction of a model is rarely complete. Instead, it is repeatedly updated as fresh experimental data becomes available. In such settings, the temporal specifications synthesized from a previous version of the model can be used to assess whether the new model is qualitatively different from the older one.

As is well known there are two major classes of models to describe the dynamics of bio-pathways, namely deterministic ones based on ODEs [2] and stochastic ones [13] based on continuous-time Markov chains (CTMCs). In this paper, we shall focus on ODEs based models. In both types of models many rate constants of the reactions as well as the initial concentrations will be unknown. Here we consider this to be an important but orthogonal issue. Hence for evaluating our proposed method, we consider curated models with known parameter values taken from the Biomodels database [18].

We first build a set of property templates that capture parametrized families of pathway dynamical properties. Each template is built out of a BLTL (bounded linear time temporal logic) formula but whose time bounds associated with the temporal operators are integer-valued parameters. Furthermore, the atomic propositions appearing in the formula will be of the form ($\ell \leq x \leq u$) where x is a system variable and ℓ and u are parameters that take values from the value domain of x. (These template parameters are not to be confused with the (model) parameters associated with the ODEs model). The choice of BLTL as the specification logic -and the accompanying atomic propositions- is guided by the nature of the experimental data that is usually available for our models of interest, namely signaling pathways. Here the experimental data (i.e. observations of the system states) will typically consist of finite precision and noisy measurements regarding a small subset of system variables at a finite number of discrete time points. Further, only qualitative temporal properties will be applicable.

To focus on the main issues, we restrict our attention to four templates that capture key behavioural patterns of interest such as: "the concentration of a species x starts from an initial level in the interval $[c_1, c_2]$, rises to a level $[d_1, d_2]$ within k time units and remains in this interval until t_{max}". Based on these templates we develop a synthesis framework as shown in Fig. 1. First, by assuming an initial probability (usually uniform) distribution over the initial states of the system variables, a set of trajectories is generated through numerical simulations. Next, the trajectories are presented to a pre-trained convolutional neural network to identify the template ψ that best corresponds to the trajectories. We then employ a simulated-annealing [21] based global optimization procedure to estimate the parameters of the template. Specifically, in each step of the procedure, the *value generator* instantiates from ψ a concrete property $\widehat{\psi}$. We then use statistical model checking to evaluate the quality of satisfaction of $\widehat{\psi}$. Subsequently, the *loss function* computes the loss, and reports it to the simulated-annealing procedure which then terminates, or generates a new set of values for the parameters.

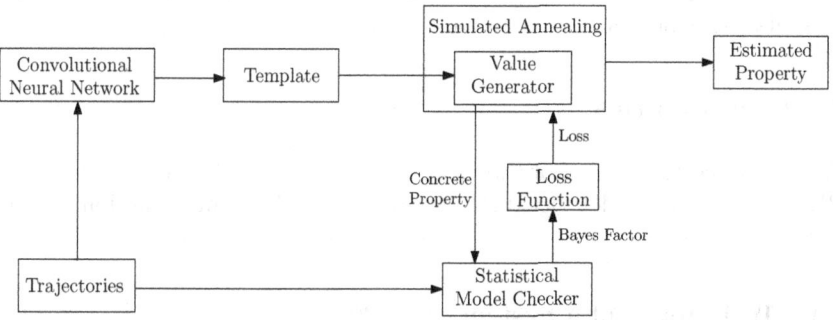

Fig. 1. Overview of the property synthesis framework

1.1 Related Work

Learning temporal logic formulas from data (or a generative model) is becoming a well explored field. The applications come from both cyber physical systems [8,15,17] and biological domains [3,7,12]. In the latter domain—which is our interest—the line of work reported in [7] is particularly relevant. The authors first learn a stochastic hybrid system from data and then use the model to generate data for learning the temporal logic formulas of interest in two steps. First, using an evolutionary algorithm, the structure (template as we call it) of the formula is learned. Then the parameters in the template are calibrated using a previously developed stochastic optimization method called the Gaussian Process Upper Confidence Bound (GP-UCB) algorithm [4]. The specification logic is bounded metric temporal logic. In our setting the model is available as a system of ODEs. We fix a set of templates in advance and train a convolution network using synthetic data not generated by the model in order to avoid bias. Then using this network and trajectories randomly generated by the model we match a template to the model. We then learn its parameters using simulated annealing combined with statistical model checking. Finally we use BLTL as our specification logic since it is a good fit for the class of models we wish to study.

An important aspect of parameter learning is to determine how well the formula instantiated by a particular choice of parameters matches the training data. Here again there is a good deal of literature on "robustness of satisfaction" [3,9,10,28]. Specifically, [28] is illustrative for ODEs based models in which a continuous notion of satisfaction is combined with an evolutionary search procedure to estimate kinetic parameters meeting temporal logic specifications. On the other hand the work reported in [3] formulates a robustness of satisfaction notion for stochastic systems and then uses this notion to optimize chosen control parameters of a stochastic system in order to maximize the robustness of satisfaction.

The paper is organized as follows. Section 2 presents the preliminaries and property templates. In Sects. 3 and 4, we explain our search procedure. Section 5 presents the experimental results and we conclude in Sect. 6.

2 Preliminaries

We introduce the basic notations we will be using in connection with ODEs, BLTL, statistical model checking. We conclude with the introduction of property templates.

2.1 Trajectories of a System of ODEs

Suppose there are n molecular species $\{x_1, x_2, \ldots, x_n\}$ in the pathway. For each species x_i, an equation of the form $\frac{dx_i}{dt} = f_i(\mathbf{x}, \Theta_i)$ describes the kinetics of the reactions that produce and consume x_i where $\mathbf{x}$ is the concentrations of the molecular species taking part in the reactions. Θ_i consists of the rate constants governing the reactions. Each x_i is a real-valued function of $t \in \mathbb{R}_+$, the set of non-negative reals. We shall realistically assume that $x_i(t)$ takes values in the interval $[L_i, U_i]$, where L_i and U_i are non-negative rationals with $L_i < U_i$. Assuming there are m reactions, we let $\Theta = \{\theta_1, \theta_2, \ldots, \theta_m\}$ be the set of rate constants. We define for each variable x_i an interval $[L_i^{init}, U_i^{init}]$ with $L_i \leq L_i^{init} < U_i^{init} \leq U_i$. We assume the value of the initial concentration of x_i to fall in this interval. We also assume the nominal value of the rate constant θ_j falls in the interval $[L_j^{init}, U_j^{init}]$ for $1 \leq j \leq m$. We set $INIT = (\prod_i [L_i^{init}, U_i^{init}]) \times (\prod_j [L_j^{init}, U_j^{init}])$. Here $INIT$ is meant to capture the variability in the initial concentrations of the variables and the rate constants across a population of cells. Further, we let $\mathbf{v}$ to range over $\prod_i [L_i^{init}, U_i^{init}]$ and $\mathbf{w}$ to range over $\prod_j [L_j^{init}, U_j^{init}]$. We define in the usual way the notion of a trajectory $\sigma_{\mathbf{v},\mathbf{w}}$ starting from $(\mathbf{v}, \mathbf{w}) \in INIT$ at time 0. We let TRJ denote the set of all finite trajectories that start in $INIT$.

As mentioned earlier we assume a probability distribution over $INIT$ and for convenience assume it to be the uniform one. The ODEs systems arising in our setting will induce vector fields that satisfy a natural continuity property. Hence one can define the probability that a trajectory starting from a randomly chosen state in $INIT$ will satisfy a given BLTL formula. Consequently one can develop a statistical model checking procedure to verify whether the system of ODEs meets the given BLTL specification with required probability [27].

2.2 Bounded Linear-Time Temporal Logic

An atomic proposition for our setting will be of the form $(L \leq x_i \leq U)$ with $L_i \leq L < U \leq U_i$ where L, U are rationals. The proposition $(L \leq x_i \leq U)$ says "the current concentration level of x_i lies in the interval $[L, U]$" and we fix a finite set of atomic propositions. BLTL formulas are then defined in the usual way.

We fix a finite set of time points $T = \{t_0 < t_1, \ldots, t_K\}$ and interpret a BLTL formulas over a trajectory σ in TRJ observed at the time points in T as usual. We say that σ is a *model* of ψ if $\sigma, t_0 \models \psi$.

For a formula ψ the statement $P_{\geq r}(\psi)$ where r $\in$ [0,1) will mean "the probability that a trajectory in TRJ is a *model* of ψ is at least r". To verify this, we consider the sequential hypothesis testing problem where the null hypothesis is $\mathcal{H}_0 : P_{\geq r}(\psi)$ and the alternate hypothesis is $\mathcal{H}_1 : P_{<r}(\psi)$. A convenient termination criterion here is the Bayes factor [16,19].

$$\mathcal{B} = \frac{Pr(d|\mathcal{H}_0)}{Pr(d|\mathcal{H}_1)} \qquad (1)$$

where d is the collection of Bernoulli random variables denoting the outcome whether a random trajectory generated by the ODE system satisfies ψ. Comparing $\mathcal{B}$ against a pre-defined threshold h, the property is accepted if $\mathcal{B}$ is larger than h and is rejected if it is less than $1/h$. Unlike the SPRT ratio test one doesn't have to specify an indifference region.

2.3 Templates

A template is a BLTL formula in which the bounds on system variable values in the atomic propositions and the integer bounds associated with the temporal operators are replaced by symbolic variables. These variables will be called *propositional variables* and *temporal variables* respectively in what follows. In addition, the template is augmented by a set of constraints. These constraints will be of the form $[u_j \leq \ell_k]$ or $[u_k \leq \ell_j]$ given two atomic propositions of the form $(\ell_j \leq x_j \leq u_j)$ and $(\ell_k \leq x_k \leq u_k)$.

Here is an example of a template:

$$p_1 \wedge F^{\leq t_1} G^{\leq t_2} p_2 \mid [u_1 \leq \ell_2]$$

$$\text{where } p_1 = (\ell_1 \leq x_1 \leq u_1) \text{ and } p_2 = (\ell_2 \leq x_1 \leq u_2).$$

This template represents the statement "value of x_1, starting from a low level (p_1) reaches within t_1 time units a high level (p_2) and stays at p_2 for at least t_2 units". $[u_1 \leq \ell_2]$ captures the constraint the level (ℓ_1, u_1) is lower than (ℓ_2, u_2).

The main idea is to search over the temporal and atomic proposition variables and use Bayes factor to measure of how well a synthesized property characterizes the observed behavior. A property with a Bayes factor larger than a given Bayes factor threshold is accepted while one with a small Bayes factor is rejected. In this initial study we consider the templates listed in Table 1.

3   Classifying Templates Using a Convolutional Neural Network

Our workflow first trains a convolution network to recognize trajectories presented to it as belonging to one of the set of templates we have fixed. It then

Table 1. Basic templates

No.	Template	Description
1	$p_1 \wedge F^{\leq t_1} G^{\leq t_2} p_2$ where $p_1 : (\ell_1 \leq x \leq u_1)$ $p_2 : (\ell_2 \leq x \leq u_2)$	Starting from the level p_1, within t_1 steps, the value of x reaches the level p_2 and stays there for at least t_2 steps. Typically describes sustained activations or deactivations. Constraints can be used to specify whether x decreases or increases from the initial level
2	$p_1 \wedge F^{\leq t_1} (p_2 \wedge F^{\leq t_2} p_3)$ where $p_1 : (\ell_1 \leq x \leq u_1)$ $p_2 : (\ell_2 \leq x \leq u_2)$ $p_3 : (\ell_3 \leq x \leq u_3)$	Starting from an initial level p_1, the value of x reaches the level p_2 within t_1 steps. Then, from p_2, x reaches a level p_3 within t_2 steps. Formulates evolution of species concentration from an initial level to a new level and then further to another new level or back to the initial level
3	$p_1 \wedge F^{\leq t_1} (p_2 \wedge F^{\leq t_2} G^{\leq t_3} p_3)$ where $p_1 : (\ell_1 \leq x \leq u_1)$ $p_2 : (\ell_2 \leq x \leq u_2)$ $p_3 : (\ell_3 \leq x \leq u_3)$	Similar to Template 2, the value of x starts from the level p_1, reaches the level p_2 within t_1 steps. Then within the next t_2 steps, reaches a level p_3 and stays in p_3 for at least t_3 steps. Characterizes transient or sustained activations, can be extended to formulate bistability
4	$p_1 \wedge F^{\leq t_1} (p_2 \wedge F^{\leq t_2} (p_3 \wedge F^{\leq t_1} (p_4)))$ where $p_1 : (\ell_1 \leq x \leq u_1)$ $p_2 : (\ell_2 \leq x \leq u_2)$ $p_3 : (\ell_3 \leq x \leq u_3)$ $p_4 : (\ell_4 \leq x \leq u_4)$	Starting from an initial level p_1, the value of x reaches the level p_2 where $(u_1 < \ell_2)$ within t_1 steps. Then, from p_2, x reaches a level p_3, $(u_3 < \ell_2)$ within t_2 steps. Further from p_3, x reaches a level p_4 where $(u_3 < \ell_4)$. Imposing constraints $[u_1 < \ell_2] \wedge [u_1 < \ell_4] \wedge [u_3 < \ell_2] \wedge [u_3 < \ell_4]$ characterizes oscillations

classifies a set of random trajectories generated by a model as belonging to one of the templates and then proceeds to synthesize a concrete property using the template.

3.1 Data Preprocessing

The evolution of a variable x is mainly reflected by changes in its value over time. We first transform the trajectories by evaluating the change in x at each time point, and computing the normalized $\Delta x(t)$ data over time as indicated by the formula below. This transformed data is then fed to the convolutional neural network for classification.

$$\Delta x(t) = \frac{x(t) - x(t-1)}{max(x) - min(x)}, \qquad (2)$$

where $max(x)$ and $min(x)$ are the maximum and minimum values of x across all the time points in the simulation.

3.2 Training and Deploying the Convolutional Neural Network

A convolutional neural network (CNN) is a type of feed forward neural network proposed in [24]. It has been successfully used to classify time series data and other features [30]. In this paper, we have adopted a standard convolutional neural network and implemented it using Tensorflow, a deep learning framework by Google [1]. There is a vast literature available including [24] on CNNs.

The CNN receives the pre-processed inputs described in Sect. 3.1 and feeds it to two convolutional and pooling layers, connected to two fully connected layers. Then it outputs to four output neurons, corresponding to the four templates. Due to space limitations we present the architecture and other details of the this CNN in the full report [31].

Our CNN is trained for the templates listed in Table 1. The training set is generated from mathematical functions found in [29]. Specifically, we selected 25 functions that conform to the four templates. For each of these functions, we generated 68 'seed' curves using different random initial parameters. Next, we transformed these into the frequency domain using Fast Fourier Transform (FFT). In the frequency domain, we perform further randomization before transforming them back into curves in time domain using the inverse FFT. We obtained 2,000 randomized curves from each seed curve. In total, 136,000 curves were used to train the CNN.

After training, the CNN is deployed to identify a template that best matches a set of trajectories randomly generated by a given model. Since neural networks take as inputs fixed-length data, the trajectories need to be re-scaled using a different sampling rate of simulation as follows. We first generate 20 trajectories. The same simulation time as given in the literature for the respective model is divided up into 200 equally spaced time-points, and sampled. The trajectories are then transformed into $\Delta x(t)$ as mentioned before in Sect. 3.1, and fed to the neural network. A simple majority across the results of classifying these 20 trajectories is used to determine the final template.

4 The Search Procedure

Given a template ψ identified from the convolutional neural network with time variables Var_T, and propositional variables Var_{AP}, we automatically mine the values of Var_T and Var_{AP} such that the concretized formula is optimal in a certain sense.

In order to reduce the search complexity, we assume the BLTL based template is given as a conjunction of component formula skeletons. We consequently optimize each conjunct in the template.

We adopt a simulated annealing based procedure presented in Algorithm 1 to estimate the parameters.

Algorithm 1. optimizeProperty

Input : Template ψ
Output: Synthesized property ψ_{syn}
1 $\widehat{\psi} \leftarrow$ Initialize Var_T and Var_{AP} using random values;
2 **while _Simulated Annealing_** _decides to continue_ **do**
3 $\quad$ Compute Bayes Factor $\mathcal{B}_{\widehat{\psi}} \leftarrow$ **SMC**($\widehat{\psi}$) ;
4 $\quad$ Compute $Loss_{\widehat{\psi}} \leftarrow$ **Loss Function**($\widehat{\psi}, \mathcal{B}_{\widehat{\psi}}$);
5 $\quad$ **Simulated Annealing** $\leftarrow Loss_{\widehat{\psi}}$;
6 $\quad$ Update Var_T and Var_{AP} ;
7 **return** $\psi_{syn} \leftarrow \widehat{\psi}$ with minimum loss if exists;

We generate values for the propositional variables using the constraints specified in the propositional variables and the template constraints. Though the constraint satisfaction problem is NP-complete the constraints in our framework are simple inequalities which enables us to adopt a tree-based solution. The value intervals of a variable are parsed as a tree structure where the values of the child nodes are larger than the parent nodes.

For example, for the template $p_1 \wedge F^{\leq t_1}(p_2 \wedge F^{\leq t_2} p_3)$ suppose we have the constraints $[u_1 < \ell_2]$ and $[\ell_2 < u_3]$, together with the implicit constraints $[\ell_1 < u_1]$, $[\ell_2 < u_2]$ and $[\ell_3 < u_3]$ the tree is constructed as shown in Fig. 2. We generate values for the leaf variables (u_2 and u_3) first and then use them to bound the value range of parents (ℓ_2 and ℓ_3), recursively till the root (ℓ_1) is reached.

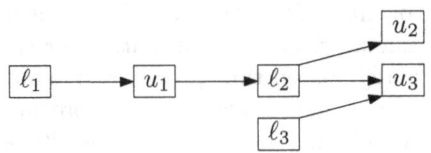

Fig. 2. Generating values for propositional variables using a tree

4.1 Loss Function

Each instantiated property $\widehat{\psi}$ is scored using a *loss function* and the score will guide the direction of the search. The score is composed out of the "loss" suffered by three factors: temporal variables, atomic propositions and the quality of satisfaction. For the temporal variables we use the intuition that if ψ_1 and ψ_2 are two instantiations such that ψ_1 implies ψ_2 then ψ_1 is to be preferred. This suggests that if $\psi_1 = F^{\leq t_1}\varphi$ and $\psi_2 = F^{\leq t_2}\varphi$ are two instantiations and $t_1 \leq t_2$ then t_1 is preferred to t_2. Similarly t_2 is preferred to t_1 if $G^{\leq t_1}\varphi$ and $G^{\leq t_2}\varphi$ are two instantiations with $t_1 \leq t_2$.

The loss component L_T of the temporal variables is given by:

$$L_T = \prod_{t_i \in \mathrm{Var}_T} \left(t_i\right)^{sgn(t_i)}$$

$$sgn(t) = \begin{cases} -1, & \text{if temporal operator of } t_i \text{ is } G \text{ or } U \\ 1, & \text{if temporal operator of } t_i \text{ is } F \end{cases}.$$

Next, we define L_{AP}, the loss function component contributed by the propositional variables. For each atomic proposition, we consider both the tightness of the value range, and how precisely it describes the behaviour of the trajectories.

For each atomic proposition ap_i, we define the tightness as $(u_i - \ell_i)/(max_i - min_i)$, the range normalized to the maximum value range of the variable in trajectories. The idea is to keep the value range as small as possible.

Besides the tightness, we also measure the fitness of the atomic propositions in $\widehat{\psi}$ to the trajectories based on the constraints. Essentially for each constraint of the form $u_j < \ell_k$ attached to the atomic propositions ap_j and ap_k, the estimated levels of ap_j is expected to be lower than ap_k. This information is also used to optimize $\widehat{\psi}$. To this end, we compute the mean value of ap_i as $(\frac{\ell_i + u_i}{2})$. The weight w_i associated with each ap_i is evaluated as follows. We first initialize the set of weights W_{AP} for all the atomic propositions in $\widehat{\psi}$ to 0. Then for each constraint $u_j < \ell_k$, we decrease w_j by 1 and increase w_k by 1. The fitness of an atomic proposition is thus $\left(\frac{\ell_i + u_i}{2}\right)^{w_i}$. Intuitively, the level of ap_j tends to be in the lower range of value space while ap_k to be in higher range.

Combining these two factors, we define the loss function component due to the propositional variables as

$$L_{AP} = \prod_{ap_i \in \mathrm{Var}_{AP}} \left(\left(\frac{u_i - \ell_i}{max_i - min_i}\right)\left(\frac{\ell_i + u_i}{2 \cdot max_i}\right)^{w_i} \right).$$

Finally, in each iteration of the simulated-annealing procedure, if the Bayes factor $\mathcal{B}_{\widehat{\psi}}$ is larger than a pre-defined threshold h (in our case it is set to 100), we apply the loss function and continue with the iterations according to the search procedure. Otherwise, the loss is set as ∞ and the current combination of parameters is rejected. The search then continues with another combination of parameters.

Thus

$$Loss_{\widehat{\psi}} = \begin{cases} L_{AP} \cdot L_T, & \mathcal{B}_{\widehat{\psi}} > h \\ \infty, & \text{otherwise} \end{cases}.$$

We use the multiplicative form of the loss function since we found that the additive form performs badly. For instance, if two temporal variables and one propositional variable appear in a formula the search gets biased towards optimizing just one the three variables while fixing a trivial value for the other two variables. Admittedly the current formulation of the loss function is just a first and preliminary step. A systematic study of the various possibilities -including other notions of quality of satisfaction- needs to be carried out in the future.

5   Experimental Evaluation

We applied our method to six bio-pathway models taken from the Biomodels database [23]. For the purposes of experimentation we fixed ±5% range around the nominal values as the initial interval of values of each species and we assumed a uniform distribution over the resulting set of initial states. Using the convolutional neural network and randomly generated trajectories using the model, the most suitable BLTL template was then identified followed by a concrete instantiation for this template to a high satisfaction probability, namely, $r \geq 0.9$.

Table 2 shows $|\mathbf{x}|$, the number of system variables and $|\Theta|$, the number of rate constants of the ODEs systems associated with the six models. The time unit for the F and G operators is 'minutes'. Furthermore, the number of time points to simulate (i.e. t_K) for each of the models was fixed using the literature of the respective models [5,6,11,14,20,25]. We next present the synthesized properties for the important species in each of the bio-pathway models. Across all the six case studies, there is a total of 13 such species.

Table 2. Characteristics of the models

Bio-pathway models	EGF-NGF	Segmentation clock	MAPK cascade	Atorvastatin	Va factor	CD95 signalling
$|x|$	32	16	8	18	30	23
$|\Theta|$	48	71	22	30	9	17

Validation. In the six case studies we present here, we compared the synthesized properties against the observed qualitative trends of species documented in [5,6,11,14,20,25]. For one of the models we provide further validation by using the synthesized properties in the context of rate constants estimation problem as explained in Sect. 5.3.

5.1   Template Recognition

We first generate 20 trajectories from the model and use these as inputs to the CNN. For each trajectory and for each species (variable) of interest the CNN returns the confidence level in classifying the trajectory to each of the four templates and the template with the highest confidence is chosen. Finally, the template with most votes from all the trajectories is chosen as the template to be the candidate for synthesizing a concrete formula.

For each of the case studies in Sect. 5.2, we observed that the CNN returns the same template overwhelmingly for all the 20 trajectories with high confidence (above 98%). This data is reported in [31].

5.2   Case Studies

EGF-NGF Pathway. The EGF-NGF signalling pathway [5] captures the differential response of PC12 cells to two growth hormones, EGF and NGF. EGF induces cell proliferation while NGF stimulates cell differentiation. It has been reported that the signal specificity is correlated with different Erk dynamics. A transient activation of Erk has been associated with cell proliferation, while a sustained activation has been linked to differentiation. The model has 32 ODEs and 48 kinetic rate parameters. We simulated this model for 60 min.

Table 3(a) shows three properties that describe the sustained activation of Erk*, bound-EGFR and C3G*, rising rapidly (within 10 min) to a high level. It has been verified from experimental data that under NGF stimulation, sustained activation of Erk* is induced by the phosphorylation of C3G. The synthesized property captures this behaviour: $([0 \leq \text{Erk}^* \leq 0] \wedge F^{\leq 5}G^{\leq 55}([477401 \leq \text{Erk}^* \leq 571121]))$ returned that the concentration level of Erk* rises from an initial level $[0 \leq \text{Erk}^* \leq 0]$ to a peak level $[477401 \leq \text{Erk}^* \leq 571121]$ and stays at that level for 50 min.

Segmentation Clock Network. Formation of segments in vertebrate embryos is controlled by coupled oscillations in the Notch, Wnt and FGF signalling pathways governed by a segmentation clock network that periodically activates the segmentation genes [11]. The model consists of 16 ODEs and 71 kinetic rate parameters. We simulated this model for 250 min.

From Table 3(b), one can find that both properties characterize the oscillation of Lunatic fringe-mRNA and cytosolic NicD, capturing the peak values. Although the search space of 11 parameters is large, the mined properties are closed to the nominal ones from literature. For example, the Lunatic fringe-mRNA property is close to the one observed in [27]:

$((([\text{Lunatic fringe mRNA} \leq 0.4]) \wedge (F^{\leq 40}([\text{Lunatic fringe mRNA} \geq \overset{\bullet}{2}.2] \wedge F^{\leq 40}([\text{Lunatic fringe mRNA} \leq 0.4] \wedge F^{\leq 40}([\text{Lunatic fringe mRNA} \geq 2.2] \wedge F^{\leq 40}([\text{Lunatic fringe mRNA} \leq 0.4])))))))$.

MAPK Cascade. From yeast to mammals, mitogen activated protein kinase (MAPK) cascades are bio-molecular networks widely involved in signal transduction of extracellular stimulus from the plasma membrane to the cytoplasm and nucleus. They play a major role in processes involving cell growth, mitogenesis, differentiation and stress responses in mammalian cells. The MAPK pathway [20] consists of three levels where the activated kinase at each level phosphorylates the kinase at the subsequent level down the cascade. It has been shown that

Table 3. Properties synthesized for the six case studies.

Simulation profile	Synthesized property
(a) EGF-NGF Pathway Model	
	$p_1 \wedge F^{\leq 5}G^{\leq 55}p_2$ $p_1 : 0 \leq \text{Erk}^* \leq 0$ $p_2 : 477401 \leq \text{Erk}^* \leq 571121$
	$p_1 \wedge F^{\leq 3}G^{\leq 57}p_2$ $p_1 : 0 \leq \text{C3G}^* \leq 0$ $p_2 : 111035 \leq \text{C3G}^* \leq 138166$
	$p_1 \wedge F^{\leq 2}G^{\leq 58}p_2$ $p_1 : 0 \leq \text{bound-EGFR} \leq 0$ $p_2 : 81639.9 \leq \text{bound-EGFR} \leq 86368.9$
(b) Segmentation Clock Network Model	
	$p_1 \wedge F^{\leq 58}(p_2 \wedge F^{\leq 23}(p_3 \wedge F^{\leq 75}(p_4)))$ $p_1 : 0.096 \leq \text{Lunatic fringe mRNA} \leq 0.102$ $p_2 : 2.42 \leq \text{Lunatic fringe mRNA} \leq 2.68$ $p_3 : 0.000 \leq \text{Lunatic fringe mRNA} \leq 0.008$ $p_4 : 1.83 \leq \text{Lunatic fringe mRNA} \leq 2.65$
	$p_1 \wedge F^{\leq 40}(p_2 \wedge F^{\leq 56}(p_3 \wedge F^{\leq 26}(p_4)))$ $p_1 : 0.199 \leq \text{cytosolic NicD} \leq 0.207$ $p_2 : 1.11 \leq \text{cytosolic NicD} \leq 1.23$ $p_3 : 0.25 \leq \text{cytosolic NicD} \leq 0.46$ $p_4 : 0.86 \leq \text{cytosolic NicD} \leq 1.03$
(c) MAPK Pathway Model	
	$p_1 \wedge F^{\leq 4}(p_2 \wedge F^{\leq 13}(p_3 \wedge F^{\leq 15}(p_4)))$ $p_1 : 9.50 \leq \text{Mos-P} \leq 10.50$ $p_2 : 81.38 \leq \text{Mos-P} \leq 88.97$ $p_3 : 0.00 \leq \text{Mos-P} \leq 5.13$ $p_4 : 44.24 \leq \text{Mos-P} \leq 68.94$
	$p_1 \wedge F^{\leq 6}(p_2 \wedge F^{\leq 24}(p_3 \wedge F^{\leq 11}(p_4)))$ $p_1 : 9.50 \leq \text{Erk2-PP} \leq 10.50$ $p_2 : 275.68 \leq \text{Erk2-PP} \leq 328.35$ $p_3 : 1.77 \leq \text{Erk2-PP} \leq 40.22$ $p_4 : 263.68 \leq \text{Erk2-PP} \leq 299.09$

Table 3. *Continued*

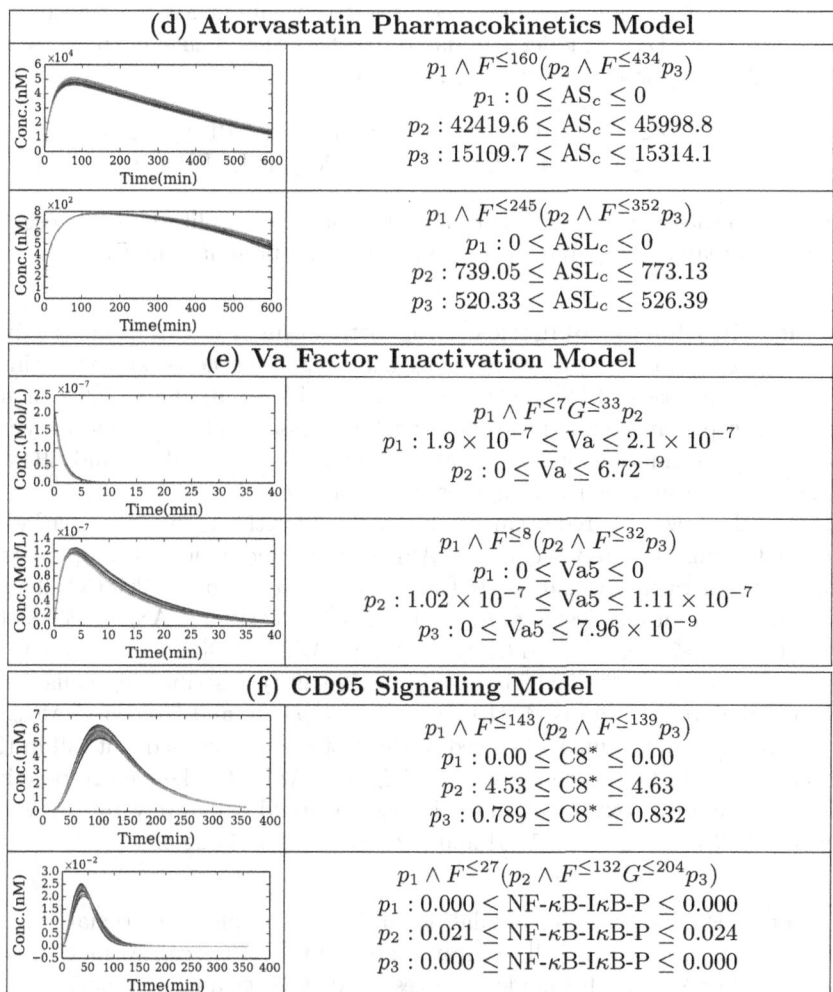

(d) Atorvastatin Pharmacokinetics Model	
	$p_1 \wedge F^{\leq 160}(p_2 \wedge F^{\leq 434}p_3)$ $p_1 : 0 \leq AS_c \leq 0$ $p_2 : 42419.6 \leq AS_c \leq 45998.8$ $p_3 : 15109.7 \leq AS_c \leq 15314.1$
	$p_1 \wedge F^{\leq 245}(p_2 \wedge F^{\leq 352}p_3)$ $p_1 : 0 \leq ASL_c \leq 0$ $p_2 : 739.05 \leq ASL_c \leq 773.13$ $p_3 : 520.33 \leq ASL_c \leq 526.39$
(e) Va Factor Inactivation Model	
	$p_1 \wedge F^{\leq 7}G^{\leq 33}p_2$ $p_1 : 1.9 \times 10^{-7} \leq Va \leq 2.1 \times 10^{-7}$ $p_2 : 0 \leq Va \leq 6.72^{-9}$
	$p_1 \wedge F^{\leq 8}(p_2 \wedge F^{\leq 32}p_3)$ $p_1 : 0 \leq Va5 \leq 0$ $p_2 : 1.02 \times 10^{-7} \leq Va5 \leq 1.11 \times 10^{-7}$ $p_3 : 0 \leq Va5 \leq 7.96 \times 10^{-9}$
(f) CD95 Signalling Model	
	$p_1 \wedge F^{\leq 143}(p_2 \wedge F^{\leq 139}p_3)$ $p_1 : 0.00 \leq C8^* \leq 0.00$ $p_2 : 4.53 \leq C8^* \leq 4.63$ $p_3 : 0.789 \leq C8^* \leq 0.832$
	$p_1 \wedge F^{\leq 27}(p_2 \wedge F^{\leq 132}G^{\leq 204}p_3)$ $p_1 : 0.000 \leq \text{NF-}\kappa\text{B-I}\kappa\text{B-P} \leq 0.000$ $p_2 : 0.021 \leq \text{NF-}\kappa\text{B-I}\kappa\text{B-P} \leq 0.024$ $p_3 : 0.000 \leq \text{NF-}\kappa\text{B-I}\kappa\text{B-P} \leq 0.000$

a negative feedback loop of MAPK cascade results in sustained oscillations in MAPK phosphorylation [20]. This ODEs model of this MAPK cascade consists of 8 species and 22 rate parameters. We simulated the model for 60 min.

Table 3(c) illustrates the properties for the two species, namely, phosphorylated Mos (Mos-P) at the initial level of the cascade, and biphosphorylated kinase Erk (Erk-PP) at the terminal level of the cascade. With the increased production of Erk-PP, the negative-feedback due to Erk-PP affects the phosphorylation of the initial level kinase, Mos. This in turn affects downstream phosphorylation of intermediate kinases, and ultimately the concentration of Erk-PP is decreased. Thus an oscillation cycle is triggered. The two properties synthesized by our method reflect this behaviour.

Experimental findings [22] indicate that "dual serine/threonine phosphorylation of SOS by Erk has been found to cooperatively inhibit MKKK phosphorylation". When the ODEs model is updated to reflect this change in the network, our method synthesized the following property:

$$[9.5 \leq \text{Erk2-PP} \leq 10.5] \wedge F^{\leq 17}([251.30 \leq \text{Erk2-PP} \leq 262.66]$$
$$\wedge F^{\leq 15}([0 \leq \text{Erk2-PP} \leq 42.74] \wedge (F^{\leq 14}[173.20 \leq \text{Erk2-PP} \leq 192.45]))).$$

From this synthesized property, one can infer that the amplitude of the oscillations has decreased compared to the nominal model presented in Table 3(c).

Atorvastatin Pharmacokinetics. Drug metabolism of statins inside the liver cells plays an important role in reducing cholesterol synthesis, and the stimulation of the uptake of LDL-cholesterol from the blood [6]. This ODEs model describes the pharmacokinetics of transport processes and metabolic enzymes in the biotransformation of atorvastatin. It consists of 18 ODEs and 30 rate parameters and the model was simulated for 600 min.

Table 3(d) shows two requirements synthesized for the atorvastatin pathway. AS (a hydrophilic hydroxyl-acid) and ASL (a very lipophilic lactone), the two forms of atorvastatin are transported into the cell and converted into different metabolites. The properties: $[0 \leq \text{AS}_c \leq 0] \wedge F^{\leq 160}([42419.6 \leq \text{AS}_c \leq 45998.8] \wedge F^{\leq 434}([15109.7 \leq \text{AS}_c \leq 15314.1]))$ and $[0 \leq \text{ASL}_c \leq 0] \wedge F^{\leq 245}([739.05 \leq \text{ASL}_c \leq 773.13] \wedge F^{\leq 352}([520.33 \leq \text{ASL}_c \leq 526.39]))$ describe this behaviour. The estimated value bounds $[42419.6 \leq \text{AS}_c \leq 45998.8]$ and $[739.05 \leq \text{ASL}_c \leq 773.13]$ are close to the peak observed in the system. The subsequent fall in the concentration due to the conversion of AS_c and ASL_c to their corresponding para- and ortho-hydroxy metabolites is also captured accurately by the value bounds $[15109.7 \leq \text{AS}_c \leq 15314.1]$ and $[520.33 \leq \text{ASL}_c \leq 526.39]$.

Va Factor Pathway. The regulation of Va factor plays a crucial role in hemostasis. As studied in [14], activated-protein-C (APC) causes inactivation of bovine factor Va and this model involves bond cleavage and dissociation of Va and its associated intermediate complexes produced in the process. The model consists of 30 ODEs and 9 kinetic rate parameters and was simulated for 20 min.

The two properties synthesized by our method characterizes the behaviour of the three species, namely Va and Va_5 are shown in Table 3(e). In particular, the properties synthesized using our method captures the rapid dissociation of Va by APC within 7 min.

CD-95 Signalling. Activation of CD-95 [25] in some situations results in cell death, and, in some other situations, induces activation of the NF-κB pathway. This has been found to be due to the cleavage of an anti-apoptotic protein, cFLIP_L and Procaspase-8. This model has 23 variables and 17 parameters and was simulated for 360 min.

The properties synthesized in Table 3(f) show the activation of Caspase-8 and the NF-κB-IκB-P by CD-95. Our method was able to mine the properties which characterize the rise and fall of the two proteins. More specifically, the third property mined for NF-κB-$I\kappa B$-P reflects transient activation within 150 min. It has been reported that CD-95 results in parallel – and not mutually-exclusive – transient activation of NF-κB and the Death Inducing Signalling Complex (DISC). This is in agreement with our findings.

5.3   Rate Constants Estimation Based on the Synthesized Properties

To further demonstrate the efficacy of the property synthesis procedure, we used the synthesized properties to estimate the unknown rate constants **w** of a pathway model in the context of the method developed in [27]. In this method both time course experimental data and known qualitative trends are encoded as BLTL formulas and the rate constant estimation problem is solved through evolutionary search combined with statistical model checking. In the present setting, we use the synthesized properties ψ_{syn} as sole inputs (i.e. no experimental data) to this estimation procedure. We then compared the quality of the rate constants obtained using our synthesized properties with the rate constants reported in the literature [25].

We applied our method to the CD-95 signalling pathway. We assumed 10 (k_2, k_3, k_5, k_6, k_7, k_{11}, k_{12}, k_{14}, k_{15}, k_{17}) out of 17 rate constants to be unknown. The inputs to the estimation procedure of [27] consists of 7 BLTL properties synthesized by our method. Figure 3 shows the simulation profiles generated using the predicted rate constant values. More precisely, 1,000 trajectories were generated using the rate constants estimated by our method and plotted against trends observed using the constants reported in [25].

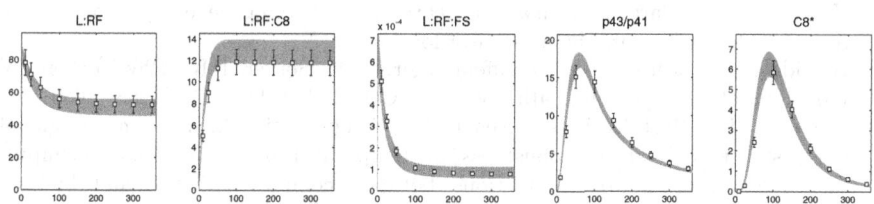

Fig. 3. Parameter estimation results for the CD-95 pathway using the synthesized properties

6   Conclusion

We have proposed an automated method to mine dynamic properties from ODEs based models of bio-pathways. Using simulated trajectories, our method first identifies a BLTL template matching their behaviour with the help of a convolutional neural network. A simulated-annealing based procedure combined with statistical model checking is then applied to this template to mine a concrete property. By checking the synthesized properties against the ones given in the literature as well as using them to do rate constants estimation of biopathways we have provided strong evidence that the mined BLTL formulas faithfully describe the behaviour of various species in our case studies.

In this preliminary study we have started with four templates. It will be useful to expand this templates library. Equally important, we have considered here only templates involving a single system variable. It will be challenging but very fruitful to learn properties that involve (at least) two system variables. This will enable for instance, to learn regulatory trends; for instance how an upstream variable representing a perturbation generates a pathway response in terms of a downstream variable.

Here we have focused on synthesizing properties for biological pathways modelled as a system of ODEs. However, our technique can be applied to ODEs systems arising in other settings as well.

To improve computational scalability, it will be important to port our current implementation to a GPU platform and exploit parallel search strategies such as parallel simulated annealing [26]. Finally it will be interesting to extend our method to the setting partial differential equations based models that capture spatial aspects of biopathways dynamics.

References

1. Abadi, M., Agarwal, A., Barham, P., Brevdo, E., Chen, Z., Citro, C., Corrado, G.S., Davis, A., Dean, J., Devin, M., Ghemawat, S., Goodfellow, I., Harp, A., Irving, G., Isard, M., Jia, Y., Jozefowicz, R., Kaiser, L., Kudlur, M., Levenberg, J., Mané, D., Monga, R., Moore, S., Murray, D., Olah, C., Schuster, M., Shlens, J., Steiner, B., Sutskever, I., Talwar, K., Tucker, P., Vanhoucke, V., Vasudevan, V., Viégas, F., Vinyals, O., Warden, P., Wattenberg, M., Wicke, M., Yu, Y., Zheng, X.: TensorFlow: large-scale machine learning on heterogeneous systems (2015). Software, tensorflow.org. http://tensorflow.org/
2. Aldridge, B.B., Burke, J.M., Lauffenburger, D.A., Sorger, P.K.: Physicochemical modelling of cell signalling pathways. Nat. Cell Biol. **8**(11), 1195–1203 (2006)
3. Bartocci, E., Bortolussi, L., Nenzi, L., Sanguinetti, G.: System design of stochastic models using robustness of temporal properties. Theor. Comput. Sci. **587**, 3–25 (2015). Interactions between computer science and biology. http://www.sciencedirect.com/science/article/pii/S0304397515002224
4. Bortolussi, L., Sanguinetti, G.: Learning and designing stochastic processes from logical constraints. In: Joshi, K., Siegle, M., Stoelinga, M., D'Argenio, P.R. (eds.) QEST 2013. LNCS, vol. 8054, pp. 89–105. Springer, Heidelberg (2013). doi:10. 1007/978-3-642-40196-1_7
5. Brown, K.S., Hill, C.C., Calero, G.A., Myers, C.R., Lee, K.H., Sethna, J.P., Cerione, R.A.: The statistical mechanics of complex signaling networks: nerve growth factor signaling. Phys. Biol. **1**(3), 184 (2004)
6. Bucher, J., Riedmaier, S., Schnabel, A., Marcus, K., Vacun, G., Weiss, T.S., Thasler, W.E., Nüssler, A.K., Zanger, U.M., Reuss, M.: A systems biology approach to dynamic modeling and inter-subject variability of statin pharmacokinetics in human hepatocytes. BMC Syst. Biol. **5**(1), 1 (2011)

7. Bufo, S., Bartocci, E., Sanguinetti, G., Borelli, M., Lucangelo, U., Bortolussi, L.: Temporal logic based monitoring of assisted ventilation in intensive care patients. In: Margaria, T., Steffen, B. (eds.) ISoLA 2014. LNCS, vol. 8803, pp. 391–403. Springer, Heidelberg (2014). doi:10.1007/978-3-662-45231-8_30
8. Chen, G., Sabato, Z., Kong, Z.: Active learning based requirement mining for cyber-physical systems. In: 2016 IEEE 55th Conference on Decision and Control (CDC), pp. 4586–4593. IEEE (2016)
9. Donzé, A., Maler, O.: Robust satisfaction of temporal logic over real-valued signals. In: Chatterjee, K., Henzinger, T.A. (eds.) FORMATS 2010. LNCS, vol. 6246, pp. 92–106. Springer, Heidelberg (2010). doi:10.1007/978-3-642-15297-9_9
10. Fainekos, G.E., Pappas, G.J.: Robustness of temporal logic specifications for continuous-time signals. Theor. Comput. Sci. 410(42), 4262–4291 (2009)
11. Goldbeter, A., Pourquié, O.: Modeling the segmentation clock as a network of coupled oscillations in the Notch, Wnt and FGF signaling pathways. J. Theor. Biol. 252(3), 574–585 (2008)
12. Grosu, R., Smolka, S.A., Corradini, F., Wasilewska, A., Entcheva, E., Bartocci, E.: Learning and detecting emergent behavior in networks of cardiac myocytes. Commun. ACM 52(3), 97–105 (2009)
13. Heath, J., Kwiatkowska, M., Norman, G., Parker, D., Tymchyshyn, O.: Probabilistic model checking of complex biological pathways. Theor. Comput. Sci. 391(3), 239–257 (2008)
14. Hockin, M.F., Cawthern, K.M., Kalafatis, M., Mann, K.G.: A model describing the inactivation of factor Va by APC: bond cleavage, fragment dissociation, and product inhibition. Biochemistry 38(21), 6918–6934 (1999)
15. Hoxha, B., Dokhanchi, A., Fainekos, G.: Mining parametric temporal logic properties in model-based design for cyber-physical systems. Int. J. Softw. Tools Technol. Transf. (2017). http://dx.doi.org/10.1007/s10009-017-0447-4
16. Jha, S.K., Clarke, E.M., Langmead, C.J., Legay, A., Platzer, A., Zuliani, P.: A Bayesian approach to model checking biological systems. In: Degano, P., Gorrieri, R. (eds.) CMSB 2009. LNCS, vol. 5688, pp. 218–234. Springer, Heidelberg (2009). doi:10.1007/978-3-642-03845-7_15
17. Jin, X., Donzé, A., Deshmukh, J.V., Seshia, S.A.: Mining requirements from closed-loop control models. IEEE Trans. Comput. Aided Des. Integr. Circ. Syst. 34(11), 1704–1717 (2015)
18. Juty, N., Ali, R., Glont, M., Keating, S., Rodriguez, N., Swat, M.J., Wimalaratne, S.M., Hermjakob, H., Le Novère, N., Laibe, C., Chelliah, V.: BioModels: content, features, functionality and use. CPT Pharmacomet. Syst. Pharmacol. 4, 55–68 (2015)
19. Kass, R.E., Raftery, A.E.: Bayes factors. J. Am. Stat. Assoc. 90(430), 773–795 (1995)
20. Kholodenko, B.N.: Negative feedback and ultrasensitivity can bring about oscillations in the mitogen-activated protein kinase cascades. Eur. J. Biochem. 267(6), 1583–1588 (2000)
21. Kirkpatrick, S., Gelatt, C.D., Vecchi, M.P., et al.: Optimization by simulated annealing. Science 220(4598), 671–680 (1983)
22. Langlois, W.J., Sasaoka, T., Saltiel, A.R., Olefsky, J.M.: Negative feedback regulation and desensitization of insulin-and epidermal growth factor-stimulated p21^{ras} activation. J. Biol. Chem. 270(43), 25320–25323 (1995)

23. Le Novere, N., Bornstein, B., Broicher, A., Courtot, M., Donizelli, M., Dharuri, H., Li, L., Sauro, H., Schilstra, M., Shapiro, B., et al.: Biomodels database: a free, centralized database of curated, published, quantitative kinetic models of biochemical and cellular systems. Nucleic Acids Res. **34**(Suppl. 1), D689–D691 (2006)

24. LeCun, Y., Bengio, Y.: The handbook of brain theory and neural networks. In: Convolutional Networks for Images, Speech, and Time Series, pp. 255–258. MIT Press, Cambridge (1998). http://dl.acm.org/citation.cfm?id=303568.303704

25. Neumann, L., Pforr, C., Beaudouin, J., Pappa, A., Fricker, N., Krammer, P.H., Lavrik, I.N., Eils, R.: Dynamics within the CD95 death-inducing signaling complex decide life and death of cells. Mol. Syst. Biol. **6**(1), 352 (2010)

26. Onbaşoğlu, E., Özdamar, L.: Parallel simulated annealing algorithms in global optimization. J. Glob. Optim. **19**(1), 27–50 (2001). http://dx.doi.org/10.1023/A:1008350810199

27. Palaniappan, S.K., Gyori, B.M., Liu, B., Hsu, D., Thiagarajan, P.S.: Statistical model checking based calibration and analysis of bio-pathway models. In: Gupta, A., Henzinger, T.A. (eds.) CMSB 2013. LNCS, vol. 8130, pp. 120–134. Springer, Heidelberg (2013). doi:10.1007/978-3-642-40708-6_10

28. Rizk, A., Batt, G., Fages, F., Soliman, S.: Continuous valuations of temporal logic specifications with applications to parameter optimization and robustness measures. Theor. Comput. Sci. **412**(26), 2827–2839 (2011). http://dx.doi.org/10.1016/j.tcs.2010.05.008

29. von Seggern, D.: CRC Standard Curves and Surfaces, 1st edn. CRC Press, Boca Raton (1993)

30. Zheng, Y., Liu, Q., Chen, E., Ge, Y., Zhao, J.L.: Time series classification using multi-channels deep convolutional neural networks. In: Li, F., Li, G., Hwang, S., Yao, B., Zhang, Z. (eds.) WAIM 2014. LNCS, vol. 8485, pp. 298–310. Springer, Cham (2014). doi:10.1007/978-3-319-08010-9_33

31. Zhou, J., Ramanathan, R., Wong, W.F., Thiagarajan, P.S.: Automated property synthesis of ODEs based bio-pathways models. http://www.comp.nus.edu.sg/~zhoujun/full_report.pdf

Tool Papers

TransferEntropyPT: An R Package to Assess Transfer Entropies via Permutation Tests

Patrick Boba and Kay Hamacher$^{(\boxtimes)}$

Technische Universität Darmstadt, Darmstadt, Germany
hamacher@bio.tu-darmstadt.de

Abstract. The package **TransferEntropyPT** provides R functions to calculate the transfer entropy (TE) [6] for time series of (binned) data. The package provides a function to assess the statistical significance of the TE using permutation tests on the sequential data of the time series. The underlying code base is written in C++ for computational efficiency and makes use of the **boost** and **OpenMP** libraries for parallelization of the data-parallel tasks in the permutation tests. In addition to p-values from hypothesis tests on independence, the package provides direct access to the percentiles themselves. An anticipatory toy model, as well as a biological network is used as show cases. Here, every time series concentrations of a single molecular species is tested and assessed against each other.

1 Introduction

A potential interdependence of two random variables can be analyzed by a variety of measures. Among the classical ones, we find Pearson's correlation coefficient or the mutual information. The Transfer Entropy (TE) introduced in [6] overcomes some of the shortcomings of these traditional measures when applied to time series data, such as only being able to identify linear correlations. To this end, TE employs conditional probabilities with respect to the specific time order of events in a time series. In recent years, TE has gained traction in many applications, e.g. the analysis of gene regulatory networks [7] or analysis procedures for magnetoencephalography in neuroscience [9]. Furthermore, Bauer et al. [1] were able to show the causal relationship between perturbations in process variables such as pressure of chemical processes. At present, there is no software package readily available to compute the TE *and* assess its significance in statistical software systems like R. The growth in the number of TE applications prompted us to adapt previous work [2] as a package explicitly for the statistical software R.

1.1 Background: Information Theory

First, we want to review some basic concepts of information theory to illustrate the issues the package addresses. The *Kullback-Leibler-Divergence* measures the amount of information needed (in bits) to get from the distribution

© Springer International Publishing AG 2017
J. Feret and H. Koeppl (Eds.): CMSB 2017, LNBI 10545, pp. 285–290, 2017.
DOI: 10.1007/978-3-319-67471-1_17

$p(Z)$ of a random variable Z from a sample space $\mathcal{Z}$ to another distribution $q(Z)$: $\mathrm{KLD}_{p|q} = \sum\limits_{z \in \mathcal{Z}} p(z) \log_2 \frac{p(z)}{q(z)}$. In the subsequent parts of this study, we use the convention "$0 \cdot \log = 0$". Note, that the $\mathrm{KLD}_{p|q}$ is therefore well defined as long as $\forall z \in \mathcal{Z} : q(z) > 0$. The *Mutual Information* (MI) is a special case of the Kullback-Leibler-Divergence which is used to quantify the information that one random variable X contains about another variable Y - and vice versa. Here, the distribution q is set to the independent joint probability of X and Y, namely $q(X,Y) := p(X) \cdot p(Y)$, while $p(X,Y)$ is the empirical joint probability of the two random variables, and $p(X)$ and $p(Y)$ the respective marginals. The MI measures how much the empirical $p(X,Y)$ deviates from the independent case. Eventually, the MI becomes a generalized correlation coefficient. Then:

$$\mathrm{MI}_{X,Y} = \sum_{x \in \mathcal{X}, y \in \mathcal{Y}} p(x,y) \log_2 \left(\frac{p(x,y)}{p(x) \cdot p(y)} \right) \tag{1}$$

using the same notation as above.

The obvious extension of the MI to address dynamics in time series of stationary processes, is a time-delayed MI (TDMI) [5]. To this end, let $\boldsymbol{X}$ and $\boldsymbol{Y}$ be two time series written in vector form. Then, the probability to find a realization $x_n \in \mathcal{X}$ at time n is $p(x_n)$. The same holds for y_n, then one can compute the MI between time-lagged $x_{n+n'}$ and y_n with lag n'. As previously discussed, however, such a TDMI is inferior to other methods of time series analysis in the detection of potential causal relations [4].

The motivation for the TE is to quantify the dependency of one process ($\{x_n\}$) on another ($\{y_n\}$) by a mutual information of *conditional* probabilities, rather than time-lagged ones.

Then, we can – in analogy to Eq. 1 – express the dependency of one variable x in relation to the other variable y in a time dependent manner by using:

$$\mathrm{TE}_{Y \to X} = \sum p(x_{n+1}^{(k)}, x_n^{(k)}, y_n^{(l)}) \log \frac{p(x_{n+1} \mid x_n^{(k)}, y_n^{(l)})}{p(x_{n+1} \mid x_n^{(k)})} \tag{2}$$

The indices k and l represent the time windows being used (time lags) for each variable creating the multi-dimensional probabilities via histogram techniques. We now detect how much information flows from Y to X by checking whether the state of X is dependent on the history of both variables – expressed in a non-trivial $p(x_{n+1} \mid x_n^{(k)}, y_n^{(l)})$ – or only depends on its own history which can be quantitatively assessed via $p(x_{n+1} \mid x_n^{(k)})$. Note, that $\mathrm{TE}_{Y \to X} \neq \mathrm{TE}_{X \to Y}$.

1.2   Statistical Significance via Permutation Tests

To test data for statistical significance under the null hypothesis of independence between the x and y measurements we employ permutation tests. Our R package **TransferEntropyPT** features the parallelized calculation of such a null model.

It is inspired by the method proposed in [8] and facilitates a shuffling within each time series to destroy potential correlations.

Each time series is randomized by shuffling the order of the data points. Neither the scale of measurements nor the overall composition, meaning the probabilities $p(x_n)$ of occurrence of individual measurements, of the vector itself is altered. Therefore the overall entropy of times series does not change, but the information transfer between both vectors and the temporal information flow of each vector is disrupted. Eventually, signatures of causality are being destroyed.

The number of necessary repetitions in the permutation test depends strongly on the sample size of the used data. Since the shuffling runs are independent from each other (data parallelism), the randomization can be easily parallelized and therefore optimized for several CPU cores (Fig. 1).

To test for statistical significance we perform the permutation test and compute from the repeatedly computed TEs the so called Z-score for the information transfer of $TE_{Y \to X}$ and $TE_{X \to Y}$ individually.

$$Z_{TE} = \frac{TE - \overline{TE}_s}{\sigma(TE_s)} \quad (3)$$

With $\overline{TE}_s$ being the mean Transfer Entropy of all shuffle runs and $\sigma(TE_s)$ the respective standard devi-

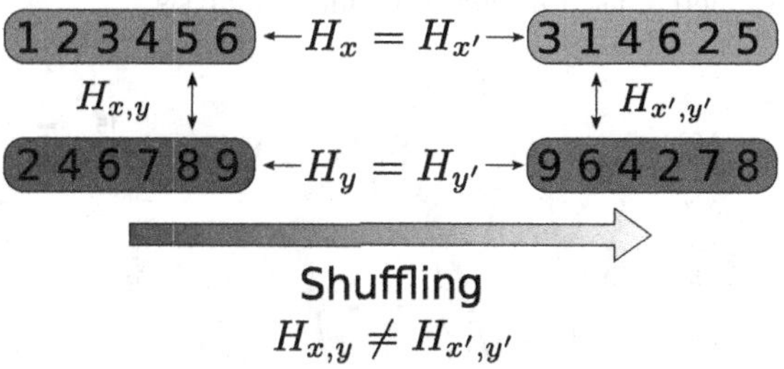

Fig. 1. Schematic of the null model. The sequential time series are shuffled to remove any causal relation between them. The entropy of each time series is kept constant, since the procedure only randomizes the order of the series, but not their respective composition.

ation. The resulting score gives the distance of the Transfer Entropy for the original, raw data in units of standard deviation from the mean of the Transfer Entropy for the shuffled data. Z can be converted to, e.g., p-values under Null-Hypothesis testing under applicable distributions of the test statistics TE.

1.3 Illustrative Application: Coupling of Clocks

In Fig. 2 we show a more involved model of two oscillators, that could implement, e.g., intracellular clocks. We have four species x, y, u, and v whose concentrations form four time series that we analyze via the tools implemented in our R package.

2 Results

Here, we show how to detect the coupling of molecular species in our toy model of Fig. 2. We created time series for the four variables u, v, x, y, adding $\mathcal{N}(\mu, \sigma_{rel} = 5\%)$ relative noise to account for sampling and measurement errors.

Applying get.te we obtain the results in Fig. 4. In Fig. 5 we show the results for varying the coupling k_{vy} between the two oscillators x, y and u, v. In Fig. 5(a)

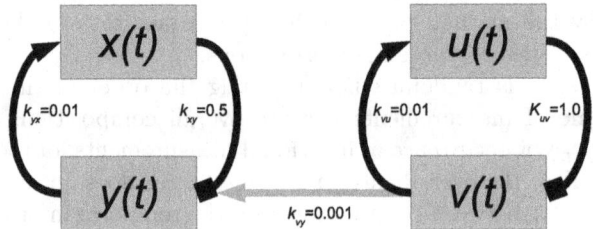

Fig. 2. A network with two coupled feedback loops that eventually implement two (almost) independent clocks. They can, however, influence each other via the interaction parametrized by k_{vy}. Note the positive feedback depicted by an arrow and the down-regulating interaction illustrated by an square-like arrow head. We show the parameters used in our subsequent analysis.

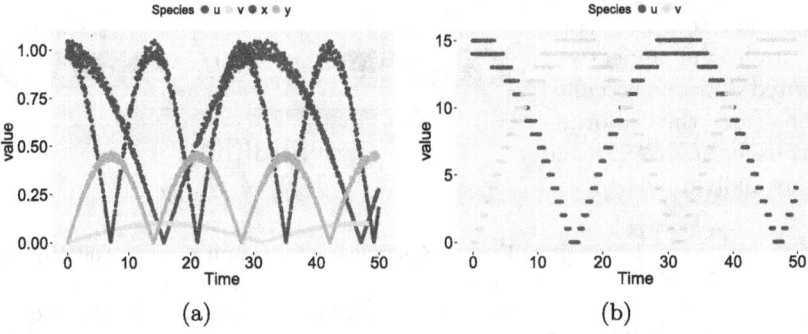

(a)   (b)

Fig. 3. (a) Sample time series for the four molecular species of the intracellular network of Fig. 2. (b) u, v after binning using `partition data`; note, that the overall scales are adjusted by this procedure.

we clearly see the difference in the TE favoring an interpretation of a coupling from v to y as the system shows by design where v and y are coupled via k_{vy}. Note, however, that under a null hypothesis testing (NHST) the p values do not meet the value of $\sim 5\%$ under the well-established Fisher rule. Thus, we would in most cases reject the TE values anyway; only for larger values of k_{vy} do we obtain significance values conforming to the threshold.

Further example applications and detailed usage instructions are discussed in the supplemental material to this paper [3].

3   Availability

The code for package **TransferEntropyPT** is made available under GPL-license. The current version can be downloaded from http://www.cbs.tu-darmstadt.de/TransferEntropy. Installation in R can be achieved via the command R CMD INSTALL TransferEntropyPT_x.y.z.tar.gz where x, y, z are the (sub)version numbers for the downloaded file.

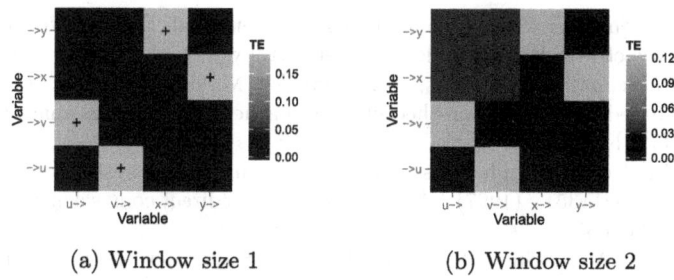

(a) Window size 1   (b) Window size 2

Fig. 4. TE for the toy molecular system in Fig. 2 and $k_{vy} = 0.001$. Note, the "+" signs indicates significant $Z > 4$ values. NHST values for TEs of other species combinations are non-significant, thus we cannot conclude on any dependency.

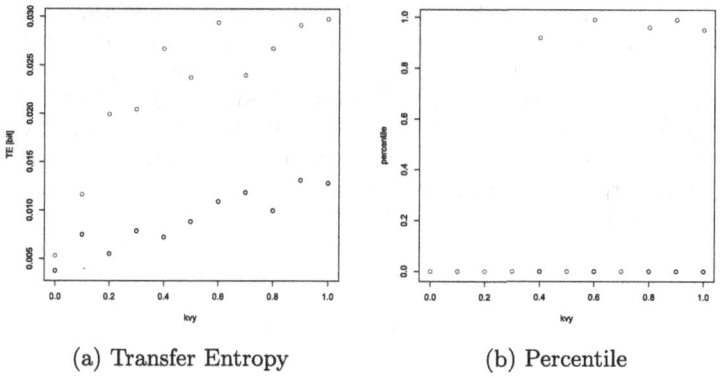

(a) Transfer Entropy   (b) Percentile

Fig. 5. (a) $TE(v \rightarrow y)$ for the toy molecular system in Fig. 2 for varying $k_{vy}1$. The black points mark $TE(y \rightarrow v)$ and the red ones show $TE(v \rightarrow y)$. (b) The percentile for the respective TE permutation test; run over 100 repetitions.

A tutorial (pdf) on how to use the package is available under the same URL.

Acknowledgments. The authors gratefully acknowledge (partial) financial support by the LOEWE projects iNAPO & compuGene of the Hessen State Ministry of Higher Education, Research and the Arts. The authors are grateful for the comments of anonymous referees that improved the manuscript and the style of presentation.

References

1. Bauer, M., Cox, J.W., Caveness, M.H.: Finding the direction of disturbance propagation in a chemical process using transfer entropy. Control Syst. **15**(1), 12–21 (2007). http://ieeexplore.ieee.org/xpls/abs_all.jsp?arnumber=4039335
2. Boba, P., Bollmann, D., Schoepe, D., Wester, N., Wiesel, J., Hamacher, K.: Efficient computation and statistical assessment of transfer entropy. Front. Comput. Phys. **3**, 10 (2015). http://journal.frontiersin.org/article/10.3389/fphy.2015.00010/abstract

3. Boba, P., Hamacher, K.: Accompanying supplemental material to this paper. https://www.cbs.tu-darmstadt.de/TransferEntropy/
4. Hlaváčková-Schindler, K., Paluš, M., Vejmelka, M., Bhattacharya, J.: Causality detection based on information-theoretic approaches in time series analysis. Physics Rep. **441**(1), 1–46 (2007)
5. Paluš, M.: Detecting phase synchronization in noisy systems. Phys. Lett. A **235**(4), 341–351 (1997). http://www.sciencedirect.com/science/article/pii/S037596019700635X
6. Schreiber, T.: Measuring information transfer. Phys. Rev. Lett. **85**(2), 461–464 (2000). http://prl.aps.org/abstract/PRL/v85/i2/p461_1
7. Tung, T.Q., Ryu, T., Lee, K.H., Lee, D.: Inferring gene regulatory networks from microarray time series data using transfer entropy. In: Twentieth IEEE International Symposium on Computing Based Medical Systems, pp. 383–388, June 2007. http://ieeexplore.ieee.org/lpdocs/epic03/wrapper.htm?arnumber=4262679
8. Weil, P., Hoffgaard, F., Hamacher, K.: Estimating sufficient statistics in co-evolutionary analysis by mutual information. Comput. Biol. Chem. **33**(6), 440–444 (2009). http://www.ncbi.nlm.nih.gov/pubmed/19910254, http://www.sciencedirect.com/science/article/pii/S1476927109001108
9. Wollstadt, P., Martínez-Zarzuela, M., Vicente, R., Díaz-Pernas, F.J., Wibral, M.: Efficient transfer entropy analysis of non-stationary neural time series. PLoS ONE **9**, 27 (2014). http://arxiv.org/abs/1401.4068

KaDE: A Tool to Compile Kappa Rules into (Reduced) ODE Models

Ferdinanda Camporesi[1,2], Jérôme Feret[1,2(✉)], and Kim Quyên Lý[1,2]

[1] INRIA, École normale supérieure, CNRS, PSL Research University,
75005 Paris, France
campores@di.ens.fr
[2] Département d'informatique de l'ÉNS,
École normale supérieure, CNRS, PSL Research University, 75005 Paris, France
feret@ens.fr, quyen@di.ens.fr

Abstract. Kappa is a formal language that can be used to model systems of biochemical interactions among proteins. It offers several semantics to describe the behaviour of Kappa models at different levels of abstraction. Each Kappa model is a set of context-free rewrite rules. One way to understand the semantics of a Kappa model is to read its rules as an implicit description of a (potentially infinite) reaction network. KaDE is interpreting this definition to compile Kappa models into reaction networks (or equivalently into sets of ordinary differential equations). KaDE uses a static analysis that identifies pairs of sites that are indistinguishable from the rules point of view, to infer backward and forward bisimulations, hence reducing the size of the underlying reaction networks without having to generate them explicitly. In this paper, we describe the main current functionalities of KaDE and we give some benchmarks on case studies.

1 The Differential Semantics of Kappa

Kappa [1] is a rule-based language which describes the behaviour of some agents that may be bound together on interaction sites. In applications to Systems Biology, agents usually abstract proteins and interaction sites specific regions on their amino acid chains. Mechanistic interactions among proteins are described by the means of rewriting rules. For instance, the rule on the left in Fig. 1 stipulates that two proteins may bind via their respective right and left sites. Graphically (we have used GKAPPA [2] to draw the rules), the shape of a protein implicitly denotes its type. The same way, sites in proteins are identified by their positions

---

This material is based upon works partially sponsored by the Defense Advanced Research Projects Agency (DARPA) and the U. S. Army Research Office under grant number W911NF-14-1-0367, and by the ITMO Plan Cancer 2014. The views, opinions, and/or findings contained in this article are those of the authors and should not be interpreted as representing the official views or policies, either expressed or implied, of DARPA, the U.S. Department of Defense, or ITMO.

J. Feret and H. Koeppl (Eds.): CMSB 2017, LNBI 10545, pp. 291–299, 2017.
DOI: 10.1007/978-3-319-67471-1_18

Fig. 1. Two rules. (left) Two proteins may bind. (right) The protein on the left may activate the right site of the protein on the right.

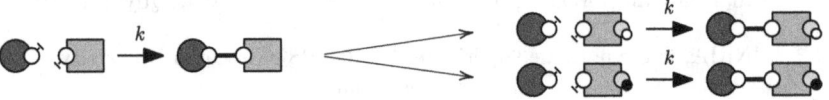

Fig. 2. From rules to reactions. The first rule in Fig. 1 is refined into two reactions according to whether or not right site of the right protein is phosphorylated.

(left, right). Sites may also carry an internal state which stands for an activation level (such as phosphorylation). In Fig. 1, the rule on the right stipulates that the bond between both proteins may activate the second one.

In a rule, the left hand side denotes some precondition, whereas the right hand side stands for a transformation. Some agents may miss some sites. This is the "Don't Care, Don't Write" convention [3]. The sites the state of which influences neither an interaction nor its kinetics are omitted. Each rule may be understood extensionally as a (finite or not) set of reactions, obtained by refining it according to its potential application contexts, until getting fully specified connected components. For instance the rule on the left in Fig. 1 may be applied with the protein on the right phosphorylated or not, as depicted in Fig. 2. In the differential semantics, rule applications preserve disconnectedness, unless specified explicitly. Thus, each connected component in the left hand side is refined separately. Agents may contain many sites and form arbitrary long chains. Thus Kappa models are usually highly combinatorial. A small number of rules may lead to a large (if not infinite) reaction networks [4,5].

The ODE semantics is defined in the following way. Each connected component in a reaction denotes an instance of a bio-molecular species. For every bio-molecular species S, a reaction $R_1 + \ldots + R_m \longrightarrow P_1 + \ldots + P_n$ gives the following contribution to the derivative of the concentration of the species S:

$$\frac{d[S]}{dt} \stackrel{\pm}{=} \sum_r \gamma(r) \cdot [r, R] \cdot \Delta(R, S) \cdot [R_1] \cdot \ldots \cdot [R_m]$$

where: 1. $\gamma(r)$ is the corrected rate of the rule r (a fraction of the rate of the rule r is taken according to a convention defining how automorphisms are taken into account); 2. $[r, R]$ is the number of different ways to induce the reaction R from the rule r; 3. and $\Delta(R, S)$ is the difference between the number of occurrences of the species S in the sequence $P_1, \ldots, P_n$ and the one in the sequence $R_1, \ldots, R_m$. We use the symbol $\stackrel{\pm}{=}$ because we totalise the contribution for each reaction R.

2   ODEs Generation with KADE

KADE generates the differential semantics of Kappa models. In command-line
mode, KADE is called with a list of Kappa files and a list of options. A rudimen-
tary graphical interface is available as well. The syntax of Kappa is described in
its reference manual [6]. KADE generates output for the numerical integration
tools MAPLE [7], MATHEMATICA [8], MATLAB [9], and OCTAVE [10], and for the
modelling standard languages DOTNET [11,12] and SBML [13]. DOTNET
is the internal format of BIONETGEN, we use it for compatibility with ERODE
[14], a tool to evaluate and reduce systems of ODEs. SBML output may be converted into LaTeX thanks to SBML2LATEX [15]. SBML is also compatible with
CELLDESIGNER [16] which provides several tools dedicated to reaction networks.

The Kappa modelling platform extends the core Kappa language with tokens,
algebraic expressions, and the possibility to allow the application of binary rules
in unary contexts. Tokens are specific continuous variables which may be consumed and produced by rules according to user-specified stoichiometric coefficients. Kappa also supports arbitrary algebraic expressions both in rate parameters and in stoichiometric coefficients. These expressions may depend on the
simulation time and on the concentration of some patterns in the current state of
the system. They permit the encoding of kinetics laws beyond mass action. This
feature is restricted to some specific backends. For instance, neither SBML, nor
DOTNET cope with non-constant stoichiometric parameters. Lastly, a rule the
left hand side of which is made of two connected components, may be provided
two rates according to whether it is applied in a binary context (each connected
component of the left hand side of the rule being embedded in two instances of
bio-molecular species), or in a unary context (both connected components being
embedded in the same instance of a bio-molecular species).

Some options let the end-user select the backend and change the name and
the repository of the output file. Some other options tune the semantics of the
model. It is also possible to truncate the ODES in order to ignore the bio-
molecular species that would have more agents than a user-specified threshold.
Three conventions exist for interpreting rate constants. In the following rule:

with the first convention (used by the simulator KASIM [6,17]), rates of rules are
not corrected; with the second one (used by the simulator SIMPLX [3]), rates are
divided by the number of automorphisms in the left hand side of rules (here 24);
the third convention (used by the simulator NFSIM [18]) accounts only for the
permutations among the agents that are undistinguishable from a mechanistic
point of view (here 2). The same issue occurs with reactions, where permutations
among identical species are considered instead of automorphisms.

KADE lets the end-user pick the convention for the rate constants of rules
(in input files) and the one for the rate constant of reactions (in output networks). BIONETGEN uses the third convention for rules and the first one for

reactions. Lastly ERODE takes the first convention for reaction rate constants in the differential setting and the second one in the stochastic one.

Some options tune the numerical integration parameters. This concerns the range for simulation time, the frequency of simulation plots, error tolerance parameters, and the size of integration steps. Moreover, the computation of the Jacobian may be disabled/enabled. It is also possible to warn numerical solvers that concentrations shall remain nonnegative.

Comparison with other tools. Both BIONETGEN and Kappa can convert rules into reactions. BIONETGEN supports compartmentalisation unlike Kappa. In BIONETGEN, equivalent sites can be specified. In contrast, KADE detects them automatically. BIONETGEN does not support tokens.

3   Equivalent Sites

Some sites may have exactly the same capabilities of interaction. This may be used to generate more compact systems of ODEs, by partitioning the set of bio-molecular species up to permutation of equivalent sites [19–21].

Consider the rules in Fig. 3. Each rule may be obtained from one another by swapping pairs of sites in agents: we say that these sites are equivalent. Equivalent sites may be used to induce forward and backward bisimulations over the stochastic and the differential semantics of Kappa [19–23].

Let us consider two sites x and y in a given kind of agent. A *set of rules* is symmetric with respect to the sites x and y if the corrected rates of every two rules that may be obtained one from the other by permuting the sites x and y in some agents, are inversely proportional to their numbers of automorphisms. The same way, a *valuation* from bio-molecular species to real numbers is symmetric with respect to the sites x and y if the images of every two bio-molecular species that can be obtained one from the other by permuting the sites x and y in some agents, are inversely proportional to their numbers of automorphisms. Lastly an *expression* over bio-molecular species is symmetric with respect to the sites x and y if and only if it takes the same values for every two symmetric valuations.

Whenever the set of rules and the initial state of the model are symmetric with respect to two sites, ignoring the difference among these sites in each bio-molecular species induces a backward bisimulation (i.e. the state of the system remains symmetric at every time [19,24]). Whenever the set of rules and each algebraic expression in rates or in stoichiometric coefficients are symmetric, ignoring the difference between these sites induces a forward bisimulation (we can define the ODEs directly over the equivalence-classes of species [19,24]).

KADE may be parameterised for detecting the forward and backward bisimulations that are induced by pairs of equivalent sites. Then, it generates the corresponding reduced ODEs without relying on the initial reaction network.

Comparison with other tools. In BIONETGEN [11,12] pairs of equivalent sites may be user-specified. In KADE, equivalent sites are inferred automatically.

Fig. 3. Sites are equivalent, if the corrected rates of the third rule is twice the corrected rate of each other rule.

The expressive power of equivalent sites in BIONETGEN and in KADE are similar. Yet in BIONETGEN equivalent sites must be equivalent in the rules, in the algebraic expressions, and in the initial state, whereas KADE may exploit pairs of sites that are equivalent in the rules and in the algebraic expressions, but not necessarily in the initial state (forward bisimulation), or that are equivalent in the rules and in the initial state, but not necessarily in the algebraic expressions (backward bisimulation). Moreover, the kinetics conventions are a bit different. As a consequence, some models require more rules to be described in Kappa and some others require more rules to be described in BIONETGEN (more details are provided in Supplementary Information [25]). From a combinatorial point of view, BIONETGEN reasons on agents with multiple occurrences of equivalent sites, which may make the detection of embeddings exponentially costly (with respect to the number of agents). In contrast, KADE quotients the set of bio-molecular species on the fly: it reasons on rigid site graphs for which the detection of embeddings is at worst quadratic [26,27].

ERODE [14] is a tool for lumping systems of ODEs. In particular, it offers some primitives to discover the best forward bisimulation (resp. the best uniform backward bisimulation) induced by an equivalence relation over the bio-molecular species of a reaction network [28,29]. ERODE can capture more forward bisimulations than KADE since equivalent sites can induce only a particular kind of equivalence relations over species. ERODE and KADE are incomparable on backward bisimulations: on the first hand, KADE focuses on equivalence among sites, but on the second hand, ERODE focuses on uniform bisimulation which means that it cannot assign weights to bio-molecular species. For instance, ERODE cannot express the backward bisimulation that gathers every kind of dimer in the example of Fig. 3 since the dimer made of a protein bound on its top site to the bottom site of another protein is twice abundant as the dimer made of two proteins bound together on their top sites (whenever the initial state and the rate constants are such that sites x and y are equivalent). As far as computation cost is concerned, ERODE works on a fully expanded description of the system (either a reaction networks, or an ODE system), which may be impossible to compute for large models. KADE discovers equivalent sites directly on the set of rules. Another difference is that KADE applies on uninterpreted parameters (KADE reductions remain valid if the value of rate parameters is modified) whereas ERODE can compute bisimulations only over fully instantiated networks.

On fully instantiated networks, KADE and ERODE may be combined. Firstly, KADE may quickly detect equivalent sites and generate reduced networks accordingly. Then ERODE may look for further reductions. When focusing on forward bisimulation, ERODE also provides a proof that final reductions are optimal.

4   Benchmarks

We test the reduction power and the time efficiency of our framework on three families of models offering various conditions about the ratio of the number of Kappa rules to the number of reactions and about the ratio of the number of different bio-molecular species configurations to the number of their equivalence classes. In KaDE, the computation time for generating networks (or ODEs) depends mainly on the number of rules and the number of equivalence classes of bio-molecular species configurations. The data-structure described in [17] is used to generate reactions efficiently. More examples, including most of the BioNet-Gen test suite, are provided in Supplementary Information [25].

model	sites	rules	species		reactions	
			original	reduced	original	reduced
kinase/phosphatase	n	$6n$	$2 + 4^n$	$2 + \binom{n+3}{3}$	$6n4^{n-1}$	$2n\binom{n+2}{2}$
multiple phosphorylation	n	$n2^n$	2^n	$n + 1$	$n2^n$	$2n$
mult. phosphoryl. with counter	n	$2n^2$	2^n	$n + 1$	$n2^n$	$2n$

Fig. 4. Key attributes of our models with respect to the parameter n.

The first family involves a kinase, a phosphatase, and a target protein. The target protein has n sites (n is left as a parameter). The kinase may bind and unbind to each non-phosphorylated site of the target protein. The kinase may phosphorylate a site when releasing it. Conversely, the phosphatase may bind and unbind to each phosphorylated site of the target protein. The phosphatase may also dephosphorylate a site when releasing it. We assume that every site has the same mechanistic properties and that the rate of reactions does not depend on the state of the other sites in the target protein.

The second and third families of models are inspired by the protein Kai. This protein plays a crucial role in the control of the circadian clock oscillations. We consider a protein with n sites (n is left as a parameter) which may each be phosphorylated, or not. The kinase and the phosphatase are not described explicitly. We assume that the rate constants of phosphoralylation (resp. dephosphorylation) of a site in a protein depend on the number of sites that are already phosphorylated in this protein. In the third family of models, a trick suggested by Pierre Boutillier is used to reduce drastically the number of rules that are required to describe the models. We use a fictitious site that is bound to a chain of fictitious proteins the length of which encodes the number of phosphorylated sites. When a site is phosphorylated, a new protein is inserted in the chain and removed when a site gets dephosphorylated. Thus the phosphorylation level of a protein can be checked by looking at the length of this chain, without having to enumerate the different combinations fot the sites that are phosphorylated.

In Fig. 4, we give the number of rules, species and reactions, for each family of models for the parameter n ranging from 1 to 10, as well as the number of reactions and species when equivalent sites are considered. In Fig. 5, we compare

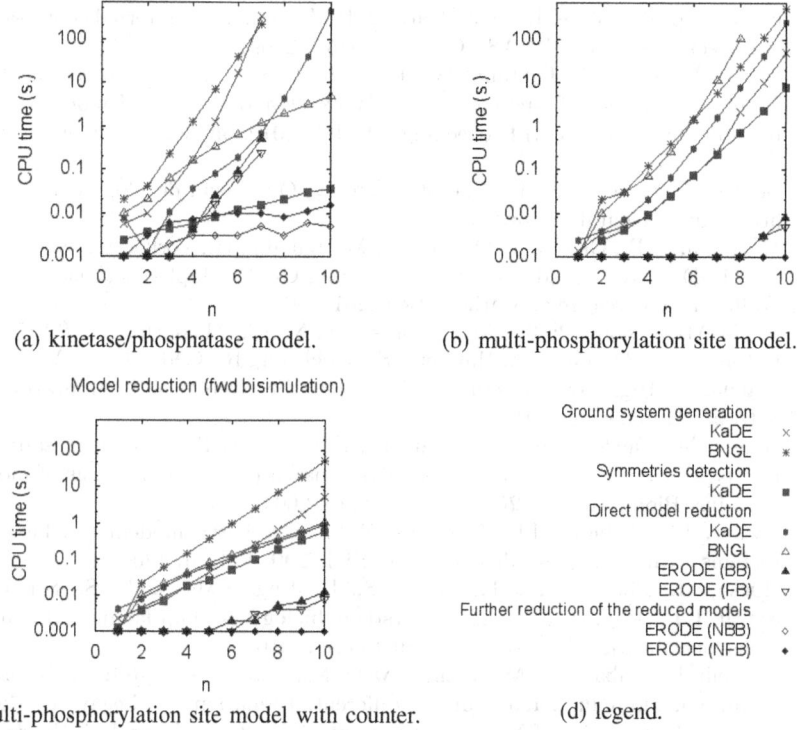

(a) kinetase/phosphatase model.

(b) multi-phosphorylation site model.

(c) multi-phosphorylation site model with counter.

(d) legend.

Fig. 5. Comparison between the time performances of KaDE, BioNetGen, and ERODE, on a MacBookPro with a 2,8 GHz Intel Core i7 CPU and a 16 Go 1600 MHz DDR3 memory and with a 10 minutes time-out.

the computation time to generate the original and the reduced networks with BioNetGen and KaDE. The generation of reduced models with KaDE (which does not require explicit annotation of equivalent sites) is much faster than the one of the unreduced networks. KaDE and BioNetGen generate exactly the same reduced networks. Lastly, we apply the fast version of ERODE of the bisimulation inference algorithm [29] on the original networks and the complete version on the reduced ones [28]. But we found not further reduction this way. In [25], we observe as good results on the BioNetGen test suite.

References

1. Danos, V., Laneve, C.: Formal molecular biology. TCS **325**(1), 69–110 (2004)
2. Feret, J.: Gkappa: a library to generate site graphs with graphviz. https://github.com/Kappa-Dev/GKappa
3. Danos, V., Feret, J., Fontana, W., Krivine, J.: Scalable simulation of cellular signaling networks. In: Shao, Z. (ed.) APLAS 2007. LNCS, vol. 4807, pp. 139–157. Springer, Heidelberg (2007). doi:10.1007/978-3-540-76637-7_10

4. Feret, J., Danos, V., Krivine, J., Harmer, R., Fontana, W.: Internal coarse-graining of molecular systems. PNAS **106**, 6453–6458 (2009)
5. Danos, V., Feret, J., Fontana, W., Harmer, R., Krivine, J.: Abstracting the differential semantics of rule-based models: exact and automated model reduction. In: Jouannaud, J.P. (ed.) Proceedings of LICS 2010, pp. 362–381. IEEE Computer Society (2010)
6. Boutillier, P., Feret, J., Krivine, J., Kim Lý, Q.: Kasim development homepage. http://dev.executableknowledge.org
7. Monagan, M.B., Geddes, K.O., Heal, K.M., Labahn, G., Vorkoetter, S.M., McCarron, J., DeMarco, P.: Maple 10 Programming Guide. Maplesoft (2005)
8. Wolfram Research, Inc.: Mathematica (2017)
9. MATLAB version 9.2: The MathWorks Inc., Natick, Massachusetts (2017)
10. Eaton, J.W., Bateman, D., Hauberg, S., Wehbring, R.: GNU Octave Version 4.0.0 Manual: A High-Level Interactive Language for Numerical Computations. Free Software Foundation (2015)
11. Blinov, M., Faeder, J.R., Goldstein, B., Hlavacek, W.S.: Bionetgen: software for rule-based modeling of signal transduction based on the interactions of molecular domains. Bioinformatics **20**(17), 3289–3291 (2004)
12. Faeder, J.R., Blinov, M.L., Hlavacek, W.S.: Rule-based modeling of biochemical systems with bionetgen. Methods Mol. Biol. **500**, 113–167 (2009)
13. Hucka, M., Bergmann, F.T., Hoops, S., Keating, S.M., Sahle, S., Schaff, J.C., Smith, L.P., Wilkinson, D.J.: The systems biology markup language (sbml): language specification for level 3 version 1 core (2010)
14. Cardelli, L., Tribastone, M., Tschaikowski, M., Vandin, A.: ERODE: a tool for the evaluation and reduction of ordinary differential equations. In: Legay, A., Margaria, T. (eds.) TACAS 2017. LNCS, vol. 10206, pp. 310–328. Springer, Heidelberg (2017). doi:10.1007/978-3-662-54580-5_19
15. Dräger, A., Planatscher, H., Wouamba, D.M., Schröder, A., Hucka, M., Endler, L., Golebiewski, M., Müller, W., Zell, A.: SBML2LATEX: conversion of SBML files into human-readable reports. Bioinformatics **25**(11), 1455–1456 (2009)
16. Funahashi, A., Matsuoka, Y., Jouraku, A., Morohashi, M., Kikuchi, N., Kitano, H.: Celldesigner 3.5: A versatile modeling tool for biochemical networks. Proc. IEEE **96**, 1254–1265 (2008)
17. Boutillier, P., Ehrhard, T., Krivine, J.: Incremental update for graph rewriting. In: Yang, H. (ed.) ESOP 2017. LNCS, vol. 10201, pp. 201–228. Springer, Heidelberg (2017). doi:10.1007/978-3-662-54434-1_8
18. Sneddon, M.W., Faeder, J.R., Emonet, T.: Efficient modeling, simulation and coarse-graining of biological complexity with nfsim. Nat. Meth. **8**, 177–183 (2011)
19. Camporesi, F., Feret, J.: Formal reduction for rule-based models. ENTCS **276**, 29–59 (2011). Proc. MFPS XXVII
20. Camporesi, F., Feret, J., Koeppl, H., Petrov, T.: Combining model reductions. ENTCS **265**, 73–96 (2010). Proc. MFPS XXVI
21. Feret, J.: An algebraic approach for inferring and using symmetries in rule-based models. ENTCS **316**, 45–65 (2015). Proc. SASB 2014
22. Buchholz, P.: Bisimulation relations for weighted automata. Theor. Comput. Sci. **393**(1–3), 109–123 (2008)
23. Feret, J., Koeppl, H., Petrov, T.: Stochastic fragments: A framework for the exact reduction of the stochastic semantics of rule-based models. Int. J. Softw. Inform. **7**(4), 527–604 (2013)
24. Buchholz, P.: Exact and ordinary lumpability in finite Markov chains. J. Appl. Probab. **31**(1), 59–75 (1994)

25. Camporesi, F., Feret, J., Lý, K.Q.: KADE: a tool to compile kappa rules into (reduced) ode models: Supplementary information. http://www.di.ens.fr/~feret/CMSB2017-tool-paper/
26. Petrov, T., Feret, J., Koeppl, H.: Reconstructing species-based dynamics from reduced stochastic rule-based models. In: Laroque, C., Himmelspach, J., Pasupathy, R., Rose, O., Uhrmacher, A.M. (eds.) Proceedings of WSC 2012, WSC (2012)
27. Oury, N., Pedersen, M., Petersen, R.L.: Canonical labelling of site graphs. In Petre, I. (ed.) Proceedings of CompMod 2013, EPTCS, vol. 116, pp. 13–28 (2013)
28. Cardelli, L., Tribastone, M., Tschaikowski, M., Vandin, A.: Forward and backward bisimulations for chemical reaction networks. In: Aceto, L., de Frutos-Escrig, D. (eds.) Proceedings of CONCUR 2015, vol. 42, pp. 226–239. LIPIcs., Schloss Dagstuhl (2015)
29. Cardelli, L., Tribastone, M., Tschaikowski, M., Vandin, A.: Efficient syntax-driven lumping of differential equations. In: Chechik, M., Raskin, J.-F. (eds.) TACAS 2016. LNCS, vol. 9636, pp. 93–111. Springer, Heidelberg (2016). doi:10.1007/978-3-662-49674-9_6

Database of Dynamic Signatures Generated by Regulatory Networks (DSGRN)

Bree Cummins[1], Tomas Gedeon[1(✉)], Shaun Harker[2], and Konstantin Mischaikow[2]

[1] Department of Mathematical Sciences,
Montana State University, Bozeman, MT 59715, USA
gedeon@math.montana.edu
[2] Department of Mathematics,
Rutgers, The State University of New Jersey, New Brunswick, USA

Abstract. We present a computational tool DSGRN for exploring network dynamics across the global parameter space for switching model representations of regulatory networks. This tool provides a finite partition of parameter space such that for each region in this partition a global description of the dynamical behavior of a network is given via a directed acyclic graph called a Morse graph. Using this method, parameter regimes or entire networks may be rejected as viable models for representing the underlying regulatory mechanisms.

1 Introduction

An important challenge in systems biology today is the lack of robust tools that can translate static network information into actionable information about a network's dynamics. It is the dynamics of the network that ultimately correlates with the cellular state and determines its phenotype. Lack of understanding of all potential dynamics that a network structure supports is one of the reasons for the apparent lack of correspondence between genotype and phenotype.

We have developed an efficient mathematically rigorous computational toolbox, called Dynamic Signatures Generated by Regulatory Networks (DSGRN), that computes the range of dynamic behaviors supported by a given network. DSGRN is based on a new mathematical framework for nonlinear dynamics, which moves away from consideration of individual solutions at particular parameter values. In networks with 5–10 nodes with 30–50 parameters any random sampling of parameters and initial conditions will only cover a negligible portion of possible dynamical behaviors. Furthermore, comparing individual solutions to experimental data, which typically carries significant uncertainty, cannot be used to reject potential models because many nearby parameters and initial conditions will produce solutions that fit the data equally well. The main applications of our tool have been to **(1)** describe coarse dynamics for the entire parameter space for a given network, allowing exploration and quantification of different dynamic signatures supported by the network architecture, and **(2)** compare

J. Feret and H. Koeppl (Eds.): CMSB 2017, LNBI 10545, pp. 300–308, 2017.
DOI: 10.1007/978-3-319-67471-1_19

dynamics across hundreds of networks with data allowing rigorous exploration of the space of networks. To illustrate this aspect, in Sect. 3 we examine 4994 networks that are a perturbations of a transcriptional network underlying cell cycle progression in yeast. We evaluate each network by the prevalence of stable oscillatory behavior in the parameter space and doing so we find those networks that most robustly exhibit oscillatory behavior.

2 Database for Dynamics

The current state of modeling gene network dynamics is characterized by a trade-off between the model's ability to quantitatively match the experimental data, and the need for a large number of kinetic parameters to parameterize the model [1–3]. A popular modeling approach uses Boolean networks, where each protein, ligand or mRNA is assumed to have two states (ON and OFF), and the discrete time evolution of the states is based on logic-like update functions [4–6]. The highly constrained character of the states and the update rules allows relatively easy parameterization of the model from data, but it limits the power of generalization and typically results in a poor quantitative match with data. In contrast, properly parameterized ordinary differential equation models can provide a good quantitative match and are easily generalized [7,8], but we lack first principle methods to select proper nonlinearities, and the parameters are usually poorly constrained, or unknown.

Our approach is derived from Conley theory [9–11] and the computations we perform allow us to identify trapping regions. As a consequence we are able to combinatorialize the approximation of the dynamics and the parameter representation of the system, while retaining the capabilities of providing rigorous descriptions of the dynamics without explicit knowledge of the nonlinearities [12]. Furthermore, we obtain computational efficiencies similar to those of Boolean nets while preserving the quantitative richness of ODEs.

It should be noted that there are a variety of techniques that have been developed and implemented that are similar in spirit, but vary in focus and detail. A complete review is beyond the scope of this note, but we remark on the following examples. The focus of [13] is on minimizing models that exhibit a particular transition system over a finite set of states. The goals of [14] a similar in spirit, but make use of piecewise linear nonlinearities. As a consequence the decomposition of parameter space is more involved and dealt with via a hierarchical decomposition [15]. Perhaps closest to our approach in the context of this work is [16] in which the dynamics and parameterization makes use of [17]. With regard to [17] our approach allows for broader state space decompositions and less restrictions on parameters. However, [16] focuses on detailed matching of dynamics to time series and thereby providing more powerful tools for model and parameter rejection.

Switching Systems of Regulatory Networks. A regulatory network involving N genes is often modeled by a system of ordinary differential equations

$$\dot{x}_i = -\gamma_i x_i + f_i(x), \quad i = 1, \ldots, N, \tag{1}$$

where x_i denotes the concentration level of protein associated with gene i, where each gene i is associated to a node in the regulatory network. The nonlinearity f_i is meant to capture how production of one gene is regulated by other genes, but in practice it is impossible to derive f_i from first principles. Biologists use a phenomenological choice of f_i; the default is usually to express f_i in terms of Hill functions. We build upon ideas of Glass and Kaufmann [18,19] and consider a particularly simple form of (1) called a *switching system*;

$$\dot{x}_i = -\gamma_i x_i + \Lambda_i(x), \quad i = 1, \ldots, N, \tag{2}$$

where Λ_i is defined as sums and products of piecewise constant functions of the variables x_j.

The parameters of the switching systems are directly relatable to the parameters of Hill function based models and include decay rates γ_i, and for each regulating edge $j \to i$ in the regulatory network, there are three parameters: a threshold value $\theta_{i,j}$ of x_j, at which the piecewise constant function Λ_i changes values, and $l_{i,j} < u_{i,j}$, the two values of Λ_i in a neighborhood of $\theta_{i,j}$. We note that because Λ_i is not continuous, classical solutions to (2) are not guaranteed to exist. This does not hinder our ability to use switching systems to combinatorialize the dynamics of a network. Furthermore, in our perspective the switching system, rather than being a model on its own right, is only a computational model to understand the dynamics of the unknown biologically relevant model that has the form of (1) where the nonlinearities are sufficiently smooth to guarantee existence and uniqueness of solutions.

A general mathematically precise exposition of how switching systems naturally admit discretization of the phase space, dynamics, and the parameter space is nontrivial and has been developed in [20]. In this contribution, our emphasis is on description of DSGRN as a computational tool and we only briefly describe the underlying theory.

Combinatorial Dynamics. The threshold parameters θ, which denote the locations of the abrupt changes in Λ_i define a decomposition of phase space into domains, and in each domain system (2) is readily solvable. Furthermore, by representing domains as nodes of a State Transition Graph (STG) we can unambiguously assign edges between nodes that represent the directions of all solutions between the domains. To be more formal we let $\mathcal{X}$ denote the set of nodes of STG and to emphasize that we are interested in dynamics we represent the directed graph as a multivalued combinatorial map $\mathcal{F}: \mathcal{X} \rightrightarrows \mathcal{X}$ where a node $\xi' \in \mathcal{F}(\xi)$ if and only if there is an edge $\xi \to \xi'$ in the directed graph.

Observe that $\mathcal{F}$ depends on the choice of parameter values. A key insight of [20] is that $\mathcal{F}$ is locally constant, that the boundaries of regions where $\mathcal{F}$ is constant are semi-algebraic sets, and that the decomposition of parameter space into regions of constancy is explicitly computable. This decomposition of high dimensional parameter space is codified via a *Parameter Graph* $\mathcal{PG}$. The nodes of $\mathcal{PG}$ correspond to regions of parameter space with identical $\mathcal{F}$, and edges correspond to co-dimension 1 boundaries between the regions which capture the geometry of the decomposition.

Storing the entire collection of multivalued maps over all parameter ranges is prohibitive. We use the concept of a *Morse graph* to extract the essential recurrent dynamics information from the combinatorial map $\mathcal{F} \colon \mathcal{X} \rightrightarrows \mathcal{X}$. We use linear time algorithms to identify maximal sets of nodes in $\mathcal{X}$ mutually related by directed paths from one node to another. These maximal sets of nodes are called *Morse sets* of $\mathcal{F}$ and are identified by nodes in the Morse graph. The directed edges in the Morse graph indicate the reachability from one Morse set to another via paths in $\mathcal{F}$. Thus, minimal nodes in the Morse graph represent stable or attracting dynamics (see Fig. 1 (c), (e), (g), (i), (k)).

The parameter graph along with the associated Morse graphs provides an extremely condensed representation of the global dynamics over all of parameter space. We augment the Morse graph with labels on the Morse sets that describe the recurrent dynamics they represent. In our graphical output, described in detail in the Supplement, we denote by FP a Morse set that corresponds to a stable equilibrium (Fixed Point). FP OFF means that every protein concentration is below its lowest threshold, and FP ON means that every variable is above its lowest threshold. We use FC (Full Cycle) to denote a Morse set in which each coordinate crosses at least one threshold (this must happen an even number of times).

The output of the DSGRN software is a SQL database, which allows the user to query for particular types of dynamics and/or for the dynamics in different regions of parameter space. Although the DSGRN database is computed using switching systems (2), the Morse graph structures are valid for smooth systems taking the form (1) under the assumption that f_i and Λ_i are sufficiently close. Furthermore, explicit expressions for what sufficiently close means can be obtained [12]. Thus the DSGRN framework allows us to make mathematically rigorous statements about the global dynamics of regulatory networks even if the nonlinearities are not explicitly known.

3   Examples

(1) Characterizing the dynamics of a network over global parameter space. To illustrate the range of dynamics being detected by DSGRN, we show five STGs and the associated Morse graphs from the two-dimensional network shown in Fig. 1 (a). These STGs and associated Morse graphs arise from five of the 120 regions in parameter space for this system.

The network graph in Fig. 1 (a) implies that X has two thresholds and Y has one, decomposing phase space into six domains. We label each domain by a pair of integers denoting the locations of X and Y compared to their thresholds.

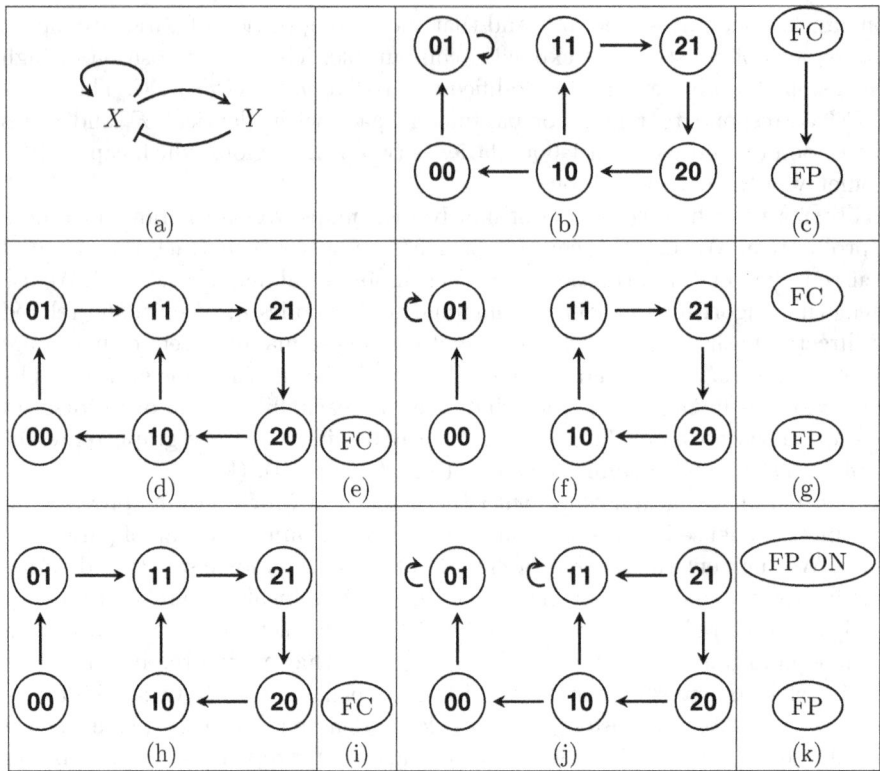

Fig. 1. (a) Two dimensional network. The pairs (b)-(c), (d)-(e), (f)-(g), (h)-(i) and (j)-(k) provide examples of STG (first panel in each pair) and their associated augmented Morse graphs (second panel in each pair). See beginning of Sect. 3 for description of labeling of nodes in STG (panels (b), (d), (f), (h), (j)) See latter part of Sect. 2 for a description of labeling of nodes in Morse graphs (panels (c), (e), (g), (i), (k)).

For example, 00 is the domain where both X and Y are below their respective lowest thresholds, and 21 is where they are above their respective highest thresholds. We use these domain labels to represent the nodes in STG, and the arrows between them represent the flow between domains in phase space (Figs. 1 (b), (d), (f), (h), (j)). From STGs, we calculate the corresponding Morse graph for each parameter, shown in Figs. 1 (c), (e), (g), (i), (k).

(2) Comparing dynamics across networks. We use DSGRN to find a network that exhibits robustly cyclic dynamics in a neighborhood of a given network. The network in Fig. 2 (a, top) is a potential network driving cell cycle progression in yeast [21]. The backbone of the network, which includes all molecules except CdH1, is a subnetwork of Fig. 4 (c) in [21]. This backbone network shows highly prevalent oscillatory dynamics in form of a Morse graph with stable FC. Experimental evidence showed that in the network where all cyclins are knocked out, the system does not oscillate, but approaches an equilibrium.

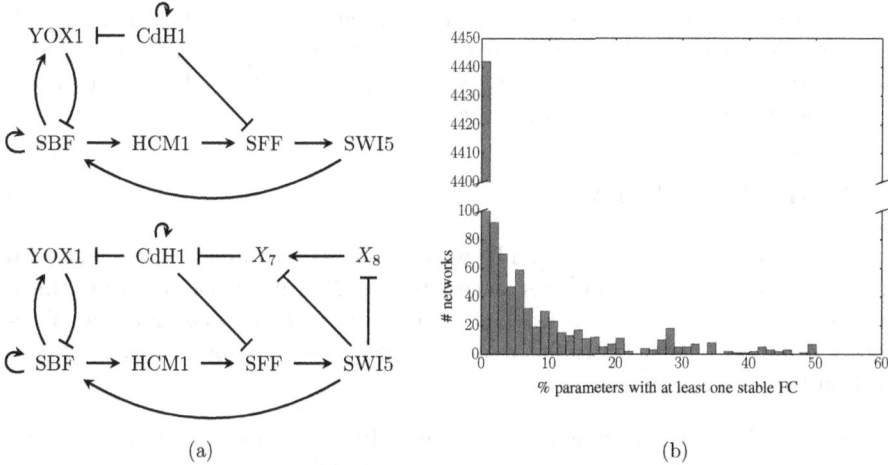

(a)   (b)

Fig. 2. A perturbation study of the network in (a, top). The histogram in (b) counts the number of nearby networks with a certain percentage of parameters that exhibit at least one stable FC. The network in (a, bottom) is one of the top 23 networks that exhibit more than 40% stable FC.

This observation showed that the backbone network is not the network that remains after cyclin knockout. However, when we include CDH1 in the network, analysis via DSGRN shows that the network in Fig. 2 (a, top) with CDH1 exhibits no stable full cycles (FCs) across all of parameter space, thus recapitulating the cyclin knockout phenotype. We now address the next question on how the unknown knocked-out cyclins impinge on the network in Fig. 2 (a, top) in such a way that their inclusion will cause the full network to oscillate.

To address the question what changes to the network will produce robust stable FC, we take the network in Fig. 2 (a, top) and add parsimonious numbers of randomly chosen nodes and/or edges to the network to sample nearby networks in the space of all networks. We analyzed 4994 such networks, seeking stable FCs using DSGRN. The resulting histogram is shown in Fig. 2 (b) with the number of networks plotted against the percentage of parameters that exhibit at least one stable FC. Most of the networks (all but 814) show no stable FCs at all. We assume that a high percentage of parameters exhibiting at least one stable FC is a reasonable proxy for robustness of full cycle oscillations. Therefore, we hypothesize that the 23 networks exhibiting at least 40% of parameters with stable FCs are the best networks in this parsimonious neighborhood of the original network, and are most likely to represent real regulatory mechanisms in the cell cycle. One of these top 23 networks is shown in Fig. 2 (a, bottom). Our method suggests that adding nodes X_7 and X_8 with the appropriate edges depicted in Fig. 2 (a, bottom) will produce a robustly oscillating network. This suggestion can be used as a guide for experimentalists to find molecular actors that fulfill these roles. Taking a different point of view, unchecked progression through cell cycle is one of hallmarks of cancer. A stable cycle oscillation can be considered a representation of such unchecked progression through the cell cycle. While the original network does not exhibit this behavior, the extended network does. From this perspective, X_7 and X_8 can be viewed as a prediction of new oncogenes, or product of oncogenes.

4   Code Availability

DSGRN is publicly available [22]. The website contains complete documentation including installation instructions, access to the GitHub repository, a graphical interface that allows user to construct the network to be analyzed, and a collection of databases for networks that have already been computed. DSGRN can work in two different modalities.

1. **Exploration of the parameter space.** In this mode, which is described in the Documentation directory on the DSGRN website [22], the program computes an SQL database from a network file. (See the Documentation on Network Specification Files for the format of input network files.) Included with this software is a command line tool `dsgrn` that accesses meta-data about the network directly from the network file without computing the whole database. Many precomputed databases can be viewed in the Databases directory on the DSGRN website [22]. The output of each database is presented as a collection of Morse graphs in descending order based on the number of parameter nodes where this Morse graph is observed. Selected filters are available that allow the user to limit the types of dynamics that are visible. To view a database that is not pre-computed, go to https://dsgrn.com and request an account. Follow the documentation on that page and also see the tutorial Bistable Repressilator Jupyter notebook in the top level Tutorial directory in the DSGRN GitHub repository (the repository is linked from [22]). These documents explain how to compute a database on the server and view it on a personal website.

2. **Computation at selected parameter nodes.** The installation described in the Documentation directory on the DSGRN website [22] downloads, but does not install, a package of Python tools that can be used to compute the State Transition Graph and other details about the dynamics at parameter nodes of the parameter graph. The parameters of interest can be chosen from examining Morse graphs in step 1. The installation instructions for the Python package and its dependencies are described in the README file in DSGRN/software/Python/ on GitHub. Furthermore, Jupyter Notebooks with tutorials on the python tools (DSGRN_Getting_Started) and on preprogrammed SQL queries wrapped in python (QueryTutorial) can be found in the GitHub repository in DSGRN/software/Python/doc.

There are two main limitations on the size of the network that can be computed. One limitation is the type of node in the network; currently, nodes with up to 3 inputs and 5 outputs can be computed, with selected higher orders available. The reason is that the structure of the parameter graph corresponding to each type of node has to be precomputed using Cylindrical Algebraic Decomposition (CAD). The current set of computed input-output logic files can be found at /usr/local/share/DSGRN/logic/ in the local installation of DSGRN. The interpretation of the logic files are given in the Documentation linked from the DSGRN website [22]. The second limitation is the size of the parameter graph. This size is a product of parameter graphs of all the nodes and grows fast with the type of the node, as well as number of nodes. The command line tool dsgrn can compute the size of the parameter graph based on the network structure, so that the user can check its size (see Documentation, section High-Level API) before committing to the calculation.

Acknowledgment. The work of T. G. was partially supported by NSF grants DMS-1226213 DMS-1361240 and DARPA D12AP200025. B. C. was supported by DARPA D12AP200025. The work of S. H. and K. M. was partially supported by grants NSF-DMS-1125174, 1248071, 1521771 and a DARPA contract HR0011-16-2-0033.

References

1. Goncalves, E., et al.: Bridging the layers: towards integration of signal transduction, regulation and metabolism into mathematical models. Mol. BioSyst. **9**(7), 1576–1583 (2013)
2. Heatha, A., Kavria, L.: Computational challenges in Systems Biology. Comput. Sci. Rev. **3**, 1–17 (2009)
3. Karlebach, G., Shamir, R.: Modelling and analysis of gene regulatory networks. Nature **9**, 770–780 (2008)
4. Bornholt, S.: Boolean network models of cellular regulation: prospects and limitations. J. R. Soc. Interface **5**, 134–150 (2008)
5. Saadatpour, A., Reka, A.: Boolean modeling of biological regulatory networks: A methodology tutorial. Methods **62**, 3–12 (2013)
6. Thomas, R.: Regulatory networks seen as asynchronous automata: a logical description. J. Theor. Biol. **153**, 1–23 (1991)
7. Tyson, J.J., Novak, B.: In: Dekker, A.M.W.V. (ed.) Handbook of Systems Biology. Academic Press, San Diego (2013)
8. Chen, K., et al.: Integrative analysis of cell cycle control in budding yeast. Mol. Biol. Cell **15**, 3841–3862 (2004)
9. Conley, C.: Isolated Invariant Sets and the Morse Index. American Mathematical Society, Providence (1978). ISBN: 9780821888834
10. Mischaikow, K., Mrozek, M.: Handbook of dynamical systems, vol. 2, pp. 393–460. North-Holland, Amsterdam (2002). doi:10.1016/S1874-575X(02)80030-3. http://dx.doi.org.proxy.libraries.rutgers.edu/10.1016/S1874-575X(02)80030-3
11. Kalies, W.D., Mischaikow, K., VanderVorst, R.C.A.M.: An algorithmic approach to chain recurrence. Found. Comput. Math. **5**, 409–449 (2005). ISSN: 1615-3375
12. Gedeon, T., Harker, S., Kokubu, H., Mischaikow, K., Oka, H.: Global dynamics for steep sigmoidal nonlinearities in two dimensions. Physica D **339**, 18–38 (2017)
13. Streck, A., Lorenz, T., Siebert, H.: Minimization and equivalence in multi-valued logical models of regulatory networks. Natural Comput. **14**, 555–566 (2015). ISSN: 1572-9796
14. Batt, G., Belta, C., Weiss, R.: Temporal logic analysis of gene networks under parameter uncertainty. IEEE Trans. Autom. Control **53**, 215–229 (2008)

15. Bogomolov, S., Schilling, C., Bartocci, E., Batt, G., Kong, H., Grosu, R.: Abstraction-based parameter synthesis for multiaffine systems. In: Piterman, N. (ed.) HVC 2015. LNCS, vol. 9434, pp. 19–35. Springer, Cham (2015). doi:10.1007/978-3-319-26287-1_2. ISBN: 978-3-319-26287-1, http://dx.doi.org/10.1007/978-3-319-26287-1_2

16. Klarner, H., Streck, A., Šafránek, D., Kolčák, J., Siebert, H.: Parameter identification and model ranking of thomas networks. In: Gilbert, D., Heiner, M. (eds.) CMSB 2012. LNCS, pp. 207–226. Springer, Heidelberg (2012). doi:10.1007/978-3-642-33636-2_13

17. Chaouiya, C., Remy, E., Mossé, B., Thieffry, D.: Qualitative analysis of regulatory graphs: a computational tool based on a discrete formal framework. In: Benvenuti, L., De Santis, A., Farina, L. (eds.) Positive Systems, vol. 294. Springer, Heidelberg (2004). doi:10.1007/978-3-540-44928-7_17. http://dx.doi.org/10.1007/978-3-540-44928-7_17

18. Glass, L., Kauman, S.A.: Co-operative components, spatial localization and oscillatory cellular dynamics. J. Theor. Biol. **34**, 219–237 (1972)

19. Glass, L., Kauman, S.A.: The logical analysis of continuous, non-linear biochemical control networks. J. Theor. Biol. **39**, 103–29 (1973)

20. Cummins, B., Gedeon, T., Harker, S., Mischaikow, K., Mok, K.: Combinatorial representation of parameter space for switching systems. SIAM J. Appl. Dyn. Syst. **15**, 2176–2212 (2016)

21. Orlando, D.A., et al.: Global control of cell-cycle transcription by coupled CDK and network oscillators. Nature **453**, 944–947 (2008)

22. Harker, S.: Dynamic Signatures Generated by Regulatory Networks (2015). http://chomp.rutgers.edu/Projects/DSGRN/

Pint: A Static Analyzer for Transient Dynamics of Qualitative Networks with IPython Interface

Loïc Paulevé[(⊠)]

CNRS, LRI UMR 8623, Univ. Paris-Sud – CNRS,
Université Paris-Saclay, 91405 Orsay, France
loic.pauleve@lri.fr

Abstract. The software Pint is devoted to the scalable analysis of the traces of automata networks, which encompass Boolean and discrete networks. Pint implements formal approximations of transient reachability-related properties, including mutation prediction and model reduction.

Pint is distributed with command line tools, as well as a Python module *pypint*. The latter provides a seamless integration with the Jupyter IPython notebook web interface, which allows to easily save, reuse, reproduce, and share workflows of model analysis.

Pint can address networks with hundreds to thousands interacting components, which are typically intractable with standard approaches.

1 Introduction

The computational analysis of the qualitative dynamics of biological networks faces the state space explosion problem, limiting the tractability of detailed models. Many studies have to use reduced models which often lose important properties and may lead to approximative results.

Pint provides formal and scalable analysis for the transient discrete dynamics (traces/trajectories) of automata networks, which subsume Boolean and multi-valued networks. Pint implements an abstract interpretation of traces based on a static analysis of causality of transitions. It results in over- and under-approximations of PSPACE-complete problems by $P \cdot \exp(k - 1)$ and $NP \cdot \exp(k - 1)$ problems, where k is the number of qualitative levels of network nodes (2 for Boolean networks). Pint then relies on Boolean constraint satisfaction (SAT) and Answer-Set Programming (ASP, [3]) for their efficient resolution.

Besides simple transient reachability analysis (from state s_0 there exists a succession of transitions leading, even briefly, to a state satisfying a given property), Pint features include the prediction of mutations to control the reachability properties, the identification of *bifurcation* transitions responsible for differentiation processes, and model reduction which preserves transient reachability properties. For each case, returned results have formal guarantees on their

This work was supported by ANR-FNR project "AlgoReCell" (ANR-16-CE12-0034) and by CNRS PEPS INS2I 2017 project "FoRCe".

© Springer International Publishing AG 2017
J. Feret and H. Koeppl (Eds.): CMSB 2017, LNBI 10545, pp. 309–316, 2017.
DOI: 10.1007/978-3-319-67471-1_20

correctness (under-approximations, satisfying sufficient conditions) or completeness (over-approximations, satisfying necessary conditions).

Most of PINT analysis can typically handle networks with several hundreds of components. PINT also provides interfaces with exact model-checkers, such as NuSMV [5], ITS [16] and MOLE [25], taking advantage of implemented static model reduction to enhance their tractability on large models. Usual explicit reachable state graph analysis are also available, although other tools dedicated to Boolean or multi-valued networks already provide them, e.g., [10,14,18].

User Interfaces. PINT can be invoked either using command line executables, suited for batch deployments, or through a programmable python interface. Moreover, its embedding in the Jupyter IPython notebook allows a user-friendly web interface to ease the management of models and calls to PINT. Jupyter notebooks provide a convenient environment for editing, saving, sharing, and reproducing model analysis workflows. It is a common framework for data-oriented bioinformatics tools [2,6,12], and has promising suitability for computational systems biology, where reproducibility is very important as well.

Distribution. PINT is written mainly with the OCaml programming language and is actively developed since 2011. It is distributed under the free software licence CeCILL, and is available at http://loicpauleve.name/pint where binary packages are provided for Ubuntu Linux and Mac OS X.

The Docker[1] image pauleve/pint provides a ready-to-use PINT environment for usual operating systems (Windows, Mac OS X, Linux), and notably the Jupyter web interface. Such a kind of distribution becomes standard for providing accessible and reproducible analyses in bioinformatics, e.g., *BioContainers* [15].

2   Input Model

PINT takes as input *asynchronous automata networks* specified in plain text. Automata networks are sets of finite-state machines having local transitions conditioned by the state of other automata in the network. The global state space of the network is the produce of the local states of individual automata, and transitions are applied non-deterministically.

Figure 1 shows an example of automata networks with its plain text representation in PINT format. By convention, the file names end with .an.

Automata networks are expressive enough to encode the asynchronous semantics of Boolean and multi-valued networks. The main difference with these latter frameworks is the explicit specification of local transitions for each automaton (node) of the network, compared to a function-centred specification for Boolean and multi-valued networks [7,19].

PINT can automatically convert models expressed as Boolean or multi-valued networks using the `pint-import` command or `pypint.load()` python function. Most of the conversions are performed using GINsim [10], enabling the support

---

[1] http://docker.com.

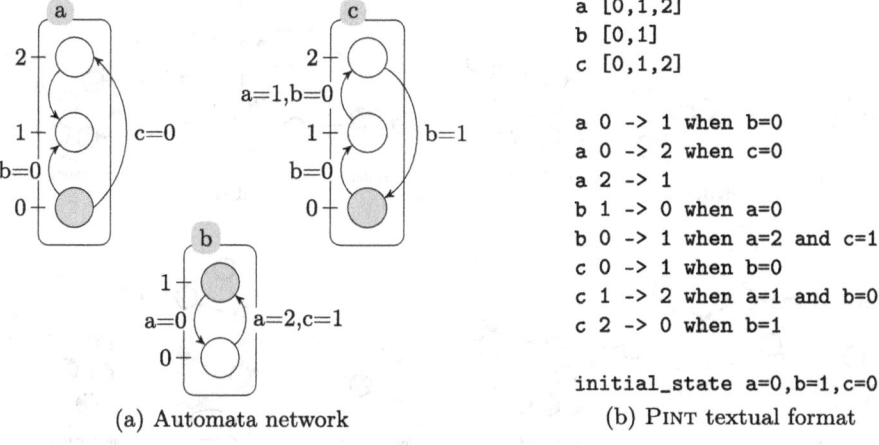

a [0,1,2]
b [0,1]
c [0,1,2]

a 0 -> 1 when b=0
a 0 -> 2 when c=0
a 2 -> 1
b 1 -> 0 when a=0
b 0 -> 1 when a=2 and c=1
c 0 -> 1 when b=0
c 1 -> 2 when a=1 and b=0
c 2 -> 0 when b=1

initial_state a=0,b=1,c=0

(a) Automata network   (b) PINT textual format

Fig. 1. (a) graphical representation of an automata network: automata are labelled boxes and their local states by circles where ticks are their identifier within the automaton. The initial state is composed of the local states in gray. A local transition is a directed edge between two local states of an automaton. Transitions can be labelled with states of other automata which are necessary to trigger the transition. (b) equivalent PINT plain text representation

for SBML-qual, GINsim, as well as various text formats. Models can be directly imported from URLs and from CellCollective database [13]. BIOCHAM reaction networks are also supported, following their Boolean semantics [4].

3 Main Features and Benchmarks

The main originality of PINT resides in the static analysis for transient reachability properties: such an approach avoids building the reachable state transition graph, neither explicitly nor symbolically. Therefore, the analysis aims at being tractable on large networks, at the price of giving possibly incomplete results.

We present the related features, illustrated in Fig. 2, with benchmarks to support their tractability on large biological networks. Computation times have been obtained on an Intel® Core™ i7-4770 3.40GHz CPU with 16GiB RAM.

Reachability analysis: formal approximation and model reduction — Given an initial state, a usual problem is to determine the existence of a sequence of transitions which leads to the activation or de-activation of key components (e.g., transcription factors) or to a particular attractor. Reachability verification is a PSPACE-complete problem and its resolution often explodes on large networks. PINT implements over- and under-approximation of reachability [9,21] which allow tackling large models, although being potentially inconclusive when the over-approximation is satisfied but not the under-approximation. In such cases, one should fall back to classical model-checking. To that aim, the *goal-oriented reduction* [19] identifies transitions that do not contribute to the goal

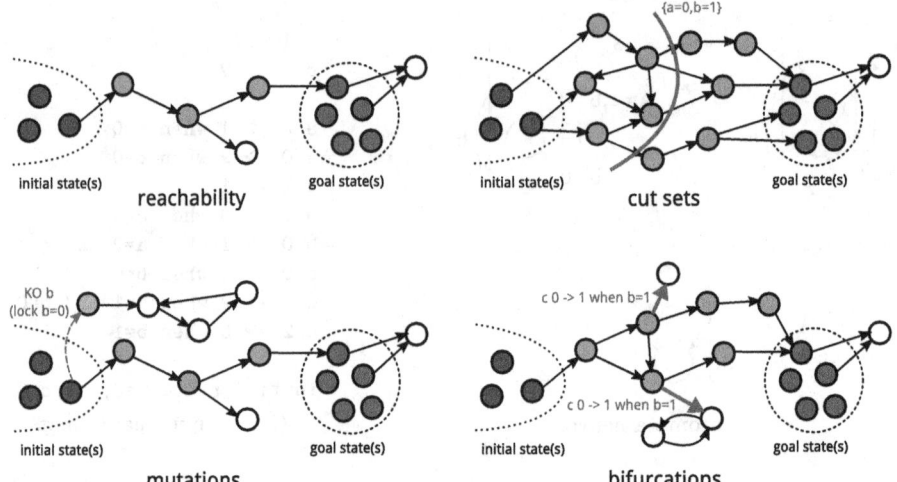

Fig. 2. Illustration of main features of PINT related to the transient reachability of a set of goal states from (a set of) initial state(s). Circles represent global states of the network and plain arrows dynamical transitions. Gray (resp. white) states are states which are (resp. are not) connected to a goal state.

reachability, and hence can be removed prior to the reachability analysis. This model reduction preserves *all* minimal traces to the goal, and can enhance greatly the tractability of model-checking. See Table 1 and [19] for benchmarks.

Prediction of mutations for controlling reachability — Given an initial state and a goal state of interest, PINT provides several methods to control the transient reachability of the goal.

The most scalable approach identifies cut sets of all the paths of transitions leading to the goal. A cut sets consists in one or several local states of automata which are necessary for the goal reachability: if one prevents the transitions involving these local states, the goal is disconnected from the initial state. PINT provides extremely scalable under-approximation of cut sets [20], which is tractable on Boolean networks with thousands of nodes (Table 2). Cut sets can thus be implemented as mutations which lock automata to its initial local state.

An alternative approach relies on a combination of static analysis and SAT solving and allows to directly infer mutations (gain or loss of function) which prevent the goal reachability. Whereas less scalable than cut set computations, it provides in general complementary solutions to cut sets, notably by identifying mutations which modify the initial state of the network.

Identification of bifurcation transitions — PINT implements static analysis for identifying so-called *bifurcation transitions* [8] after which the systems loses the capability to reach a given goal. Bifurcation transitions correspond to local transitions of the automata network which turn out to be important decision steps during differentiation processes. They can be fully identified by

Table 1. Benchmark[t] of goal reachability verification with two exact methods (NuSMV and ITS-REACH) and PINT, before (normal font) and after (bold font) goal-oriented model reduction; |T| is the number of local transitions in automata networks; |state| is the number of reachable global states, when computable. KO indicates an out-of-memory/time computation. In all cases PINT is conclusive.

Model (|nodes|)	|T|	|states|	Verification of goal reachability		
			NuSMV (EF g)	ITS-REACH	PINT
TCell-d (101) [1]	381	$\approx 2.4 \cdot 10^8$	2 s 40 Mb	0.5 s 26 Mb	0.02 s
profile 1	**0**	**1**			
TCell-d (101)	381	KO	KO	960 s 1.6 Gb	4.5 s
profile 2	**221**	**75,947,684**	**470 s 270 Mb**	**15 s 160 Mb**	
RBE2F (370) [22]	742	KO	KO	KO	0.2 s
	56	**2,350,494**	**3 s 37 Mb**	**4 s 13 Mb**	
MAPK (309) [24]	1251	KO	KO	KO	48 s
	429	**KO**	**KO**	**KO**	

Scripts and models available at http://loicpauleve.name/pint-benchmarks.tbz2

Table 2. Performance[t] of cut sets and mutations under-approximations with PINT depending on the maximal cardinality of returned sets.

Goal	TCell-d (101)		Egf-r (104) [23]		MAPK (309)		PID (10,229) [20]	
	FOXP3=1		AP1=1		ERK-PP=1		SNAIL=1	
3-cut sets	0.06 s	35	0.02 s	34	0.06 s	24	1.2 s	7
4-cut sets	0.10 s	101	0.02 s	34	0.1 s	48	5 s	37
6-cut sets	0.60 s	495	0.03 s	34	1 s	60	10 m	907
3-mutations	0.30 s	15	0.30 s	20	5 s	222	50 m	7
4-mutations	0.30 s	15	0.30 s	22	10 s	1896	50 m	67
6-mutations	0.30 s	15	0.30 s	22	KO		50 m	367

Scripts and models available at http://loicpauleve.name/pint-benchmarks.tbz2

Table 3. Performance[t] (Scripts and models available at http://loicpauleve.name/pint-benchmarks.tbz2) of exact and approximated *identification of bifurcation transitions* with NuSMV and PINT, respectively; $|t_b|$ is the number of identified bifurcation transitions.

	|T|	|states|	goal	NuSMV		PINT	
				$|t_b|$	time	$|t_b|$	time
EGF/TNF (28) [17]	53	3968	NFkB = 0	5	0.2 s	2	0.1 s
MAPK (53) [11]	173	KO	Proliferation = 1	KO		13	40 s
TCell-d (101)	381	KO	FOXP3 = 1	KO		4	58 s

Scripts and models available at http://loicpauleve.name/pint-benchmarks.tbz2

model-checking, but the static analysis in PINT allows tackling larger models, at the price of returning incomplete results (Table 3).

4   Integration with Jupyter IPython Web Notebook

Jupyter (http://jupyter.org) provides an interactive web interface for creating documents, named *notebooks*, which contain code, equations, and formatted texts. A notebook typically describes a full workflow of analysis, both with textual explanations and the full code and parameters to reproduce the results. It is a very popular framework in data science, including in bioinformatics [6,12]. A notebook is a single file which can be easily modified, shared, re-executed, and visualized online. For instance, the companion quick tutorial is available at http://nbviewer.jupyter.org/github/pauleve/pint/blob/master/notebook/quick-tutorial.ipynb.

The *pypint* module provides custom integration within the Jupyter IPython notebook, with custom menus and actions for loading models and executing PINT commands, as well as direct visualization of data structures. See Fig. 3 and the companion quick tutorial for a preview.

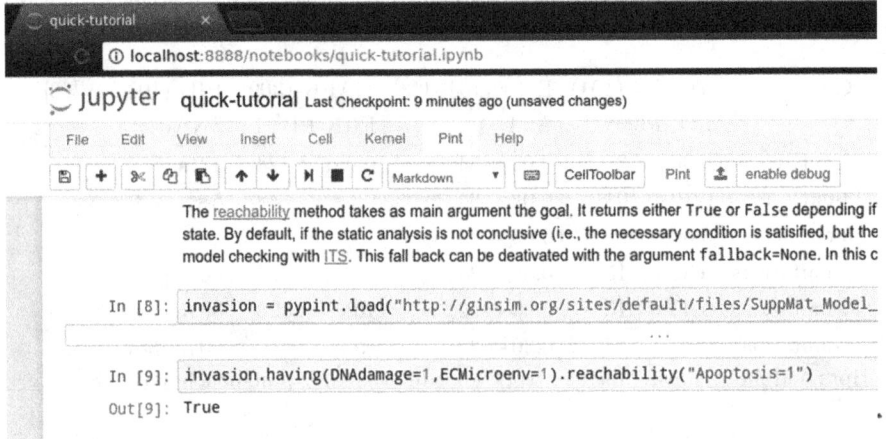

Fig. 3. Screen capture of Jupyter web interface running *pypint* in a notebook.

5   Conclusion

In this paper, we presented the prominent features of PINT on the static analysis for transient reachability of automata networks, from property verification to inference, which are tractable on large biological networks. PINT also implements classical state transition graph analysis, from fixpoint computation (using SAT solving) to explicit state space exploration, with a limited scalability. A tour of features is given at https://loicpauleve.name/pint/doc/#Tutorial.

In the next major release, we plan to add full support for synchronized local transitions, i.e., transitions that modify simultaneously the state of several automata. This improvement will allow to import any safe (1-bounded) Petri nets, broadening the class of supported dynamical models.

References

1. Abou-Jaoudé, W., Monteiro, P.T., Naldi, A., Grandclaudon, M., Soumelis, V., Chaouiya, C., Thieffry, D.: Model checking to assess t-helper cell plasticity. In: Front. Bioeng. Biotechnol. **2**, January 2015
2. Antao, T.: Bioinformatics with Python cookbook. Packt Publishing Ltd., Birmingham (2015)
3. Baral, C.: Knowledge Representation, Reasoning and Declarative Problem Solving. Cambridge University Press, New York (2003)
4. Calzone, L., Fages, F., Soliman, S.: Biocham: an environment for modeling biological systems and formalizing experimental knowledge. Bioinformatics **22**(14), 1805–1807 (2006)
5. Cimatti, A., Clarke, E., Giunchiglia, E., Giunchiglia, F., Pistore, M., Roveri, M., Sebastiani, R., Tacchella, A.: NuSMV 2: an opensource tool for symbolic model checking. In: Brinksma, E., Larsen, K.G. (eds.) CAV 2002. LNCS, vol. 2404, pp. 359–364. Springer, Heidelberg (2002). doi:10.1007/3-540-45657-0_29
6. Cock, P.J.A., Antao, T., Chang, J.T., Chapman, B.A., Cox, C.J., Dalke, A., Friedberg, I., Hamelryck, T., Kauff, F., Wilczynski, B., de Hoon, M.J.L.: Biopython: freely available python tools for computational molecular biology and bioinformatics. Bioinformatics **25**(11), 1422–1423 (2009)
7. Fages, F., Martinez, T., Rosenblueth, D.A., Soliman, S.: Influence systems vs reaction systems. In: Bartocci, E., Lio, P., Paoletti, N. (eds.) CMSB 2016. LNCS, vol. 9859, pp. 98–115. Springer, Cham (2016). doi:10.1007/978-3-319-45177-0_7
8. Fitime, L.F., Roux, O., Guziolowski, C., Paulevé, L.: Identification of bifurcations in biological regulatory networks using answer-set programming. In: Constraint-Based Methods for Bioinformatics Workshop (2016)
9. Folschette, M., Paulevé, L., Magnin, M., Roux, O.: Sufficient conditions for reachability in automata networks with priorities. Theor. Comput. Sci. **608**, 66–83 (2015). Part 1, From Computer Science to Biology and Back
10. Gonzalez, A.G., Naldi, A., Sánchez, L., Thieffry, D., Chaouiya, C.: Ginsim: A software suite for the qualitative modelling, simulation and analysis of regulatory networks. Biosystems **84**(2), 91–100 (2006). Dynamical Modeling of Biological Regulatory Networks
11. Grieco, L., Calzone, L., Bernard-Pierrot, I., Radvanyi, F., Kahn-Perlès, B., Thieffry, D.: Integrative modelling of the influence of MAPK network on cancer cell fate decision. PLoS Comput. Biol. **9**(10), e1003286 (2013)
12. Grunberg, R., Nilges, M., Leckner, J.: Biskit – a software platform for structural bioinformatics. Bioinformatics **23**(6), 769–770 (2007)
13. Helikar, T., Kowal, B., McClenathan, S., Bruckner, M., Rowley, T., Madrahimov, A., Wicks, B., Shrestha, M., Limbu, K., Rogers, J.A.: The cell collective: toward an open and collaborative approach to systems biology. BMC Syst. Biol. **6**(1), 96 (2012)
14. Klarner, H., Streck, A., Siebert, H.: PyBoolNet: a python package for the generation, analysis and visualization of boolean networks. Bioinformatics **33**, 770–772 (2016)

15. Leprevost, F.V., et al.: Biocontainers: an open-source and community-driven framework for software standardization. Bioinform. (Oxford Engl.) **33**, 2580–2582 (2017)
16. LIP6/Move. Its tools. http://ddd.lip6.fr/itstools.php
17. MacNamara, A., Terfve, C., Henriques, D., Bernabé, B.P., Saez-Rodriguez, J.: State-time spectrum of signal transduction logic models. Phys. Biol. **9**(4), 45003 (2012)
18. Mussel, C., Hopfensitz, M., Kestler, H.A.: BoolNet - an R package for generation, reconstruction and analysis of boolean networks. Bioinformatics **26**(10), 1378–1380 (2010)
19. Paulevé, L.: Goal-oriented reduction of automata networks. In: Bartocci, E., Lio, P., Paoletti, N. (eds.) CMSB 2016. LNCS, vol. 9859, pp. 252–272. Springer, Cham (2016). doi:10.1007/978-3-319-45177-0_16
20. Paulevé, L., Andrieux, G., Koeppl, H.: Under-approximating cut sets for reachability in large scale automata networks. In: Sharygina, N., Veith, H. (eds.) Computer Aided Verification. LNCS, vol. 8044, pp. 69–84. Springer, Heidelberg (2013)
21. Paulevé, L., Magnin, M., Roux, O.: Static analysis of biological regulatory networks dynamics using abstract interpretation. Math. Struct. Comput.Sci. **22**(4), 651–685 (2012)
22. Rougny, A., Froidevaux, C., Calzone, L., Paulevé, L.: Qualitative dynamics semantics for SBGN process description. BMC Syst. Biol. **10**(1), 1–24 (2016)
23. Samaga, R., Saez-Rodriguez, J., Alexopoulos, L.G., Sorger, P.K., Klamt, S.: The logic of EGFR/ERBB signaling: theoretical properties and analysis of high-throughput data. PLoS Comput. Biol. **5**(8), e1000438 (2009)
24. Schoeberl, B., Eichler-Jonsson, C., Gilles, E.D., Müller, G.: Computational modeling of the dynamics of the map kinase cascade activated by surface and internalized egf receptors. Nature Biotechnol. **20**(4), 370–375 (2002)
25. Schwoon, S.: Mole. http://www.lsv.ens-cachan.fr/~schwoon/tools/mole/

Posters

Discrete Bifurcation Analysis with Pithya

Nikola Beneš, Luboš Brim, Martin Demko, Matej Hajnal, Samuel Pastva,
and David Šafránek$^{(\boxtimes)}$

Systems Biology Laboratory, Faculty of Informatics, Masaryk University,
Botanická 68a, 602 00 Brno, Czech Republic
{xbenes3,brim,xdemko,xhajnal,xpastva,safranek}@fi.muni.cz

Bifurcation analysis is a central task of the analysis of parameterised high-dimensional dynamical systems that undergo transitions as parameters are changed. To characterise such transitions for models with many unknown parameters is a major challenge for complex, hence more realistic, models in systems biology. Its difficulty rises exponentially with the number of model components.

The classical numerical and analytical methods for bifurcation analysis are typically limited to a small number of independent system parameters. To address this limitation we have developed a novel approach to bifurcation analysis, called *discrete bifurcation analysis*, that is based on a suitable discrete abstraction of the given system and employs model checking for discovering critical parameter values, referred to as bifurcation points, for which various kinds of behaviour (equilibrium, cycling) appear or disappear. To describe such behaviour patterns, called *phase portraits*, we use a hybrid version of a CTL logic augmented with direction formulae.

Technically, our approach is grounded in a novel method of parameter synthesis from temporal logic formulae using symbolic model checking and implemented in a *new high-performance tool Pithya*[1] [1]. Pithya itself implements state-of-the-art parameter synthesis methods. For a given ODE model, it allows to visually explore model behaviour with respect to different parameter values. Moreover, Pithya automatically synthesises parameter values satisfying a given property. Such property can specify various behaviour constraints, e.g., maximal reachable concentration, time ordering of events, characteristics of steady states, the presence of limit cycles, etc. The results can be visualised and explored in a graphical user interface.

We demonstrate the method on a case study taken from biology describing the interaction of the tumour suppressor protein pRB and the central transcription factor $E2F1$ [3]. This system represents an important mechanism of a *biological switch* governing the transition from G_1 to S phase in the mammalian cell cycle. In the G_1-phase the cell makes an important decision. In high concentration levels, $E2F1$ activates the phase transition. In low concentration of $E2F1$, the transition to S-phase is rejected and the cell avoids division.

This work has been supported by the Czech National Infrastructure grant LM2015055 and by the Czech Science Foundation grant GA15-11089S.
[1] http://biodivine.fi.muni.cz/pithya/.

J. Feret and H. Koeppl (Eds.): CMSB 2017, LNBI 10545, pp. 319–320, 2017.
DOI: 10.1007/978-3-319-67471-1

References

1. Beneš, N., Brim, L., Demko, M., Pastva, S., Šafránek, D.: Pithya: a parallel tool for parameter synthesis of piecewise multi-affine dynamical systems. In: Computer Aided Verification, CAV 2017. LNCS, Springer-Verlag (2017, to appear)
2. Beneš, N., Brim, L., Demko, M., Pastva, S., Šafránek, D.: A model checking approach to discrete bifurcation analysis. In: Fitzgerald, J., Heitmeyer, C., Gnesi, S., Philippou, A. (eds.) FM 2016. LNCS, vol. 9995, pp. 85–101. Springer, Cham (2016). doi:10.1007/978-3-319-48989-6_6
3. Swat, M., Kell, A., Herzel, H.: Bifurcation analysis of the regulatory modules of the mammalian G1/S transition. Bioinformatics **20**(10), 1506–1511 (2004). Oxford University Press

Discrete Stochastic Graph Dynamics for the Nuclear Architecture of Mouse Meiotic Prophase Spermatocytes

(Extended Abstract)

Julio López Fenner[1](✉), Aude Maignan[2], Rachid Echahed[3], and Soledad Berríos[4]

[1] Univ. de La Frontera, Temuco, Chile
julio.lopez@ufrontera.cl
[2] Univ. Grenoble-Alpes, CNRS, Grenoble INP, LJK, Grenoble, France
[3] Univ. Grenoble-Alpes, CNRS, Grenoble INP, LIG, Grenoble, France
[4] ICBM, Univ. de Chile, Santiago, Chile

We propose a simple parallel stochastic dynamics for understanding random association cluster formations of $2n = 40$ *Mus musculus domesticus* bivalents during pachytene in early prophase and provide statistically optimized parameters for ensuring adequate fitting of the model with available experimental data [2]. This work represents a continuation of the discrete dynamical approach started in [2, 3] while modeling randomness for chromosome associations in $2n = 40$ - *Mus m. domesticus* spermatocytes. We focus on pachytene in prophase I (see [1]).

During pachytene, at the prophase stage of meiosis, the homologous chromosomes synapse along a proteinacious structure, called synaptonemal complex (SC), thus enabling recombination between them, a process that produces genetic variation. The synapsed chromosomes are called bivalents and are found to be attached to the nuclear envelope by both their ends, being able to move or glide upon the internal surface of it. Chromosomal bivalent's associations are said to be given by intersecting domains of *constitutive pericentromeric heterochromatin*(CPCH's), which are known to create rich dynamic and diverse scenarios via the participating elements. These are triggered by the corresponding intersection domains of CPCH located at the short arms of each bivalent, but also by the associated convergence of the rest of the constituent chromatin along them. These structures are revealed by means of squashes (or spreads), in which the nuclear envelope is removed and the spermatocyte's nucleus content is projected to a flat surface.

Data from 400 pachytene spermatocyte spreads of $2n = 40$ *Mus domesticus* treated by immunocytochemical techniques taken from [2] is used for contrasting theoretical results: we model the spermatocyte's nucleous as an (almost) six regular graph, which ensure maximal connectivity for the nodes. They represent the positions of the bivalents attached to the nuclear envelope.

© Springer International Publishing AG 2017
J. Feret and H. Koeppl (Eds.): CMSB 2017, LNBI 10545, pp. 321–323, 2017.
DOI: 10.1007/978-3-319-67471-1

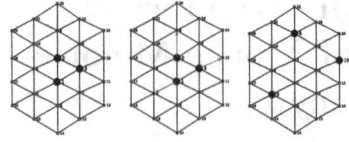

Fig. 1. Parallel evolution in G. Red lines indicate direction of displacement (Color figure online).

Upon this discrete surface, the SC evolution follows a parallel rewriting rule as in Fig. 1. Here, the bivalent's structures are represented as attributes of the vertices: The synaptonemal complex SC attached to the envelope and a random neighborhood of vertices for the CPCH. Pathwise connected domains of overlapping CPCH are considered to build an association cluster between the corresponding bivalents and we describe and analyze their statical and dynamical distribution.

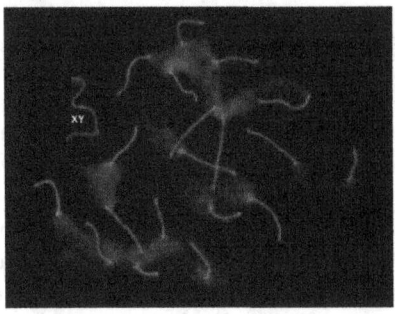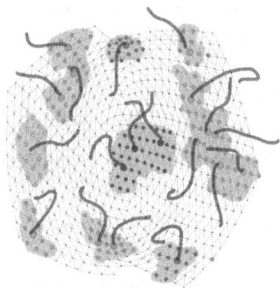

Fig. 2. Squash v/s model: An immunocytochemichally treated nucleous with an artist conception of its mathematical counterpart.

The model can now be used for interrogating different phenomena associated to the superposition of chromatin domains of the bivalents during pachytene, as well as providing a theoretical description of the kind of randomness involved in these phenomena (Fig. 2).

Undoubtedly, a model-theoretical approach to the general principles behind bivalent's associations in prophase meiotic nuclei, as well as precising the type of randomness being at play at this stage could bring us also a step closer to a better understanding of the different chromosome combinations present in the gametes. Since these associations and combinations persist until the meiotic divisions, the chromosomal associations as described here necessarily leave some imprint in the chromosomal sets passed on to gametes and hence their importance to evolution.

Acknowledgements. This work has been partially supported by Chilean MINEDUC Grant MECE-SUP 2016-2017 and LabEx PERSYVAL-Lab (ANR-11-LABX-0025-01) funded by the French program Investissement d'avenir.

References

1. Berrios, S.: Nuclear architecture of mouse spermatocytes: chromosome topology, heterochromatin, and nucleolus. Cytogenet. Genome Res. **151**(2), 61–71 (2017)
2. Berríos, S., Manterola, M., Prieto, Z., López-Fenner, J., Page, J., Fernández-Donoso, R.: Model of chromosome associations in Mus domesticus spermatocytes. Biol. Res. **43**(3), 275–285 (2010)
3. López-Fenner, J., Berríos, S., Manieu, C., Page, J., Fernández-Donoso, R.: Bivalent associations in Mus domesticus 2n = 40 spermatocytes: are they random? Bull. Math. Biol. **76**(8), 1941–1952 (2014)

Non-disjoint Clustered Representation for Distributions over a Population of Cells

Matthieu Pichené[1]($\boxtimes$), Sucheendra Palaniappan[2], Eric Fabre[1], and Blaise Genest[3]

[1] Inria, Team SUMO, Rennes, France
matthieu.pichene@inria.fr
[2] The Systems Biology Institute, Tokyo, Japan
[3] CNRS, IRISA, Rennes, France

1 Motivation

We consider a large homogenous population of cells, where each cell is governed by the same complex biological pathway. A good modeling of the inherent variability of biological species is of crucial importance to the understanding of how the population evolves. In this work, we handle this variability by considering multivariate distributions, where each species is a random variable. Usually, the number of species in a pathway -and thus the number of variables- is high. This appealing approach thus quickly faces the curse of dimensionality: representing *exactly* the distribution of a large number of variables is intractable.

To make this approach tractable, we explore different techniques to *approximate* the original joint distribution by meaningful and tractable ones. The idea is to consider families of joint probability distributions on large sets of random variables that admit a compact representation, and then select within this family the one that best approximates the desired intractable one. Natural measures of approximation accuracy can be derived from information theory. We compare several representations over distributions of populations of cells obtained from several *fine-grained* models of pathways (e.g. ODEs). We also explore the interest of such approximate distributions for approximate inference algorithms [1, 2] for *coarse-grained* abstractions of biological pathways [3].

2 Results

Our approximation scheme is to drop most correlations between variables. Indeed, when many variables are conditionally independent, the multivariate distribution can be compactly represented. The key is to keep the most relevant correlations, evaluated using the *mutual information (MI)* between two variables.

The simplest approximation is called *fully factored (FF)*, and assumes that all the variables are independent. It leads to very compact representation and fast computations, but it also leads to fairly inaccurate results as correlations between variables are entirely lost, even for highly correlated species (MI = 0.6).

© Springer International Publishing AG 2017
J. Feret and H. Koeppl (Eds.): CMSB 2017, LNBI 10545, pp. 324–326, 2017.
DOI: 10.1007/978-3-319-67471-1

Alternately, one can preserve a few of the strongest correlations, selected using MI, giving rise to a set of *disjoint clusters* of variables. For efficiency reason, we used clusters of size two. This model was able to capture some of the most significant correlations between pairs of variables (representing around 30% of the total MI), but dropped significant ones (MI = 0.2).

A better trade-off between accuracy and tractability was obtained by using *non-disjoint* clusters of two variables, structured as a tree, called the *tree-clustered approximation (TCA)*. The approximated joint distribution is fully determined by the marginals over each selected cluster of 2 variables. This gives a compact representation (<800 values in our experiments). Further, any marginal over k out of n total variables can be computed with time complexity $O(nv^{k+1})$, where each variable can take v possible values. Last, a tractable algorithm [4] allows to compute the best approximation of any distribution by a tree of clusters. TCA succeeded in capturing most correlations between pairs of variables (representing around 70% of the total MI), losing no significant ones (MI < 0.1).

Regarding inference, FF, disjoint clusters and TCA were compared to *Hybrid FF (HFF)* [2]. In short, HFF preserves a small number of joint probabilities of high value (called spikes), plus an FF representation of the remaining of the distribution. The more spikes, the more accurate the approximation, and the slower HFF inference. Overall, TCA is very accurate, while HFF generates sizable errors, even with numerous spikes (32k). Further, TCA is faster than HFF, even with few spikes (3k). FF and disjoint-clusters are even faster (1 to 2 order of magnitudes) than TCA, but the accuracy of both remains problematic.

3   Perspectives

We now aim at modeling and studying a tissue, made of tens of thousands of cells. In this context, capturing the inherent variability of the population of cells is crucial. In order to study multi-scale systems in a tractable way, we advocate a two-step approach: Firstly, abstract the low level model of the pathway of a single cell into a stochastic discrete abstraction, e.g. using [3]. Secondly, use a model of the tissue, which does not explicitly represent every cell but qualitatively explains how the *population* evolves. In this way, one need not explicitly represent the concentration of each of the tens of thousands of cells, but rather only keep one probability distribution.

Acknowledgement. This work was partially supported by ANR-13-BS02-0011-01 STOCH-MC.

326   M. Pichené et al.

References

1. Murphy, K., Weiss, Y.: The factored frontier algorithm for approximate inference in DBNs. In: UAI 2001, pp. 378–385. Morgan Kaufmann (2001)
2. Palaniappan, S.K., Akshay, S., Liu, B., Genest, B., Thiagarajan, P.S.: A hybrid factored frontier algorithm for dynamic Bayesian networks with a biopathways application. TCBB **9**(5), 1352–1365. IEEE/ACM
3. Palaniappan, S.K., Bertaux, F., Pichené, M., Fabre, E., Batt, G., Genest, B.: Abstracting the dynamics of biological pathways using information theory: a case study of apoptosis pathway. Bioinformatics **33**(13), 1980–1986 (2017). OUP
4. Chow, C.K., Liu, C.N.: Approximating discrete probability distributions with dependence tree. IEEE ToIT **14**(3), 462–467 (1968). IEEE

Effects of the Dynamics of the Steps in Transcription Initiation on the Asymmetry of the Distribution of Time Intervals Between Consecutive RNA Productions

Sofia Startceva[1], Vinodh Kumar Kandavalli[1], Ari Visa[2], and Andre S. Ribeiro[1(✉)]

[1] Laboratory of Biosystem Dynamics, BioMediTech Institute
and Faculty of Biomedical Sciences and Engineering,
Tampere University of Technology, 33101 Tampere, Finland
andre.ribeiro@tut.fi
[2] Signal Processing Unit, Faculty of Computing and Electrical Engineering,
Tampere University of Technology, 33101 Tampere, Finland

Abstract. Asymmetries in the distribution of time intervals between consecutive RNA productions from a gene can play a critical role in, e.g., allowing/preventing the RNA and, thus, protein numbers to cross thresholds involved in gene network decision making. Here, we use a stochastic, multi-step model of transcription initiation, with all rate constants empirically validated, and explore how the kinetics of its steps affect the temporal asymmetries in RNA production, as measured by the skewness of the distribution of intervals between consecutive RNA productions in individual cells. From the model, first, we show that this skewness differs widely with the mean fraction of time that the RNA polymerase spends in the steps preceding open complex formation, while being independent of the mean transcription rate. Next, we provide empirical validation of these results, using qPCR and live, time-lapse, single-molecule RNA microscopy measurements of the transcription kinetics of multiple promoters. We conclude that the skewness in RNA production kinetics is subject to regulation by the kinetics of the steps in transcription initiation and, thus, evolvable.

Keywords: Transcription initiation · Asymmetries in RNA production · Stochastic models · Single-RNA measurements

Gene expression regulation in bacteria occurs mostly in transcription initiation [1]. In *Escherichia coli*, this process is sequential [2], starting with an RNA polymerase (R) binding to an active promoter (P_{ON}) and forming a closed complex (RP_{cc}). Next, the open complex (RP_{oc}) forms. Relevantly, the subsequent steps of RNA elongation [3], termination, and RNA and R release are much faster. Thus, dynamically, transcription can be approximately modeled as:

$$R + P_{ON} \xrightarrow{k_{cc}} RP_{cc} \xrightarrow{k_{oc}} RP_{oc} \xrightarrow{\infty} P_{ON} + R + RNA \tag{1}$$

© Springer International Publishing AG 2017
J. Feret and H. Koeppl (Eds.): CMSB 2017, LNBI 10545, pp. 327–329, 2017.
DOI: 10.1007/978-3-319-67471-1

Here, RNA production kinetics is controlled by k_{cc} and k_{oc}. The probability density function (pdf) of the distribution of intervals between transcription events is the convolution of their pdfs: $f_{\Delta t}(t) = \frac{k_{cc} \cdot k_{oc}}{k_{oc} - k_{cc}} \left(e^{-k_{cc} \cdot t} - e^{-k_{oc} \cdot t}\right)$. To measure asymmetries in this distribution, we use skewness, $S = \frac{m_3}{m_2^{3/2}}$, where $m_r = \frac{1}{n} \Sigma (x_i - \bar{x})^r$ [4]. We estimate the sample skewness $S_s = \frac{\sqrt{n(n-1)}}{n-2} \cdot S$, where n is the sample number [5]. To obtain confidence boundaries for S_s we use non-parametric bootstraps as in [6].

In (1), k_{cc} is the inverse of the mean time for R to bind the promoter and complete a closed complex (τ_{cc}), while k_{oc} is the inverse of the mean time for an open complex to form (τ_{oc}). The mean time between transcription events: $\Delta t = \tau_{cc} + \tau_{oc}$.

To validate the model predictions of skewness, we collected empirical data for Δt and $\tau_{cc}/\Delta t$ for various promoters (P_{TetA}, P_{BAD}, $P_{Lac-ara-1}$, and $P_{Lac-ara-1}$ under oxidative stress) [7–9] (Fig. 1). Next, given the mean Δt of each promoter, we varied $\tau_{cc}/\Delta t$ (from 0 to 1) while maintaining Δt constant. Then, for each value of $\tau_{cc}/\Delta t$, we calculated S from the pdf of the distribution of intervals between transcription events (solid line, Fig. 1). Interestingly, we observed that S is independent of the mean value of Δt. Finally, from Fig. 1, we find that the model predictions of S fit the empirical data.

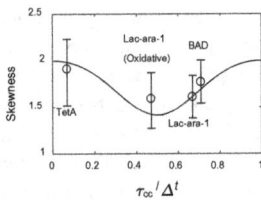

Fig. 1. Predicted skewness of Δt distributions with given $\tau_{cc}/\Delta t$ (solid line) and sample skewness of the empirical Δt distributions (with 95% confidence intervals) for the studied promoters. For each promoter, 100 or more Δt intervals were extracted from a total of 100 or more cells.

Importantly, as S is tunable by τ_{cc} and τ_{oc}, which are sequence dependent and subject to regulation, we expect it to be evolvable and adaptive to environment shifts.

References

1. Kaern, M., et al.: Stochasticity in gene expression: from theories to phenotypes. Nat. Rev. Genet. **6**, 451–464 (2005)
2. McClure, W.R.: Mechanism and control of transcription initiation in prokaryotes. Annu. Rev. Biochem. **54**, 171–204 (1985)
3. Uptain, S.M., et al.: Basic mechanisms of transcript elongation and its regulation. Annu. Rev. Biochem. **66**, 117–172 (1997)
4. MacGillivray, H.L.: Skewness and asymmetry: measures and orderings. Ann. Stat. **14**, 994–1011 (1986)

5. Joanes, D.N., Gill, C.A.: Comparing measures of sample skewness and kurtosis. R. Stat. Soc. **47**, 183–189 (1998)
6. Carpenter, J., Bithell, J.: Bootstrap confidence intervals: when, which, what? a practical guide for medical statisticians. Stat. Med. **19**, 1141–1164 (2000)
7. Muthukrishnan, A.B., et al.: In Vivo transcription kinetics of a synthetic gene uninvolved in stress-response pathways in stressed escherichia coli cells. PLoS ONE **9**, e109005 (2014)
8. Lloyd-Price, J., et al.: Dissecting the stochastic transcription initiation process in live escherichia coli. DNA Res. **23**, 203–214 (2016)
9. Kandavalli, V.K., et al.: Effects of σ factor competition are promoter initiation kinetics dependent. Biochim. Biophys. Acta - Gene Regul. Mech. **1859**, 1281–1288 (2016)

Author Index

Printed in the United States
By Bookmasters